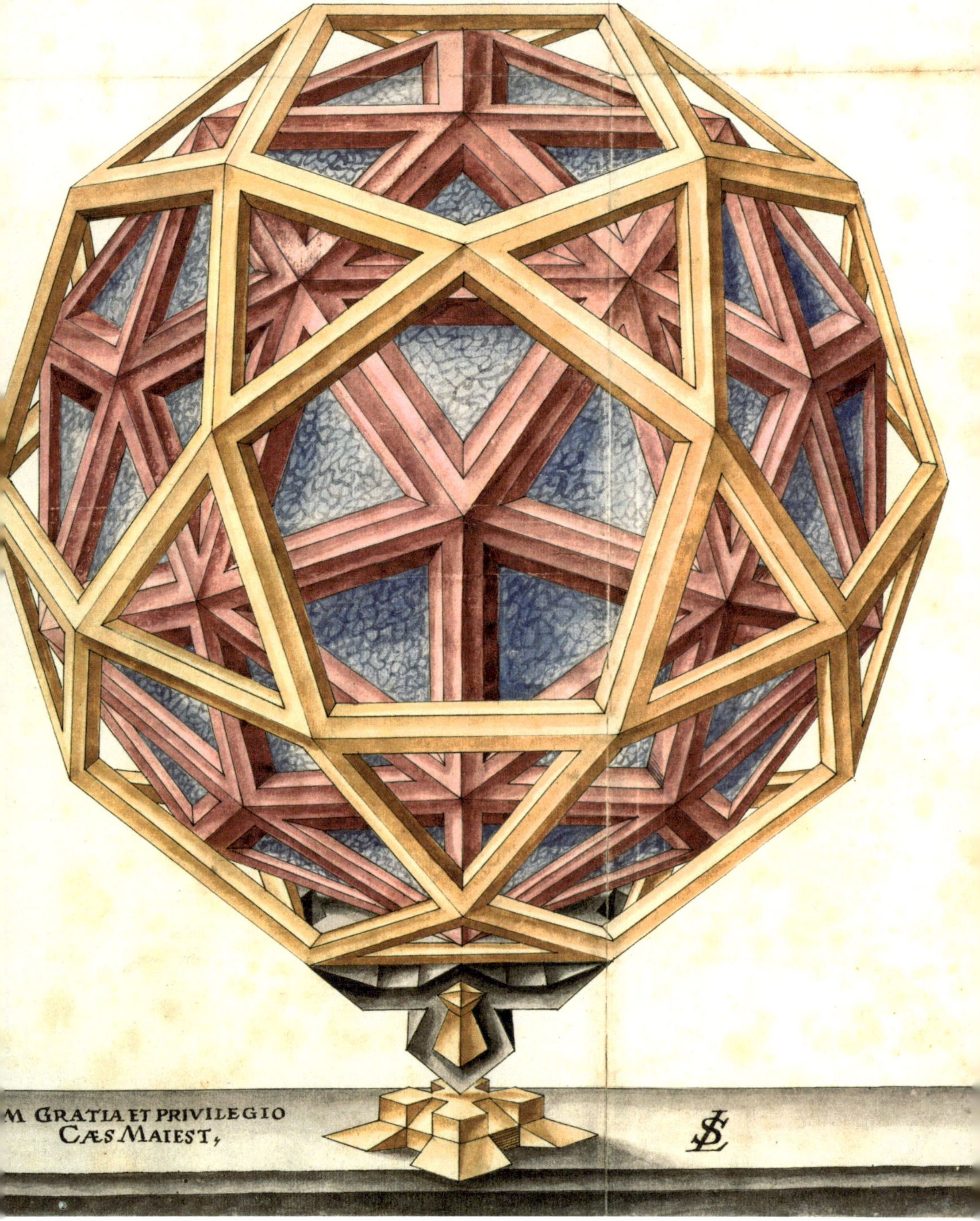
M GRATIA ET PRIVILEGIO
CÆS MAIEST,

THE MIT PRESS
Cambridge, Massachusetts / London, England

THE POLYHEDRISTS
ART AND GEOMETRY IN THE LONG SIXTEENTH CENTURY

NOAM ANDREWS

to Nimrod

Rhombicosidodecahedra 10

Corpora Regulata et Irregulata 28

Instruction in Measurement 64

Geometria and the Lehrbuch 100

Lover of the Art of Perspective 138

Cut Apart in Many Ways 168

The Violence of Whimsy 206

Epilogue: Corpora Irregulata et Regulata 242

Notes *254*

Bibliography *276*

Index *290*

Acknowledgments *297*

R H O M B I

D E C A

COSIDO-

EDRA

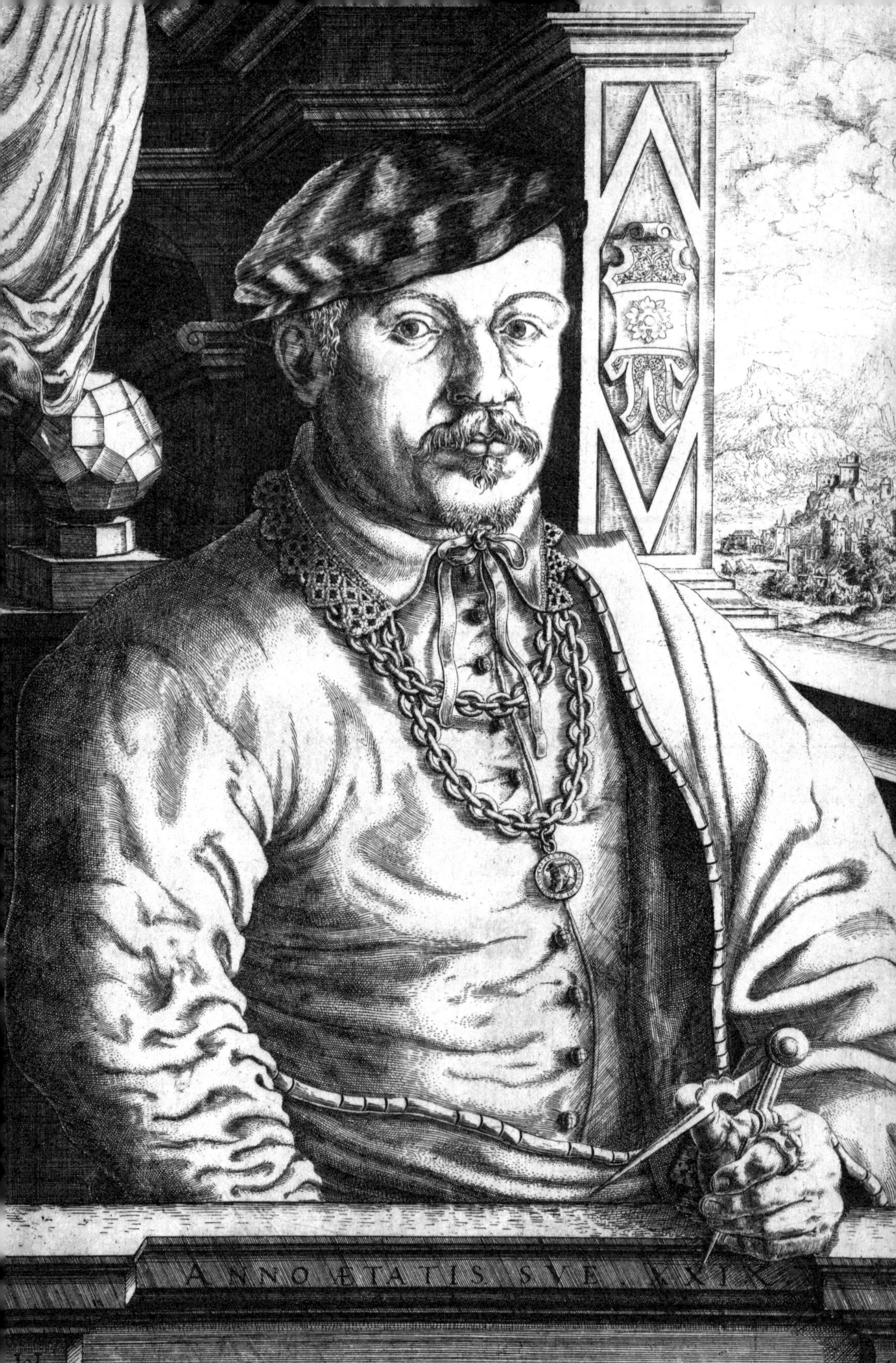
ANNO ÆTATIS SVE. XXIX
IsL

Resting on a plinth behind the fleshy figure of the German scholar and humanist Oswald von Eck (before 1539–1573) lurks a model of a rhombicosidodecahedron partially obscured by the drooping corner of phantom drapery. Languidly gripping a compass in his left hand, the view onto the Bavarian countryside visible through an open window, von Eck cuts a rotund and wobbly figure, his cap jauntily placed, his garments rumpled if neither wrinkled nor pleated nor creased. The portrait presents a slightly different sort of man than one might expect in a Renaissance staging of geometrical mastery, a man whose complacent posture and timid manière do little to assuage the viewer's fear that von Eck's faith in his abilities may be misplaced. Charting a terrain untouched by compass and ruler, the rounded edges and rambling textures of his body bear disjunctively upon the planarity of the room's architecture—the window frame above his left shoulder, the banister upon which his hand is rested, and the rear entablature fading into darkness. Indeed, it is the polyhedron, its outlandishness heightened by von Eck's nonchalance, that shoulders the bulk of the iconographic burden.

That a rhombicosidodecahedron would be exhibited with such an apparent lack of fanfare might at first seem odd. After all, such a multifaceted geometry could not have been directly traced back to Euclid's *Elements* (ca. 300 BC), the omnipresent textbook of classical mathematics that had formed the basis of all instruction in geometry for nearly two millennia. Euclid's description was limited to the five Platonic or "regular" solids (*corpora regulata/regularia*)—the pyramid (tetrahedron), the cube (hexahedron), octahedron, dodecahedron, and the icosahedron—convex polyhedra which only appeared toward the end of *Elements* and fell far beyond the curricular boundaries of the typical liberal arts education (*quadrivium*). So called because 1
they had been described by Plato in *Timaeus* (ca. 360 BC), the Platonic solids were touted in premodern European culture as the epitome of geometrical knowledge. Hence the audacity of presenting "irregular" solids or bodies (*corpora irregulata/irregularia*) without acknowledging their uniqueness: by the mid-sixteenth century the cast of solids exceeded Plato and Euclid, becoming at once commonplace enough not to attract undue attention and stable enough as icons to serve as universal signifiers of geometrical knowledge in portraiture, much like the Platonic solids themselves. 2

Hanns Lautensack (ca. 1520–1564/1566). *Oswald von Eck*. 1553. The Metropolitan Museum of Art, New York, Gift of Felix M. Warburg and his family, 1941 (41.1.143).

Deemed *ungeregulirt* as per Albrecht Dürer, or *irregulata* by the artist Lorentz Stöer, the irregular bodies appear as "irregulari corpore" in the work of Charles de Bovelles, "de corporibus irregularibus" in Piero della Francesca, "i corpi irregulari" in Daniele Barbaro,
3 and as stereometric volumes in Leonardo's Codex Forster. But the concept of what it meant to be irregular, as early modern geometrical irregularity was defined, centered at least initially on what is known today as the "semiregular" or Archimedean solids, namely the set of highly symmetrical solids that can be derived through consistent modification of the five original Platonic solids. The pursuit of the Archimedean solids, a finite set of thirteen geometries to be puzzled over and "discovered," as the astronomer Johannes Kepler would in *Harmonices mundi libri V* (1619), did not ultimately challenge the overall predominance of regularity as an organizing structure motivating geometrical visualization. Despite the occasional foray into the asymmetrical (Dürer's famous truncated rhomboid from *Melencolia I*), early *corpora* all subscribed to the rule of the regular, namely that their sides were all the same length and symmetrically
4 placed about a common center.

The advent of an irregular body in von Eck's parlor sprang from a sustained period of exponentially developing skills in drawing, perspective, and measurement that had transformed the Platonic solids into a new and distinct genre of "mathematical art." Surfacing at the dawn of printing as early visualizations of Euclid's propositions in *Elements*, the Platonic solids were adopted initially as pedagogical objects by makers eager to refine their design aptitude. Almost immediately, polyhedral models most often made from paper or wood appeared in workshops. As "perspective tools" used to set out the baselines of a solid in perspective that could then be embellished by hand, they were also ephemeral artifacts bound to their own precise reproducibility and destined to be discarded. Their attendance at the drafting table, preserved in polyhedral drawings by Leonardo da Vinci and Parmigianino, Dürer and Carpaccio, speaks to the new practical utility of the Platonic solids for artists and the visual possibilities of geometrical knowledge as an explicit artistic subject. Piero della Francesca and Luca Pacioli published influential treatises on the solids, *Libellus de quinque corporibus regularibus* (fifteenth century)

Parmigianino (1503–1540). Two studies of dodecahedra. Museo di Capodimonte, Naples (1366 verso). Su concessione del Ministero per i Beni e le Attività Culturali e per il Turismo—Museo e Real Bosco di Capodimonte.

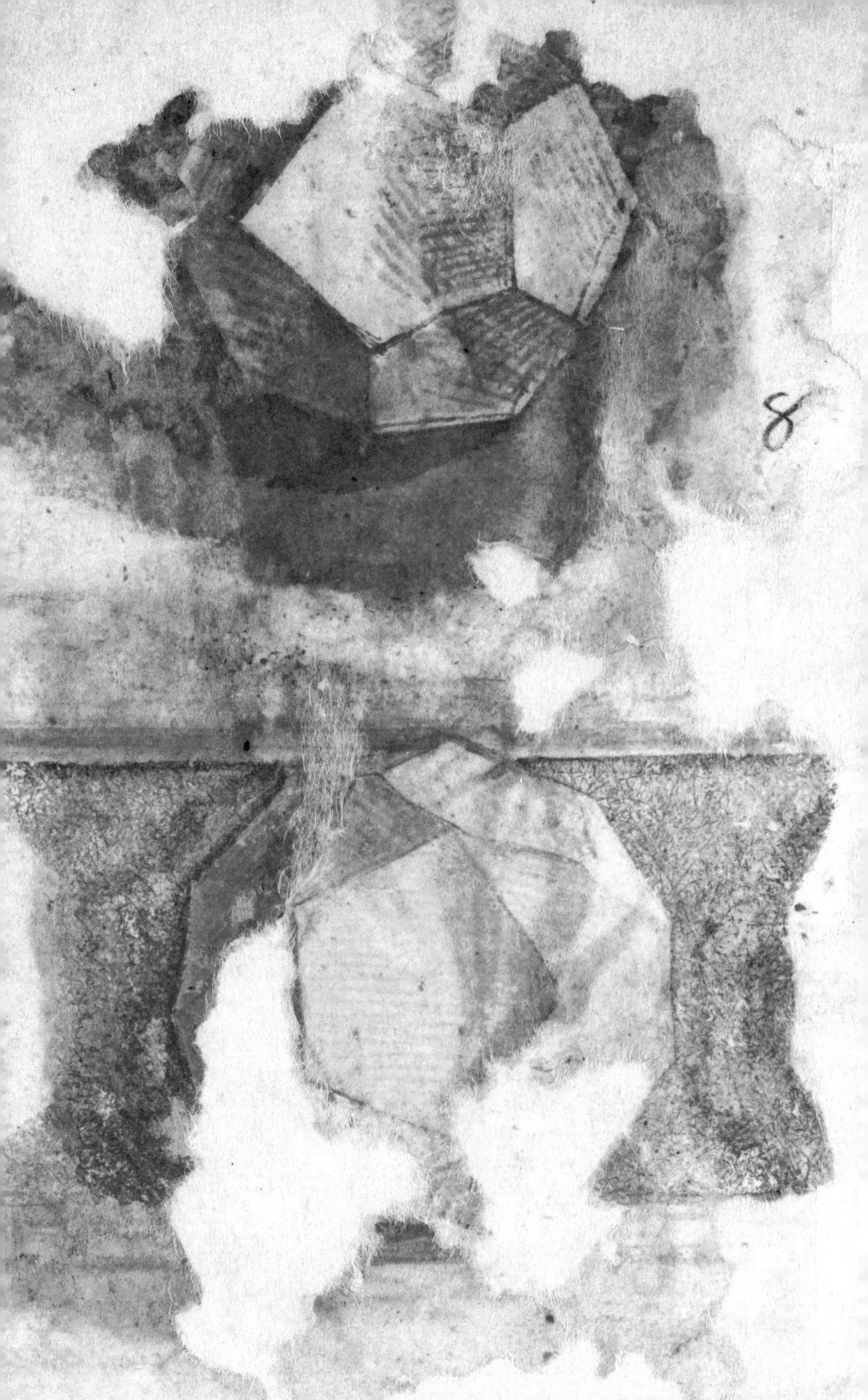
8

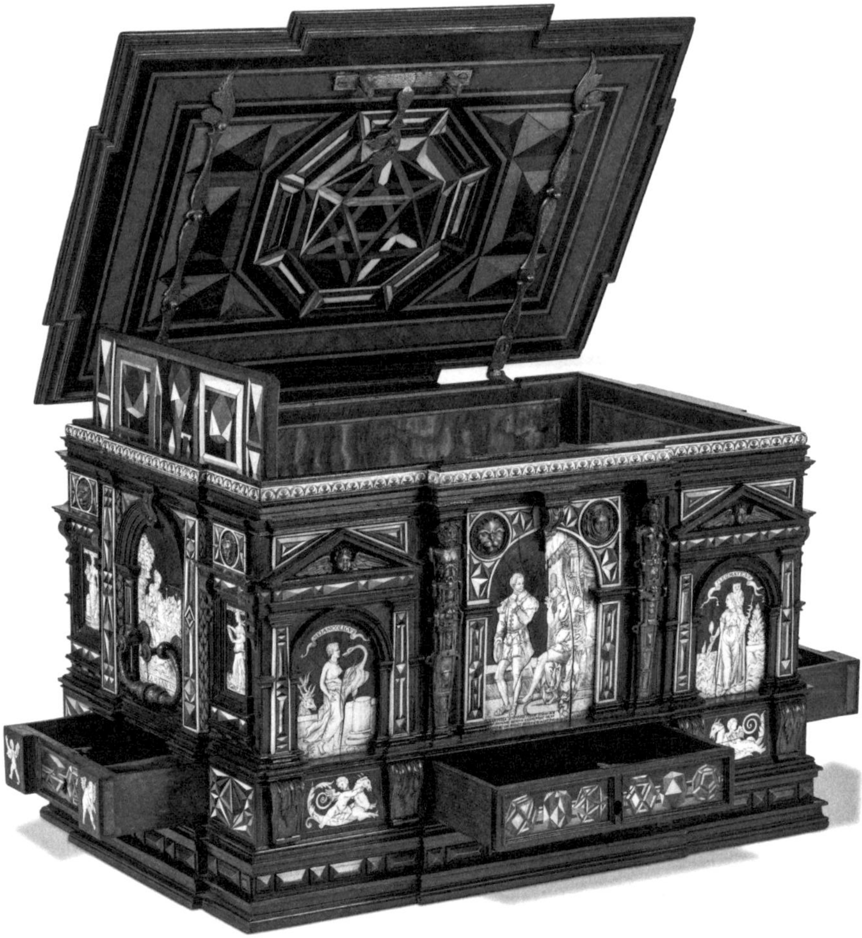

Master of Perspective. Court casket of Newbattle Abbey. Nuremberg, 1565.
Photo: Courtesy of Kunstkammer Ltd., London.

and *De divina proportione* (manuscript ca. 1498, first printed edition 1509), respectively, the latter including technical renderings of the solids variously attributed to Leonardo or his circle. Mastery of three-dimensional geometry, pushed beyond what could be described through lengthy textual description, began to redefine what it meant for artisans to hone their craft, and knowledge of the solids and their spatial properties became readily available for consumption through purchase or copying, a privilege hitherto reserved for only the most advanced mathematicians and natural philosophers.

The materialization of the solids initiated a stripping of their bookish, Neoplatonic trappings, generating a self-consciously tactile approach to geometry that would ignite a century of novel geometrical form and form-making strategies. Through a massive groundswell of technical investment, the solids became the single most recognizable emblems of perspectival measurement, transitioning into a unique category of *Meisterstücke* (masterpieces) that could take the form of diplomatic gifts or ornamental surfaces for furniture. These developments did not take place within the world of academic publishing and in turn proffered an alternative system of value, tied less to the prestige of Euclid and the classical world than to the modern artist. Adopted first to hone precision measurement and perspective skills in the "middle sciences" or "mixed mathematics," so called because they eluded Aristotle's taxonomy of knowledge, the solids emerged from the Renaissance workshops of European artists and artisans, many of whom were quick to grasp the potential of polyhedra as a basis for exploring the near limitlessness of three-dimensional abstraction. 5

One only has to look at the devotion poured into the court casket, dated 1565, by the Master of Perspective, formerly owned by the Marquesses of Lothian at Newbattle Abbey in Scotland, to glimpse the exclusive company the solids enjoyed. Inlaid with mother-of-pearl, ebony, engraved bone, and ivory, the casket is distinguished by a cluster of prominent skeletal bodies bordered by a thick band of scrolling ivory *Rollwerk* ornament. Regular and irregular polyhedra appear on the interior of the lid, on the fronts of each drawer, and ringing the lower edge of the base, while the outer sides of the casket are adorned with historical tableaux after copper plates by Heinrich Aldegrever (1502–1538) and personifications of the four temperaments (the sanguine, phlegmatic, choleric, and melancholic) and the four

virtues (faith, hope, fortitude, and prudence), all based on engravings
6 by Virgil Solis. The adjacencies of these representations, riffing on the longstanding association of the Platonic solids with the elements (fire, earth, wind, and water), speaks to the loose network that could be recalled to provide a pleasing frisson of classicism in a *Kunstkammer* object. In the gap between the figural and the abstract is the casket itself and its masterful layering of two semantic systems whose progressive decoupling by artisans, occasionally revived to stupendous aesthetic effect, could now only be resolved at the level of the object.

Situated at the nexus of print matter and drawing, goldsmithing, intarsia, and ivory, the Renaissance afterlife of the solids uncovers the influence of the decorative arts and artisanal experimentation with geometry on the formulation of Renaissance mathematical knowledge and its engagement with materials and concepts of materiality. As highly covetable objects of "mathematical wonder" serving to satiate the early modern appetite for novelty, *corpora* transformed from philosophical objects into base motifs for virtuosic displays of artisanal skill integral to the attaining of practice-based mathematical knowledge. Their numerous identities and states all existed concomitantly. Geometrical *Lehrbücher* (textbooks) filled with elementary solids remained in circulation even as salons across Europe imported cabinetry with virtuosic polyhedral motifs from Augsburg. *Elements* was assiduously taught at the same time as demand flourished for books of geometrical curiosities. Private notebooks catalogue the intense copying critical to the transmission of polyhedral knowledge. Furniture displaying regular and irregular solids were exported across the continent, presenting the fruits of geometrical experimentation to wealthy patrons intent on collecting and displaying the new and wondrous. Though less common than the ubiquitous cubic dial, the polyhedral sundial, a rarified type of scientific
7 instrument made in wood, brass, or paper, matured in this period. As a case in point, Friedrich I, Duke of Württemberg, received a rhombicuboctahedron sundial in 1596 from Ludwig Hohenfeld decorated with opulently illustrated scenes from Ovid's *Metamorphoses*
8 on gold gilt paper. Polyhedra, then, did not serve solely pedagogical purposes nor were they opportunities for invention alone. Whether as workshop tools or *Kunstkammer* artifacts, polyhedra were always intended to decelerate the experience of seeing—used as much to

teach artisans to see with accuracy and to translate their visions onto paper with precision as to entrap viewers and patrons in highly condensed displays of otherworldly, geometrical skill.

If one intervention is to be singled out for its pivotal importance, it would have to be Dürer's late treatise on measurement, *Underweysung der Messung* (1525). A new baseline for the burgeoning contributions to geometrical practice by Renaissance artisans, Dürer's treatise demonstrated how the study of geometry within the context of the practice of art was necessary for the indoctrination of conceptual rigor and the precise utility of painterly technique. 9
To this end, *Underweysung* synthesized the internal structure and principles of *Elements* with select reference to other classical mathematicians (such as Ptolemy) alongside German and Italian sources, though Dürer's most radical maneuver, one that would profoundly shift the terrain of geometry, was the decomposition of the solids into "unfolded" polyhedral nets. Operating on the regular bodies, and inventing new irregular bodies to operate on, made possible a critical differentiation between the page-long, textual definitions of polyhedra in *Elements* and their pictorial representation. For Dürer and his followers, the solids became quantities that could be worked by hand and eye, responsive to the rules and tastes of the physical world through drawing and modeling.

Dürer's graphic-geometrical innovations provided license for *corpora* to morph into a range of untamable and to an extent unexpected polyhedral shapes resembling, without necessarily being, classic Platonic or Archimedean solids. At a certain undefinable mo- 10
ment, the fluid invention of new *corpora* gradually peeled off from regularity and threatened to outpace a mainstream geometrical discourse premised on structuring oppositions and translatability between solids, planes, and images. Disseminated through popular *Lehrbücher* often in part based upon *Underweysung*, open-ended and iterative geometrical experimentation flourished throughout artisanal workshops. *Irregulata* attended to the enhancement of the artistic imagination, further disseminating graphical techniques for experimenting with geometry and responding to the demand for practical training in geometry outside of the university. They were invariably represented as physical objects that borrowed the spatial characteristics of workshop models to substantiate the conceit that they

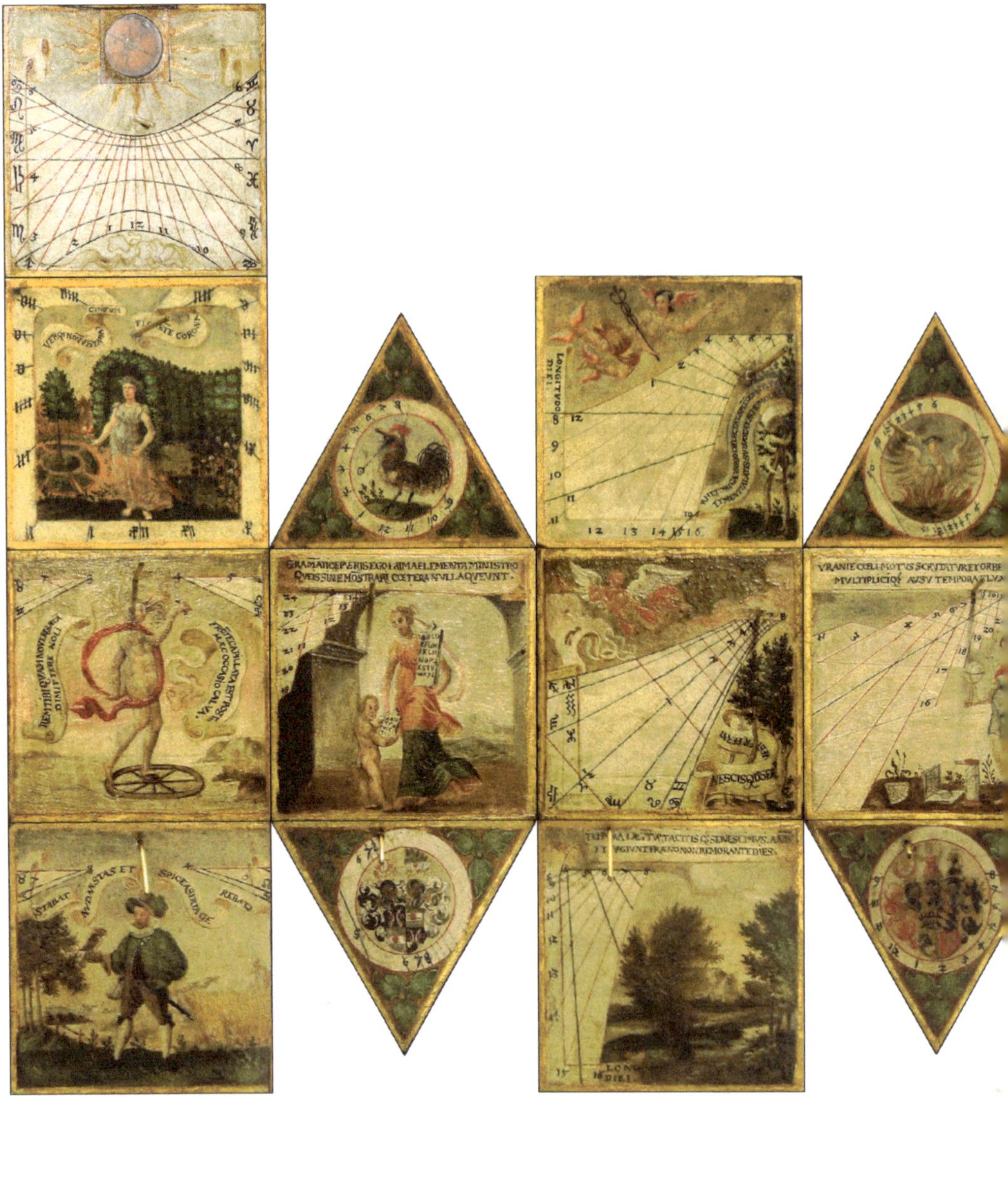

Ludwig von Hohenfeld (1576–1644). Unfolded polyhedral sundial from the collection of Duke Friedrich I. 1596. Landesmuseum Württemberg, Stuttgart (KK rosa 13). Photo: M. Paysan.

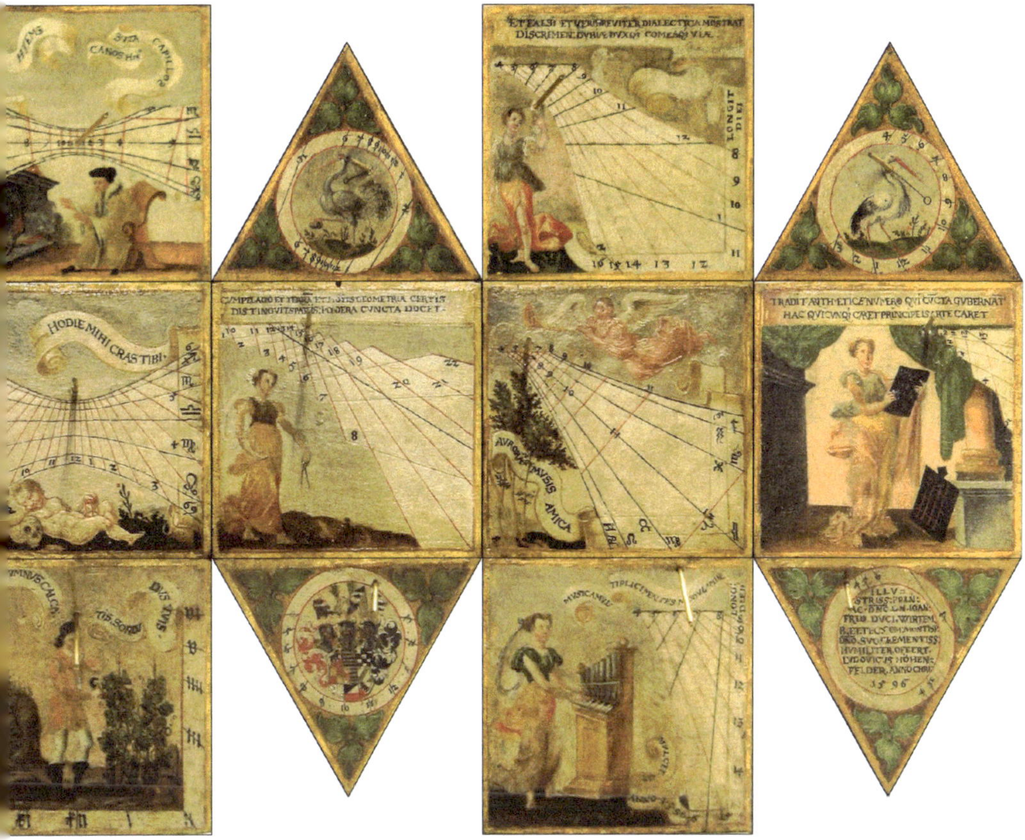

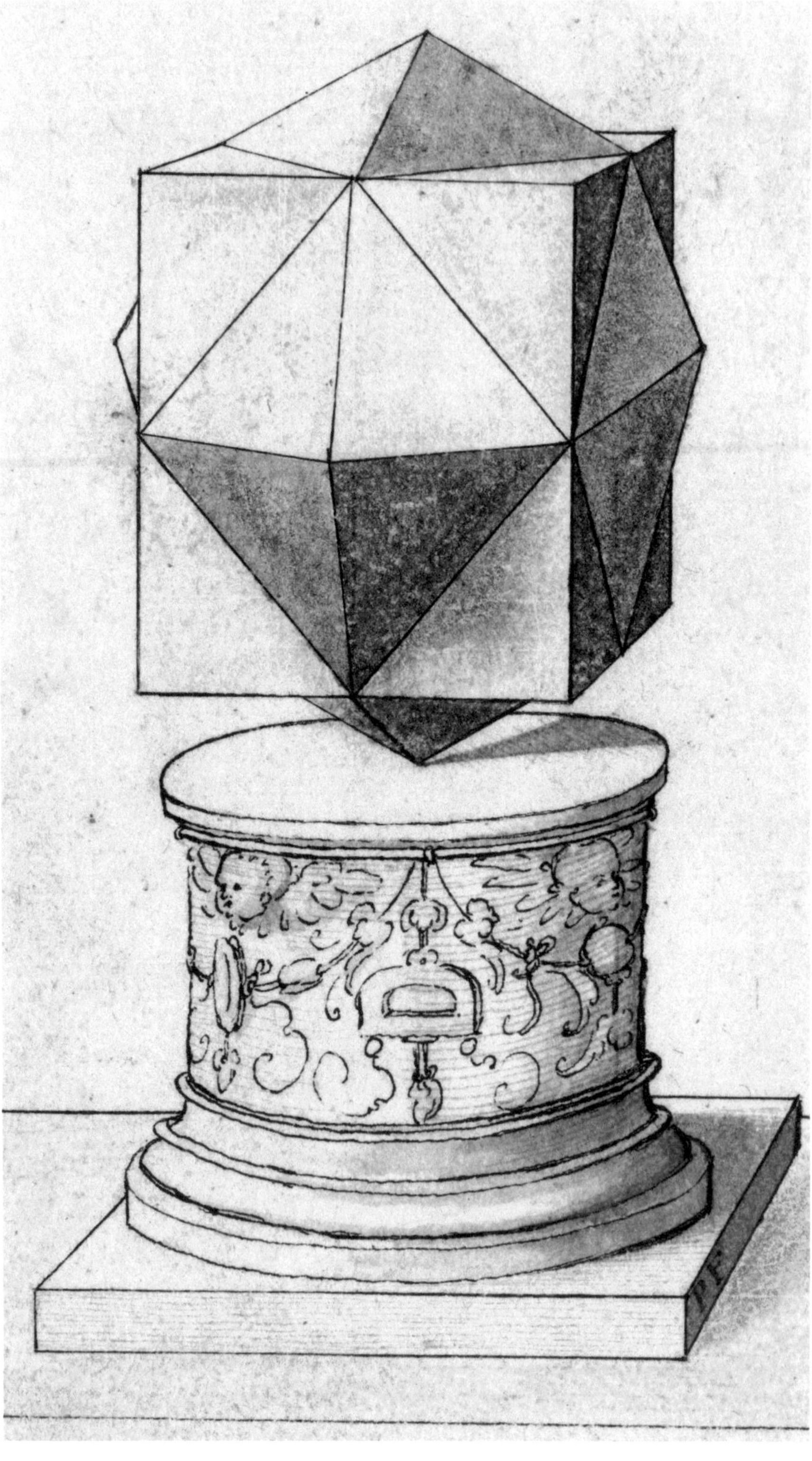

were images drawn after real objects; one of the most poignant early instances can be attributed to the Nuremberg artist Peter Flötner, who produced a small perspectival drawing of an irregular solid balanced atop an ornamental column base. The Flötner drawing heralded a synthetic universe that would refuse distinction between the real and the unreal, the built and the unbuildable. In its uncanny substitution of an alien body for the classical figure one might expect to surmount a podium, Flötner's juxtaposition unmasked geometry on the cusp of becoming a singular site of graphic testing inhabited by bodies like nothing that had been seen before. Teaching the precision measurement of *corpora* had delivered a new graphic space for the development of the early modern geometrical imagination.

Skill in representing polyhedra in three dimensions, having become central to communicating mastery of perspective for many artists, culminated in the epitome of the polyhedral print genre, *Perspectiva corporum regularium* (1568). A splendid array of printed polyhedra by the goldsmith Wenzel Jamnitzer, *Perspectiva* owed its graphic proficiency to the integral role of drawing, and the drawing of geometry, in the craft workshop, and attends to the sea change in the status of the irregular solids, which by the 1560s amounted to a catalogue of unbuildable paper projects steadily accumulating on the back of a generation of experiments by geometers. As the lavish peak of this collective work on polyhedra, in which the Platonic solids acted as a limiting condition on experimentation and polyhedral variations served as tests of innovation, *Perspectiva* solidified the fruits of artisanal research on geometry for new courtly audiences and opened up further media for the *irregulata* to inhabit.

Progressively estranged from Euclid and populated with irregular inventions born of the graphic techniques promulgated by the many perspectival *Lehrbücher*, the decorative arts proved to be fertile terrain upon which to seed and perfect a synthetic iconography of geometrical abundance hitherto restricted to pedagogical aids, drawings, and prints. In particular, German craftsmanship reached its peak in the sixteenth century, achieving levels of technical skill unparalleled in contemporary Europe. The imperial city (*Reichsstadt*) of Nuremberg became a world-renowned center of precision manufacturing, its guilds benefiting from, and exerting influence upon, the swirling artistic currents begot by the Reformation. By flooding the

Peter Flötner (1485–1546). Perspectival drawing of a column base with geometrical form. 1528. The Metropolitan Museum of Art, New York, Purchase, Jean A. Bonna Gift, 2007 (2007.223.5).

decorative art market with luxury items that could not be produced either by nature or more practically by guilds, crafts organizations, or cities bereft of skilled networks of collaborators and technological expertise, southern German cities like Nuremberg were able to respond to and invent new parameters for taste that were, for the most part, unattainable by competitors jockeying for the same courtly commissions.

The repertoire of culturally acceptable subject matter in the Reformation era expanded to feature polyhedral geometry as subject of contemplation and pleasure in the domestic interior. In opposition to the princely *studiolo* at the Palazzo Ducale in Urbino, where geometry was concretized as staid, still life objects in the style of the early *mazzocchio* (a disklike chaperon) of Paolo Uccello, the later solids were not spared the transformative pressures of fashion and commercialism that rewarded hyperembellishment. If, as classical archetypes, *corpora* had occupied a hybrid space straddling pure and applied mathematics, differing spaces and materialities could alter the valences of their signification. Thus, while workshop polyhedra in paper or wood slanted toward the representation of erudition made physical, polyhedra in Nuremberg's heady decorative art culture were adopted as mascots of the city's commitment to precision manufacturing. By 1608, when Hieronymus Braun (1566–1620), chancellery clerk of Nuremberg, presented to the city council his *Prospekt der Reichsstadt Nürnberg* (Prospect of the imperial city of Nuremberg), the first map of the city to accurately measure and record all of Nuremberg's existing buildings in three dimensions, he made sure to crown the map's dedication with two *corpora irregulata*.

The breadth of the polyhedral phenomena spanned media and geography, from the artist's workshop to the printing house, the cabinet to the court. It was not only because the nature of the bond to Euclid and the certainties of classical geometry had irrevocably shifted, though it was this too. Eccentric editions of *Elements*, such as Henry Billingsley's 1570 English translation, were printed with pop-up diagrams in an attempt to aid mathematical comprehension, containing DIY paper flaps that embedded the potential to perform the critical transition between the second and third dimensions. To learn how to represent a world of hidden geometries,

Page with interactive shapes, from *The Elements of Geometrie of the Most Auncient Philosopher Evclide of Megara*, trans. Henry Billingsley (London, 1570). Rare Book & Manuscript Library, Columbia University in the City of New York (Smith 513 1570 Eu2, fol. 314 recto).

ward narower and narower, at length ende their angles (at the heigth or toppe therof) in one point. So all their angles there ioyned together, make a ſolide angle. And for the better ſight thereof, I haue ſet here a figure wherby ye ſhall more eaſily conceiue it, the baſe of the figure is a triangle, namely, A B C, if on euery ſide of the triangle A B C, ye rayſe vp a triangle, as vpon the ſide A B, ye raiſe vp the triangle A F B, and vpon the ſide A C the triangle A F C, and vpon the ſide B C, the triangle B F C, and ſo bowing the triangles raiſed vp, that their toppes, namely, the pointes F meete and ioyne together in one point, ye ſhal eaſily and plainly ſee how theſe three ſuperficiall angles A F B B F C, C F A, ioyne and cloſe together, touching the one the other in the point F, and ſo make a ſolide angle.

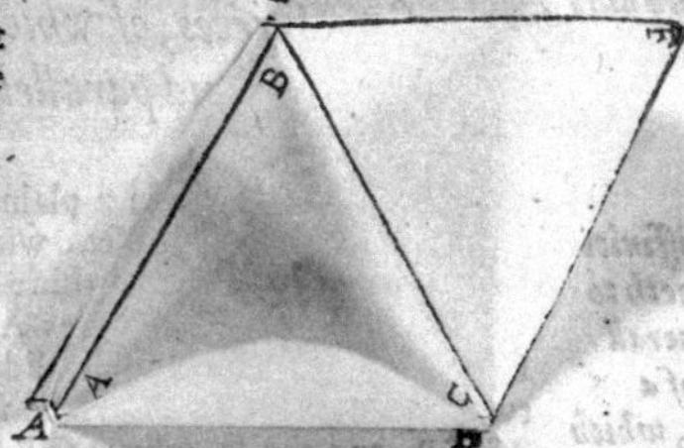

10 *A Pyramis is a ſolide figure contained vnder many playne ſuperficieces ſet vpon one playne ſuperficies, and gathered together to one point.* **Tenth diffinition.**

Two ſuperficieces rayſed vpon any ground can not make a Pyramis, forthat two ſuperficiall angles ioyned together in the toppe, cannot (as before is ſayd) make a ſolide angle. Wherfore whẽ thre, foure, fiue, or moe (how many ſoeuer) ſuperficieces are raiſed vp frõ one ſuperficies being the ground, or baſe, and euer aſcẽding diminiſh their breadth, till at the lẽgth all their angles cõcurre in one point, making there a ſolide angle: the ſolide incloſed, bounded, and terminated by theſe ſuperficieces is called a Pyramis, as ye ſee in a taper of foure ſides, and in a ſpire of a towre which containeth many ſides, either of which is a Pyramis.

And becauſe that all the ſuperficieces of euery Pyramis aſcend from one playne ſuperficies as from the baſe, and tende to one poynt, it muſt of neceſſitie come to paſſe, that all the ſuperficieces of a Pyramis are trianguler, except the baſe, which may be of any forme or figure except a circle. For if the baſe be a circle, then it aſcendeth not with ſides, or diuers ſuperficieces, but with one round ſuperficies, and hath not the name of a Pyramis, but is called (as hereafter ſhall appeare) a Cone.

Of Pyramids there are diuers kindes. For according to the varietie of the baſe is brought forth the varietie and diuerſitie of kindes of Pyramids. If the baſe of a Pyramis be a triangle, then is it called a triangled Pyramis. If the baſe be a figure of fower angles, it is called a quadrangled Pyramis. If the baſe be a Pentagon, then is it a Pentagonall or fiueangled Pyramis. And ſo forth according to the increaſe of the angles of the baſe infinitely. Although the figure of a Pyramis can not be well expreſſed in a playne ſuperficies, yet may ye ſufficiently conceaue of it both by the figure before ſet in the definition of a ſolide angle, and by the figure here ſet, if ye imagine the point A together with the lines A B, A C, and A D, to be eleuated on high. And yet that the reader may more clerely ſee the forme of a Pyramis, I haue here ſet two ſundry Pyramids which will appeare bodilike, if ye erecte the papers wherin are drawen the triangular ſides of eche Pyramis, in ſuch ſort that the pointes of the angles F of ech triangle may in euery Pyramis concurre in one point, and make a ſolide angle: one of which hath to his baſe a fower ſided figure, and the other a fiue ſided figure. The forme of a triangled Pyramis ye may before beholde in the example of a ſolide angle. And by theſe may ye conceaue of all other kindes of Pyramids.

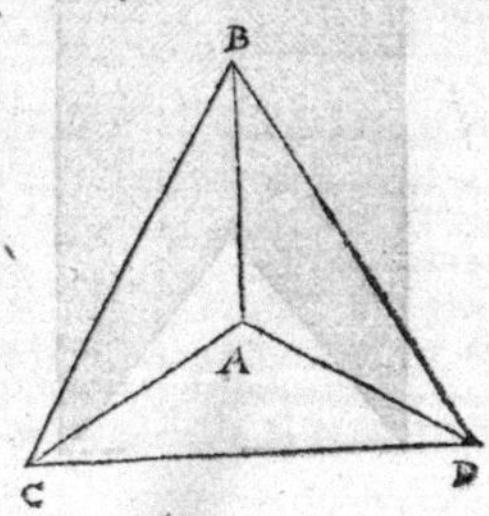

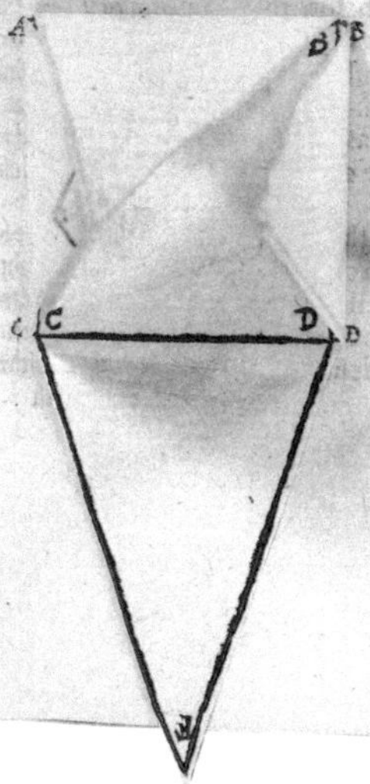

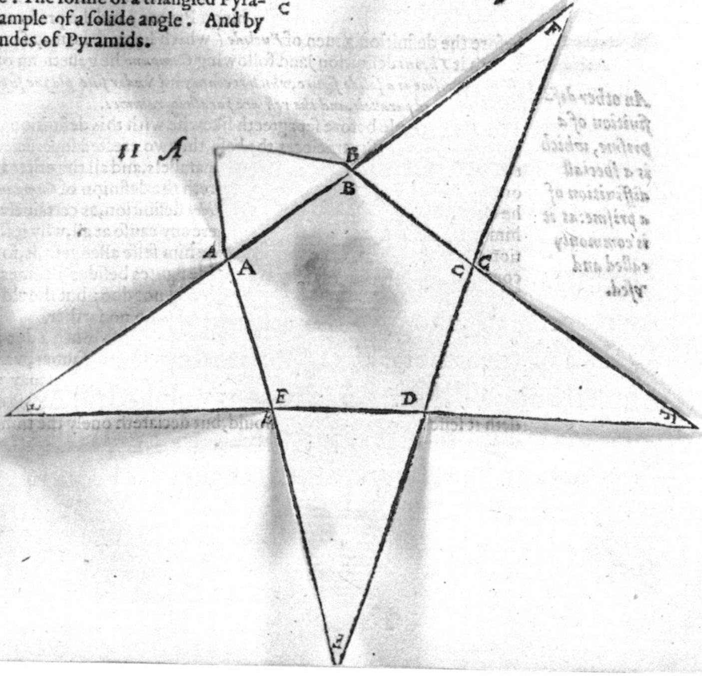

trajectories that could not be seen because they were too vast or too small, or because they were hidden in the contours of objects or the processes of conceiving of ideas, one needed the visible, the graphic, the scale of the hand, and materials that were close by and could be reused and discarded without consequence. The artist Augustin Hirschvogel lyricized the surface as charged with potentiality, a site "wherein some kinds of more beautiful bodies and spheres [lie] flattened, woven and locked together (much like the spider's web, who from her long thread weaves and constructs her dwelling enmeshed
11 together)." Through the maneuvering of these threads, the acceleration away from idealized forms was hastened; the physical and conceptual movements into and out of the third dimension gestured toward the incremental distortion of a Euclidean space increasingly under strain by its tensile connections to media. The irregularity of the *corpora*, both formally and at the level of their interaction with an imperfect materiality, fashioned a space for geometry evolving around and outward from unstable objects rather than through the hegemonic, spatial definitions of perspective.

Geometry in the sixteenth century, on the cusp of absorbing the symbolic language of algebra that would pave the way for higher-level mathematics in the centuries to come, rendering knowledge at the mathematical frontier increasingly incomprehensible to the educated layman, still owed an intimate and immediate debt to representation. It was credible to feel, as Wenzel Jamnitzer felt, that if pushed far enough *Perspectiva corporum regularium* had the potential to be at the forefront of geometrical research; not just that visualization was a vital component of his own public expression of artistic bravura but that it was also lacking from the expression and practice of mathematical knowledge. This was different from how a non-mathematician, an outsider, might instinctively subject an algebraic equation to a visual "reading" or interpretation, dwelling upon the enigmatic traces of human gesture scrawled in chalk upon a blackboard, an inscrutable code concealing years of labor behind an austere aesthetic rendered in black and white. Artistic research had leapt ahead of the scholarly, far surpassing the descriptive capacities of text through the stark newness of polyhedral inventions.

Though they reached the apex of both their visibility and importance in the late sixteenth century and would gradually come to

be seen as relics of classical mathematics in the centuries to come,
polyhedra today seem to have descended into the somewhat kitsch
realm of the trinket. Sold in museum stores, mused about on innu- 12
merable blogs, linked to on Pinterest pages, and available for purchase in various materials on Etsy or Amazon: all of these various manifestations, even if downscaled by their populism, speak to the solids as cyclical objects of desire, much as they once generated excitement in the early modern period.

Nevertheless, the casualness with which *corpora* are presented for cutting out and folding together in children's books or the ease with which corpora may be modeled on the computer, belies the radicality of their prehistory as the premier test subjects for a newly material engagement with geometry and geometrical knowledge. In the hands of artisans, the Platonic solids evolved into an intense field of experimentation, at once a new language of abstraction and a starting point for potentially limitless geometrical invention. Their transformation comprises the century-long journey of polyhedral geometry out from cloistered establishments of academic learning and their rebirth as covetable emblems communicating knowledge of geometry and perspective. In time, the exploration of the visual properties of the solids would diverge from reference to Euclid altogether and come to constitute an alternative method for the assembling of geometrical knowledge that spoke to the unique strengths and talents of a diverse array of European image and object makers. So considerable is the distance between the diagrams in early printed editions of Euclid's *Elements* and the numerous subsequent permutations of geometrical bodies that to account for their difference we must account for how classical geometry was unbound from text at the beginning of the century and distributed across Europe as irregularities by the century's end. To this end the present book is devoted.

COR

REGVL

IRREG

P O R A

A T A E T

V L A T A

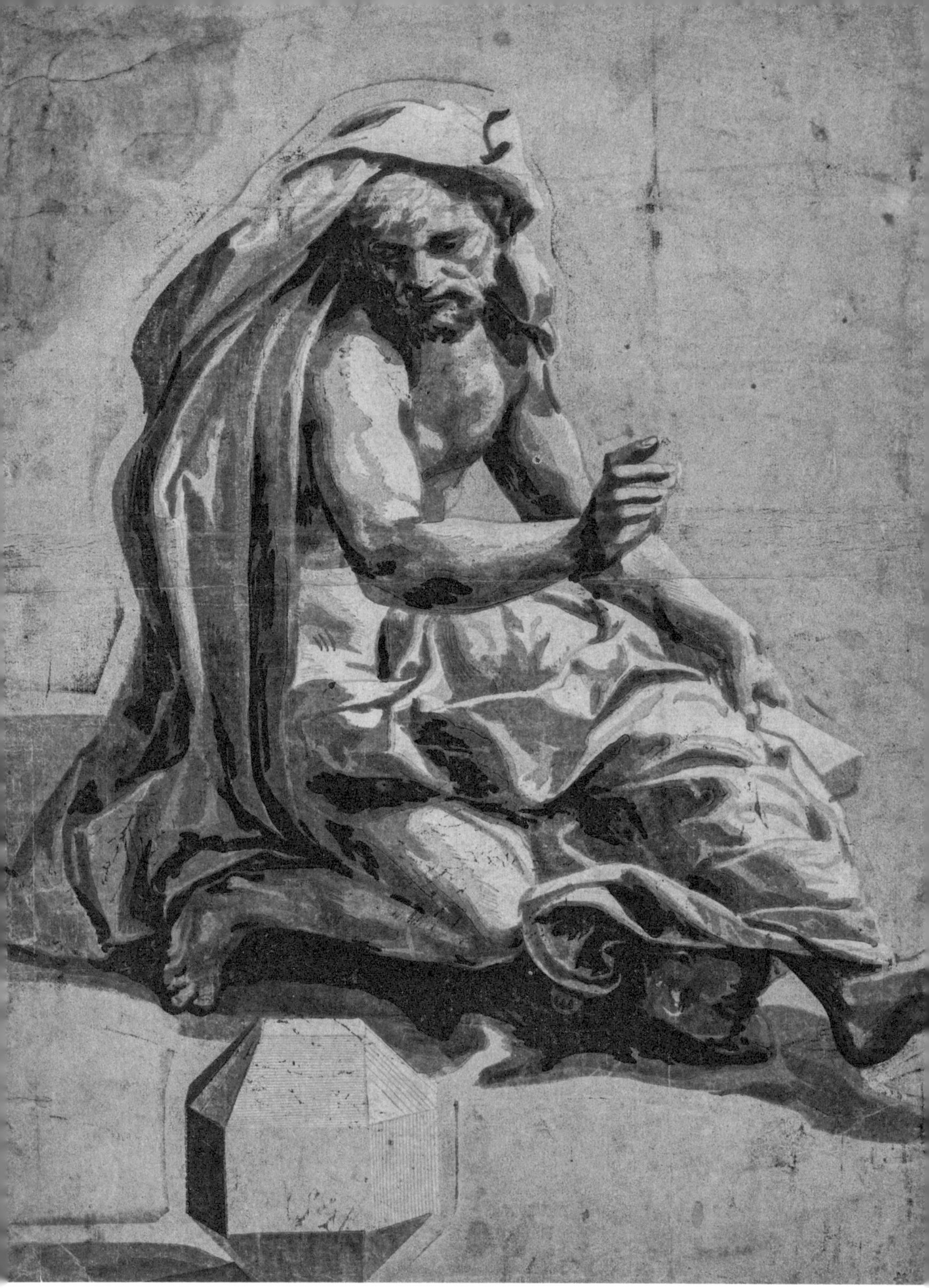

Ugo da Carpi (ca. 1480–1520/1532), after Parmigianino (1503–1540). *Archimedes* (?). Ca. 1518–1520. The Albertina Museum, Vienna (DG2002/524).

Archimedes hesitates, transfixed by the rhombicuboctahedron hovering on the edge of the page in the foreground, like a strange species never before encountered in nature. His body, rendered in layered chiaroscuro print, appears in a pose of almost vigorous, if haltingly arrested, contemplation. The polyhedron, its faces adorned with striated lines, loiters with inscrutable purpose. It has always been, and always will be, complete and outside taste, always already existent. In Ugo da Carpi's dramatic rendering, human and geometrical body are staged as a study in contrasts. At once rational, ordered, and finite, the outlines of its impression upon the page faintly visible around its outer edge, the rhombicuboctahedron is also out of this world. It stands both for itself and for the idea of itself, so much so that it appears to flicker in and out of focus, as if in excess of the real. Swathed in a cascade of wrinkled and tangled cloth still far beyond the capacity of mathematics to describe, the philosopher is captured on the cusp of apprehending the crux of knowledge made palpable as a discrete object. He grips a tabula rasa in his left hand, grappling with the decision of whether to divert his gaze in order to ensnare it, perhaps wondering whether in so doing he would run the risk of the polyhedron, and all it represented, vanishing beyond the reaches of his imagination.

Polyhedra are a spectral yet constant presence in the history of Western culture. Essential emblems of essence, they encompassed two major groups in the early modern period—the Platonic or regular solids (*corpora regulata*) and the Archimedean or semiregular solids (*corpora irregulata*), the endless potential forms derived by slicing up, truncating, or embellishing the regular solids. The rhombicuboctahedron in Ugo da Carpi's *Archimedes* is one such *corpus*. While the 1
corpora irregulata maintained a connection to their Platonic progenitors in that they were created by operating upon the *corpora regulata*, their number limited only by the imagination of their inventors, Renaissance interest in polyhedra stemmed from the Platonic solids' uniqueness. There have only ever existed five completely regular and symmetrical polyhedra. No more, no less. 2

To artists, mathematicians, and philosophers at the turn of the sixteenth century, the Platonic solids—the tetrahedron/pyramid (four triangular faces), the hexahedron/cube (six square faces), octahedron (eight triangular faces), dodecahedron (twelve pentag-

onal faces), and icosahedron (twenty triangular faces)—radiated a tantalizing promise of divine symmetry, order, and perfection that spanned from the building blocks of matter to the proportions of the human body and the structure of the universe. The maturation of perspectival geometry had only made the representation of the solids—sensible particulars through which to apprehend universal knowledge, at once everywhere and nowhere—a more pressing issue that verged into the realm of the existential. As the capacity to produce *disegni* (drawings/designs), as theorized by Giorgio Vasari (1511–1574) and others, resulted from years of artistic training required to fully transmit onto the page "an apparent expression and articulation of the conceit that one has in the mind," so the inability to image the solids threatened to signify an inability to imagine them in the first
3 place. Likewise to successfully represent polyhedra in three dimensions was to boldly venture beyond the gauze separating the world of appearances from the world of causes and to return with an artifact that could be tangibly appreciated. Imaging heretofore abstract, philosophical concepts had the potential to bring the solids into being in the only way they could be comprehended by any but the most advanced of mathematicians, if not in the sum totality of their hermetic mystery than through an image that stood for this totality.

While the history of art and architecture has traditionally couched the Renaissance in terms of achievement and genius, the history of science and mathematics has conjured an ulterior reality of clashing ideals and concepts, a period swamped in irregularity, only beginning to grapple with the measurement and definition of newly discovered lands, trading in rumors of monstrous foreign creatures, consumed by the verification of data sets and the purging of translation errors from newly printed, classical texts. The story of the early struggle to visualize the solids and their eventual transformation into a unique image typology collapses these historiographic vistas and disciplinary registers. Ciphers of Platonism steeped in Christological and humanist significations suddenly and in a matter of decades at the advent of the sixteenth century collided head on with perspective in all its novel complexity and contradiction. The force of this meeting would dislodge the solids from the text of *Elements*, Euclid's epoch-defining treatise on geometry, and reconstitute them as the quintessence of the third dimension in its most pure and unadulterated form.

The most popular, medieval version of *Elements* belonged to the mathematician Campanus of Novara (ca. 1220–1296), who had reworked earlier manuscripts of *Elements* some of which had been translated into Latin from Arabic in the twelfth century by Robert of Chester. Written and compiled around 1250, Campanus's *Elements* was reissued in 1482 by Erhard Ratdolt (1442–1528), a publisher from Augsburg working in Venice. The Ratdolt edition retains the distinction of being the first printed book of mathematics to lay out explanatory diagrams to accompany Euclid's technical terminology. Like the early publishers of Vitruvius's *De architectura*, which had been recovered in the fifteenth century as a manuscript without illustrations, Ratdolt surely would have recognized the marketing opportunity that adding images to a classical text would provide. But unlike Vitruvius, whose explicit references to architectural descriptions and drawings left ample room for artistic intervention, the Euclidean paradigm was premised upon the painstakingly detailed instructions necessary to precisely construct geometrical figures. All Ratdolt would have to do, it would seem, would be to direct a printmaker to follow these instructions to their expected conclusion.

Naturally the enterprise proved more challenging than Ratdolt had expected. In the preface and dedication to his patron, Doge Mocenigo of Venice, Ratdolt apologized for what apparently had been the book's long gestation and blamed the complexity of creating and integrating the more than four hundred woodblock prints of his 1482 edition. Not only did the images need to rise to the level of the source material in terms of quality and clarity, but the production of a precisely printed, straight line from a woodblock had revealed itself to be a deceptively difficult art, testifying to the complications of translating a geometrical drawing into print. While the engraved line can be executed with relative ease, using a burin to cut a line into a metal surface, the woodblock-printed line must be constructed and excavated mechanically, the wood incrementally excised from either side of an area until it becomes thin enough to produce a linear element.

Hence while the pyramid and the cube in Ratdolt's edition are depicted in legible axonometric (parallel projection) views, the octahedron, dodecahedron, and icosahedron struggle valiantly to be read despite the lack of linear hierarchy and the overlapping demonstration of their graphic construction. For the most part envisioned as the sort of diagrams recognizable in any modern mathematical textbook, Ratdolt's portrayal of the solids had run aground on the inhospitable shores of the third dimension. Three factors had converged to frustrate the solids' effective realization: the difficulty of comprehending the final books of Euclid, which fell far beyond the boundaries of the typical *quadrivium* curricula, the mechanical production of the woodblock prints based on a technology that was still in its infancy, and the conceptual difficulty of understanding what the geometrically composed third dimension would look like on the page.

The challenge of visualizing Euclidean principles and translating them into printed form was surely a key factor in the diagrams' subsequent graphic stability. Despite never attaining the status of true three-dimensional representations, the Ratdolt solids endured as shorthands for a nascent three-dimensionality and were recycled repeatedly in future editions of *Elements*. Among numerous examples are the Venetian humanist Bartolomeo Zamberti's (ca. 1473–1543) Latin translation of *Elements* from the original Greek, published as *Opera* (1505); an *Elements* from 1546 in Basel, *Euclidis Megarensis mathematici clarissimi Elementorum geometricorum*, replete with scaled-down versions of diagrams from the Ratdolt edition, though the diagrams are more interspersed within the text than in the 1482 original; and an *Elements* published in Venice in 1565, more than eighty years after Ratdolt's edition, that encloses identical copies of the Ratdolt dia-
4 grams. Luca Pacioli (1445–1517), the celebrated Tuscan mathematician, also relied upon the Ratdolt diagrams in his 1509 translation of *Elements* (likewise financed by Ratdolt) despite his ongoing relationship with Leonardo da Vinci, with whom he presumably collaborated on the polyhedral drawings to his manuscript *De divina*
5 *proportione* (ca. 1496–1499).

The early printed editions of *Elements* reveal a sustained period of almost slavish devotion to the 1482 edition, rather than concerted efforts to find new approaches to the visual interpretation of Euclid. Invention was limited to the relative size and placement of the Ratdolt diagrams within the text. As with other sets of "firsts" from the age of printing—the Swiss physician and naturalist Conrad Gessner's magisterial *Historia animalium* (1551–1558) comes to mind—these first diagrams of the Platonic solids likewise attained a measure

Diagram of a dodecahedron, from *Preclarissimus liber elementorum Euclidis perspicacissimi* (Venice, 1482), Book 13, Proposition 17. Boston Public Library (Q.402.22 folio).

cuius diameter.p.n.ideoq̄ȝ ⁊ spere cuius diameter.a.b. Latus aũt huius solide fi/gure dico esse lineã minorẽ.cõstat enĩ ꝙ linea. b.d .ẽ rõnalis in potentia cũ eius quadratũ sit subquincuplũ ad quadratũ linee.a.b.que posita ẽ rõnalis siue in lon gitudine siue in potentia tñ:itaq̄ȝ semidiameter atq̄ȝ semidiametri circuli.e. f.g.ẽ etiã rõnalis in potentia.nam eius semidiameter ẽ equalis.b.d. igiť ex.12.hui⁹ lat⁹ pẽtagoni equilateri huic circulo iscripti ẽ linea minor.at vero sicut i huius demon strationis processu patuit latus huius figure est quantũ latus pentagoni:ergo la/tus huius figure.20.alchaidaꝝ ẽ linea minor quẽadmodum proponitur.

Propositio .17.

Corpus duodecim basium pentagonoꝝ equilateraꝝ atq̄ȝ equiãgulariũ ab assignata spera circũscriptibile cõstitue re.eritq̄ȝ palã lat⁹ eiusdẽ corporis irrationale esse.id qd̃ residuũ diciť. ¶ Fiat cubus ſm ꝙ docet.14.huius circũscriptibile ab assignata spera:sitq̄ȝ huius cubi due superficies.a.b.⁊.a.c.ima/ginemur aũt nunc ꝙ.a.b.sit suprema superficies cubi ⁊.a.c. sit vna ex laterib⁹.sit q̄ȝ linea.a.d.cõmunis istis duabus superficiebus.diuidanť itaq̄ȝ in superficie.a.b duo opposita latera ꝑ equalia videlicet.d.b.⁊ latus ei oppositũ:⁊ puncta diuisio/nis cõtinuenť ꝑ lineã.c.f.latus quoq̄ȝ.a.d.⁊ illud qd̃ sibi opponiť in supficie.a.c. diuidanť per equalia ⁊ puncta diuisionis ꝯtinuenť linea recta cuius medietas sit g.h.sitq̄ȝ punctus.h.medius punct⁹ linee.a.d.similiter linea. e.f. diuidať ꝑ equa/lia in .k.⁊ protrahať.h.k.quãlibet igiť triũ lineaꝝ.e.k.k.f.⁊.g.h.diuide ſm pro/portionẽ.ha.me.⁊.du.ext.in trib⁹ pũctis.l.m.q.sintq̄ȝ maiores portiones eaꝝ.l. k.k.m.⁊.g.q.quas manifestũ ẽ eẽ eq̄les cũ tote linee diuise sint equales videlȝ que libet eaꝝ medietati lateris cubi.deinde a duobus punctis.l.⁊.m.erige perpendicu lares vt docet.12.vndecimi ad supficiẽ.a.b.quaꝝ vtrãq̄ȝ ponas equalẽ linee .k.l. sintq̄ȝ.l.n.⁊.m.p.similiter a puncto.q.erige perpendiculariter.q.r.ad superficieȝ a.c. quã ponas eq̄lẽ.g.q. ꝓtrahe itaq̄ȝ lineas.a.l.a.n.a.m.a.p.d.m.d.p.d.l.d.n a.r.a.q.d.r.d.q. Manifestũ est igitur ex quinta huius ꝙ due linee.k.e.⁊.e. l.po/tentialiter sunt triplũ ad lineã.k.l.ideoq̄ȝ etiã ad lineã.l.n.cũ.k.l.⁊.l. n.sint eq̄les At vero.k.e.ẽ equalis.e.a.igiť due linee.a.e.⁊.e.l.sunt potentia triplũ ad lineã.l. n.quare ex penuľ.primi.a.l.ẽ potentia tripla ad.l. n.ideoq̄ȝ per candẽ.a. n.ẽ po/tentia quadrupla ad.l.n. Cũq̄ȝ ois linea sit potentia quadrupla ad medietatẽ sui sequiť ex cõi scia ꝙ.a.n.sit dupla in longitudine ad.l.n. ⁊ q̄r.l.m.dupla est ad.l.k At.k.l.⁊.l.n.sunt equales:erit.a.n.equalis.l.m.sunt eni eaꝝ dimidia equalia .Et q̄r ex.33.primi.l.m.ẽ equalis.n.p.erit.a.n.equalis.n.p.eodẽ modo ꝓbabis tres lineas.p.d.d.r.⁊.r.a.esse equales sibi inuicẽ ⁊ duabus predictis .habem⁹ itaq̄ȝ ex his quinq̄ȝ lineis pentagonũ equilateꝝ qui ẽ.a.n.p.d.r.sed forsasse dices ipsũ nõ esse pentagonũ q̄r nõ nec forsan ẽ totus in supficie vna.qd̃ esset necessariũ ad hoc vt esset pentagonus .Qd̃ ergo sit tot⁹ in supficie vna sic habeto:producat equideȝ a puncto.k.linea.k.s.perpendicularis ad supficiẽ.a.b.que sit equalis. l.k. eritq̄ȝ ob hoc equalis vtriq̄ȝ duaꝝ.l.n.⁊.m.p.cũq̄ȝ ipsa sit equidistãs vtriq̄ȝ eaꝝ ex sexta vn decimi:ideoq̄ȝ cũ ambab⁹ in eadẽ supficie ex diõne lineaꝝ equidistantiũ necesse ẽ vt punct⁹.s.sit in linea.n.p.⁊ ꝙ diuidat eã ꝑ equalia: ꝓtrahanť igiť due linee.r.h. ⁊.h.s.sunt itaq̄ȝ duo trianguli.k.s.h.⁊.q.r.h.sup vnũ angulũ videlicet.k.h.q.cõ/stituti ⁊ ẽ proportio.k.h.ad.q.r.sicut.k.s.ad.q.h.nã vt.g.h.ad.q.r.sic.k.h.ad.q

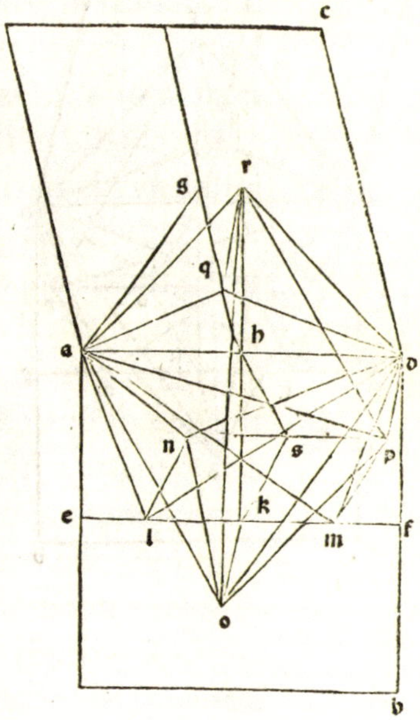

of scholarly irrefutability and *de rigueur* citing despite their pictorial
6 deficiencies. Much as the image of a rare or fantastical animal could enjoy a long afterlife in early modern book culture once committed to paper, so too were the Ratdolt solids repeated and aggregated without independent authentication.

Polyhedra had been experimented with before, most notably by Paolo Uccello (1397–1475) in *The Battle of San Romano* (ca. 1455) and his stellated dodecahedron in the marble floor of the basilica of San
7 Marco in Venice. But the first printed text to theorize their construction and manipulation in the science of perspective was Pacioli's influential print edition of *De divina proportione* (1509). Much of the theoretical apparatus in *De divina proportione* relied upon an unattributed Italian translation of Piero della Francesca's (ca. 1415–1492) treatise on the Platonic solids, *Libellus de quinque corporibus regularibus*, a fact that did not escape Vasari, who claimed that Pacioli had unscru-
8 pulously plagiarized Piero's manuscripts after his death.

Libellus was only one of several texts Piero had dedicated in part to the theme of polyhedral geometry. Unlike Euclid, Piero had designated numerical values for the solids and their sides in *Trattato d'abaco* (ca. 1450), and in *De prospectiva pingendi* (ca. 1482–1487) had taken on
9 the task of describing their perspectival construction. *Libellus* added a systematic arithmetical and geometrical exploration of the regular bodies and was among the first texts to subject truncations of the Platonic solids (*corpora irregularia*) to the analysis and measurement
10 of their sides and surfaces. The impression that the divine might be correspondingly located in the irregular lingers, if unsaid and unbeknownst to the authors. Although the first book in *De divina proportione* is dedicated to unfurling the namesake "divine proportion"—the "golden ratio"—it too details instructions on how to derive a poten-
11 tially unending number of *corpora irregulata* from the regular solids.

The first category to be explored was that of the semiregular or Archimedean solids, as per the designation of Pappus of Alexandria. Consisting of thirteen unique and uniform convex geometries composed by two or more regular polygons meeting in identical vertices, they fit neatly within the early modern, mathematical category of *corpora irregulata*, based as the Archimedean solids are upon cutting
12 solid angles from the five regular bodies. Though the term *irregulata* may be misleading to a modern reader, given that the Archimedean

solids are highly symmetrical entities, the pursuit of discovering and describing them constituted a subset of "irregular" geometrical experimentation. In addition to *De divina proportione*, exemplars of such solids often appear in books on geometry alongside geometrical constructs ill-fitting the definition of semiregular solids. Albrecht Dürer (1471–1528), Germany's most famous living artist at the turn of the sixteenth century, describes seven Archimedean solids that he deems "irregular bodies" (*ungeregulirten corporen*) in *Underweysung der Messung* (1525), his celebrated treatise on measurement, while Daniele Barbaro (1514–1570), translator of and commentator on Vitruvius, follows his discussion of the Platonic solids in *La pratica della perspettiva* (1569) with a section dedicated to the "description of the irregular bodies, which were born from the regular bodies [*Descrittione dei corpi irregulari, che nasceno dai corpi regulari*]." 13

De divina proportione was critical to the dissemination of an approach to learning three-dimensional geometry that used the Platonic solids as base experimental subjects. While the images would yield to future refinement, the book's prints ignited a broad interest in the visualizing of geometry. But *De divina proportione* also 14
reflected the perspectival glitches the solids proffered. While the two manuscripts in Milan and Geneva dazzle with vivid color drawings by Leonardo da Vinci or his circle, the woodblock-printed solids in the equivalent published edition were roughly and unconvincingly rendered with clumsy and imprecise crosshatching and awkward shading, whether as volumetric (*solidum*) or perforated (*vacuum*) bodies. The dashed lines representing indirect light on the leftmost face 15
of the *Octaedron planum solidum*, to take a representative specimen, do not read against the uneven distribution of shading across the solids' two darker faces. And the dark shading patterns printed on the solid's faces, meant to articulate the edges of the solid, in their seemingly haphazard orientations efface the very three-dimensionality the artist was trying to achieve. Similarly confused is the *Octaedron elevatum solidum*, where the orienting of contour lines conflicts with the edges of the solid. The creator of these inchoate images, operating with a limited repertoire of stock textures (dashed, solid, and blank), had evidently deployed his patterns with only the most basic knowledge of how to bring each solid to life, unable to capture the way shadows cast their presence upon a subject. In the interval between

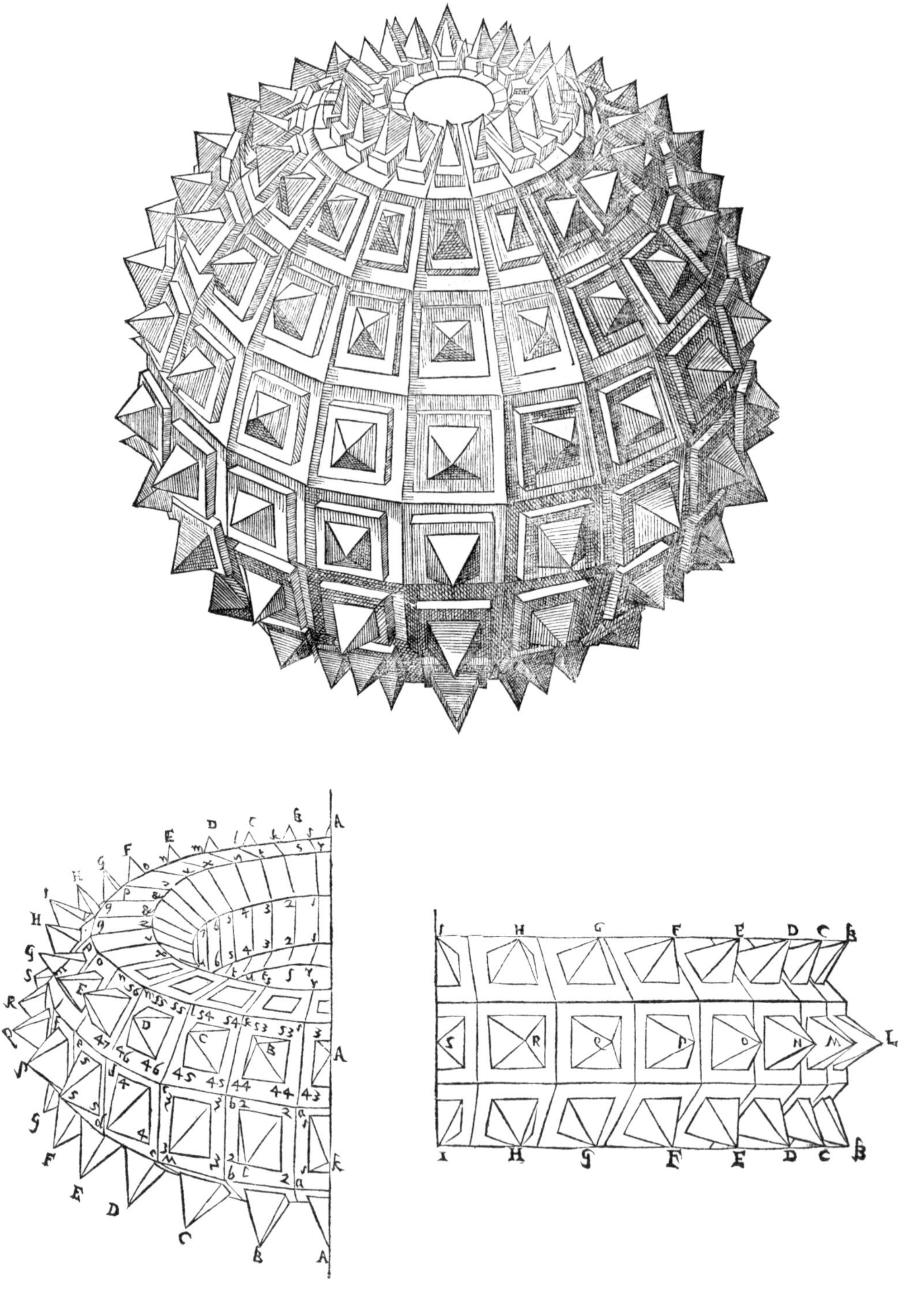

Corpora irregulata, from Daniele Barbaro, *La pratica della perspettiva* (Venice, 1569). Bibliothèque nationale de France, Paris, Département Arsenal (FOL-S-1290 [3], p. 128 and p. 162).

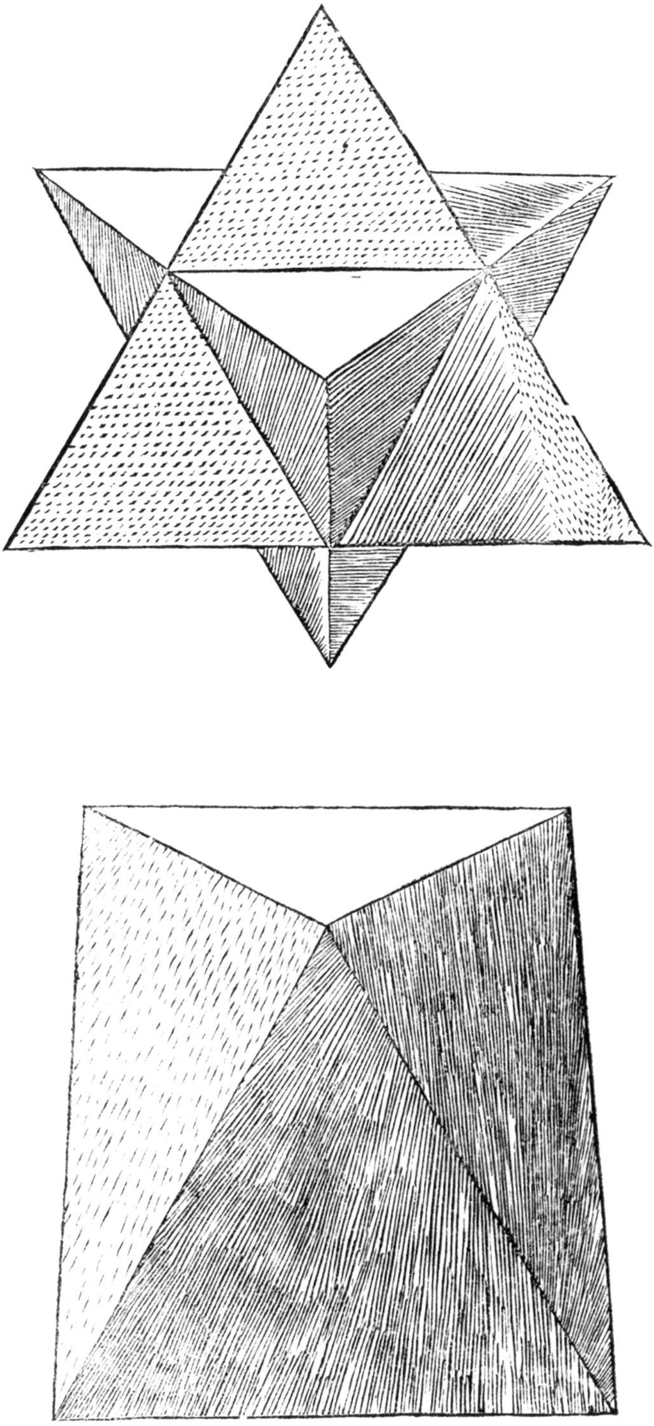

Octaedron planum solidum and *Octaedron eleuatum solidum*, from Luca Pacioli, *De divina proportione* (Venice, 1509), fols. xv and xix. History of Science Collections, University of Oklahoma Libraries.

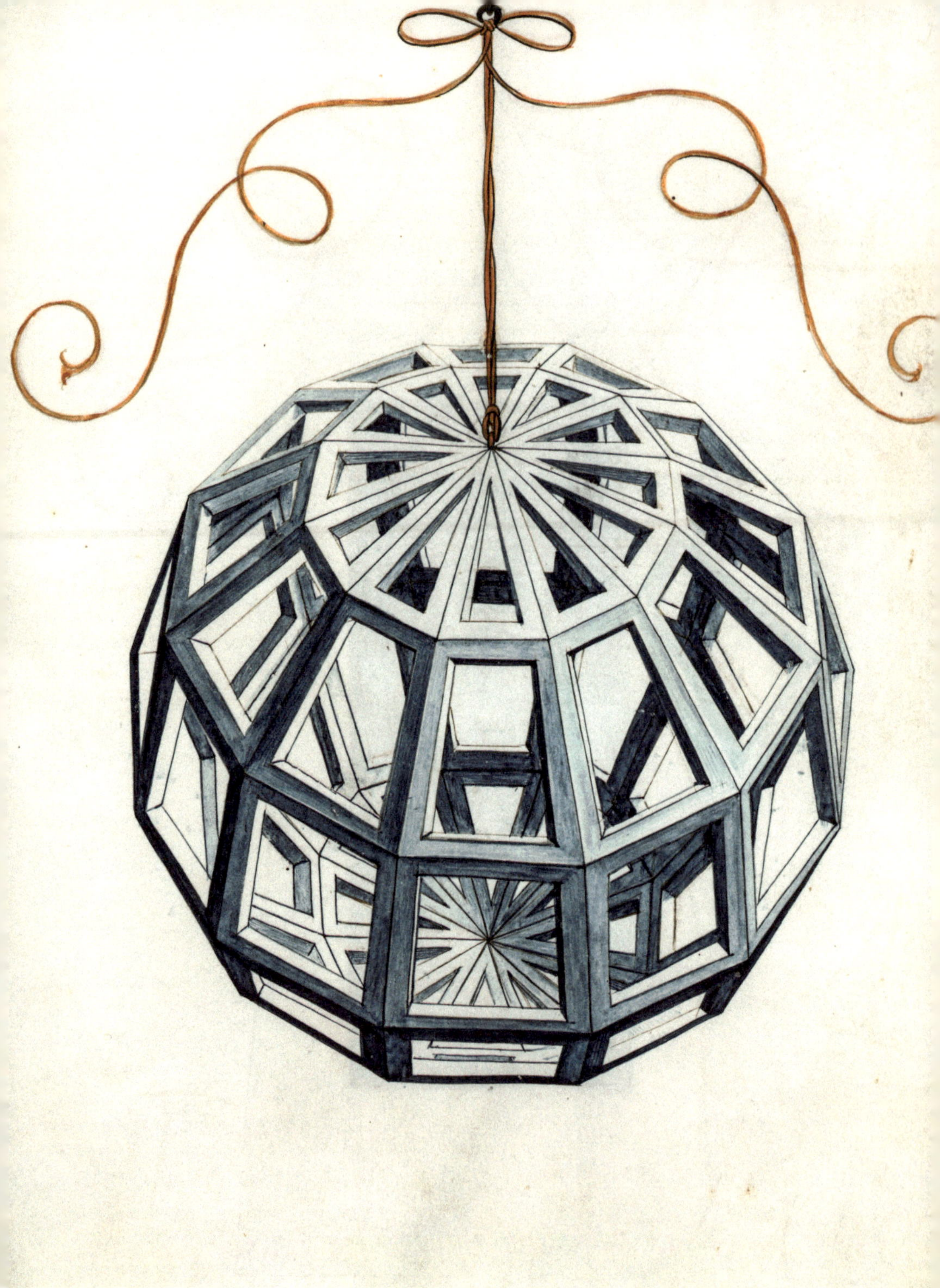

Leonardo da Vinci (1452–1519), attrib. Sphere with seventy-two sides [*Septvaginta dvarvm basivm vacvvm*], from Luca Pacioli, *De divina proportione* (manuscript completed in Milan, 1496–1498), fol. 111 recto, fig. XL. Veneranda Biblioteca Ambrosiana, Milan. Photo: ©Veneranda Biblioteca Ambrosiana/Mondadori Portfolio.

geometry and its image, the caprices of pictorial description were already divulging their indiscriminate impact on the transmission of mathematical knowledge.

Despite the voluminous literature on the history of perspective,
there has been little attention devoted to early modern polyhedra,
let alone to the excitement and the trepidation that the newly vi-
sualizable Platonic solids and their irregular cousins inspired after
De divina proportione. Hiding in plain sight within the frame of the 16
newly three-dimensional, pictorial gaze, polyhedra have mostly
been left off of history's stage, never quite aligning with our mod-
ern vision of the innovations entailed by the Renaissance and recast,
if at all, as geometrical frivolities from the inception of perspective.
And yet to do so is to ignore the central position polyhedra occupied
as both subject and problem for sixteenth-century artists and arti-
sans, for whom the studying of perspective through the represen-
tation of polyhedral bodies was *de rigueur.* In Italy alone, drawings
by Leonardo, Parmigianino, and Carpaccio, prints by Ugo da Carpi
and Caraglio, and intarsia in the dimly lit and sumptuously wood-
paneled *studiolo* of the Palazzo Ducale in Urbino by Giuliano da Maiano
(ca. 1432–1490), and in the monastery of Monte Olivetto Maggiore
near Siena and the church of Santa Maria in Organo, Verona—both by
Fra Giovanni da Verona (ca. 1457–1525)—hint at the prominence and
utility of polyhedra both as graphic studies and as physical models in
the workshop aiding the drawing of perspectival geometry. Studying
the precision measurement and construction of the solids was seen as
a necessary step in learning to access higher dimensions of knowledge,
exhibiting a tacit faith in the relationship between the mathematics
of accurate pictorial representation and true understanding of the
underlying mathematical rules governing the universe. The coinci- 17
dence of the golden section with the geometry of regular polygons
and polyhedra imbued polyhedra with a harmonic aesthetic princi-
ple present in the most perfect of human bodies and architectures,
and even in the capital letters of the Latin alphabet as enumerated by
Pacioli in the second book of *De divina proportione.* 18

The desire to depict substance, and substantiveness, was born out of the unique priorities of perspectival representation in the Italian Renaissance. But neither can the fraught solidity of the illustrations of polyhedra in *De divina proportione*, their attempts to be

seen as real objects on paper surfaces, not objects composed *of* surfaces with their own histories of making and unmaking, be fully attributed to the classical world. After all, Plato had insisted in the *Timaeus* that the five *corpora* were never, or at least were only temporarily, solid. Composed of triangular surfaces in flux that performed the state changes of matter, the solids were linked directly to the elements: fire with the tetrahedron, air with the octahedron, water with the icosahedron, the earth with the cube, and the universe or "the whole" with the dodecahedron. Elements dissolved, disintegrated, or evaporated into each other by shedding or gaining triangles. That said, there were some rules. The triangles of three of the solids—fire-tetrahedron, water-octahedron, and air-icosahedron—were held to be interchangeable, in that all three bodies could be created from equilateral component triangles. But the separate nature of the earth-cube and the universe-dodecahedron stemmed from their geometrical incompatibility, as neither could be created from
19 equilateral triangles alone.

> The second species of solid [octahedron] is formed out of the same triangles [as the pyramid], which unite as eight equilateral triangles and form one solid angle out of four plane angles, and out of six such angles the second body is completed. And the third body is made up of 120 triangular elements, forming twelve solid angles, each of them included in five plane equilateral triangles, having altogether twenty bases, each of which is an equilateral triangle. ...The isosceles triangle produced the fourth elementary figure, which is compounded of four such triangles, joining their right angles in a centre, and forming an equilateral quadrangle. Six of these united form eight solid angles, each of which is made by the combination of three plane right angles; the figure of the body thus composed is a cube, having six plane quadrangular
> 20 equilateral bases.

Once the triangles were combined into the faces of each of the regular bodies, the bodies gained elemental affinities based upon their formal characteristics. The variability within each of the four elements was due to differences in the size of the elementary triangles that
21 composed the sides of the bodies. As Marsilio Ficino (1433–1499),

philosopher and head of the Medici-sanctioned *Accademia platonica* in Florence, summarized:

> And he [Plato] considers the pyramidal shape to harmonize with fire because it is slender and better at cutting than the others are, being made of fewer triangles and therefore being the lightest of all shapes. The cube, by contrast, harmonizes with earth, being very solid and stable. The remaining shapes harmonize with the intermediate elements, for they stand at points midway between fire and earth in their relationship to motion and stillness. The dodecahedron, which has twelve faces, harmonizes with the cosmos, in his view, on account of the twelve spheres and the twelve signs of the Zodiac. 22

Though Plato did not specify the mechanics of how the triangles would assemble and reassemble themselves, he offered a geometrical explanation for the transformation of one element into another, or the mixing of two or more elements, disassembled back into their triangular surfaces. When "greater bodies are broken up, many small 23
bodies will spring up out of them and take their own proper figures; or, again, when many small bodies are dissolved into their triangles, if they become one, they will form one large mass of another kind." 24

The high visibility of polyhedra as the premier mathematical artifacts of antiquity, brought into broader cultural consciousness through the use of perspective to represent and experiment upon the Platonic solids, benefited from a resurgence of interest in the *Timaeus*. While most medieval philosophers had to work from Calcidius's Latin version (ca. 321 AD), the edition of the *Timaeus* used by humanist scholars was the new Latin translation by Ficino, which he had translated from the original Greek in its entirety for the first time. Plato 25
had been in the midst of unveiling the geometrical constitution of the four elements at the conclusion of Calcidius's abridged *Timaeus*. But the special significance the regular bodies held for Plato's elemental theory would have to wait for Ficino, who detailed Plato's geometrically modulated natural philosophy in which all "natural phenomena are based on the principles of mathematics." 26

Ficino's *Timaeus* delineated geometrical theories of the elements just as Italian geometrical-perspectival culture was coming

27 into its own. The emergence of geometry as a new if invisible force structuring the constitution of matter seemed destined to set up a clash with the sensorial imperatives of Aristotle, who in *De caelo* (350 BC) had claimed that Plato's Tetris-like accounts were ill-supported by nature's behavior and could not be borne out by empirical evidence. "The thinkers that put forth this view turn out to suggest a theory about the phenomena which, though, conflicts with the phenomena. …The advocates of the reduction to planes are so attached to their preconceived notions that they behave like discus-
28 sants defending a thesis." Nevertheless, by describing the elements as composite geometries, Plato had implied the possibility of visually representing the elements *as* geometries, or rather of drawing the component parts of the elements scaled to the size of the paper and the human hand. Renaissance artists and mathematicians exploited this angle, in essence reconstituting geometry within the realm of sensorial experience. To these early geometers, as Wenzel Jamnitzer would elaborate in *Perspectiva corporum regularium*, the augmentation and graphic combination of the solids, associated as they were with Plato and Euclid, simulated on paper the irregular and unexpected creations produced by combining the base forms of matter.

The recovered tactility and constructability of Plato's *corpora* contributed to the solids becoming sought-after subjects of pan-European perspectival culture, and imbued with prestige the practical investment in learning to construct and deconstruct polyhedral geometry. Creative investment in the visualization of polyhedral geometry spilled out from Italy, and from the early years of the sixteenth century crossed the Alps as part of the fruitful exchanges of skills and styles, motifs and prints between Italy and Germany. Jacopo de' Barbari (ca. 1460/70–before 1516), the enigmatic Italian painter and likely creator of the famous *Portrait of Luca Pacioli* (ca. 1500) with a glass rhombicuboctahedron, first moved to Nuremberg to work for Emperor Maximilian I and then stayed on in Germany in the employ of Elector Friedrich III of Saxony and Elector Joachim of Brandenburg. Albrecht Dürer, no stranger to Italian journeys, devoted himself to copying interlaced geometrical designs of knots after Leonardo (or from the Academia Leonardi Vinci), exchanged drawings with Raphael, and traveled to Venice to protest the copies
29 of his own work by Marcantonio Raimondi. And Daniele Barbaro,

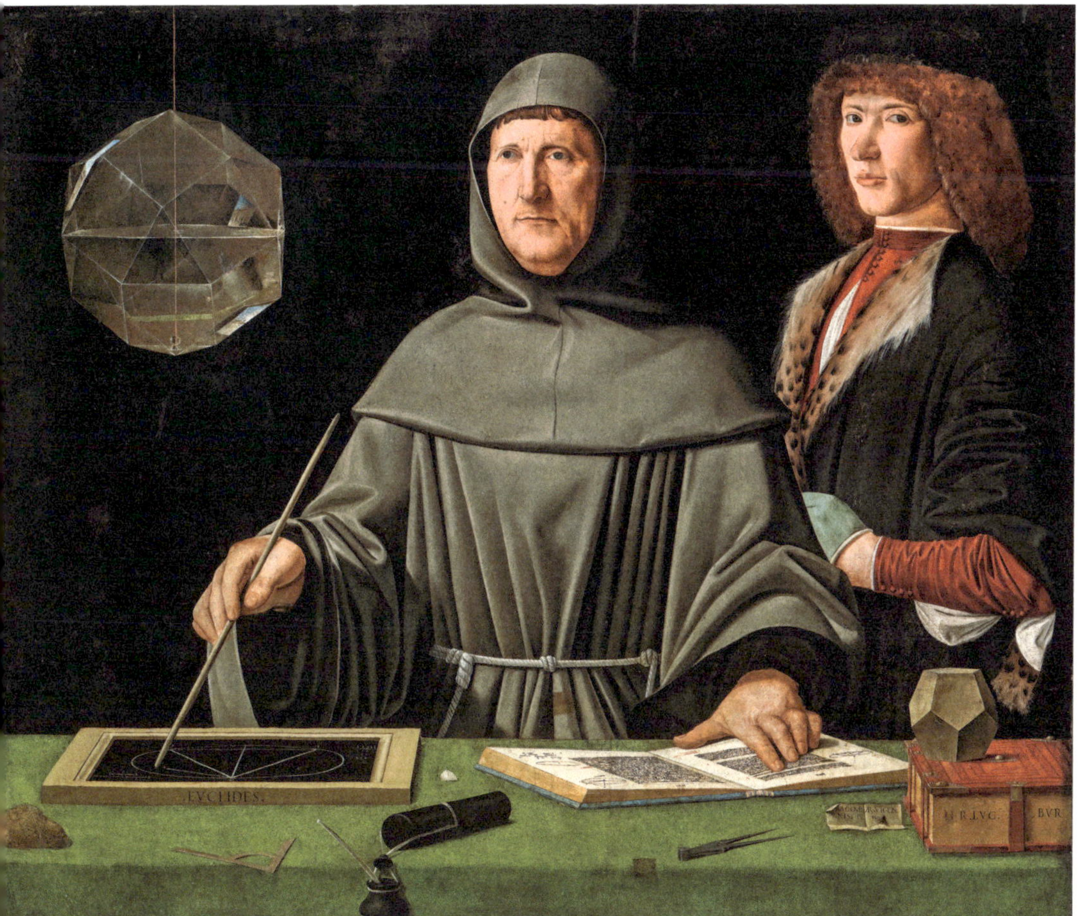

Jacopo de' Barbari (ca. 1460/1470–1516). *Portrait of Luca Pacioli.* Ca. 1500. Museo di Capodimonte, Naples, Italy (Q.58 [I.C.80 Jaco-Bar]). Su concessione del Ministero per i Beni e le Attività Culturali e per il Turismo—Museo e Real Bosco di Capodimonte.

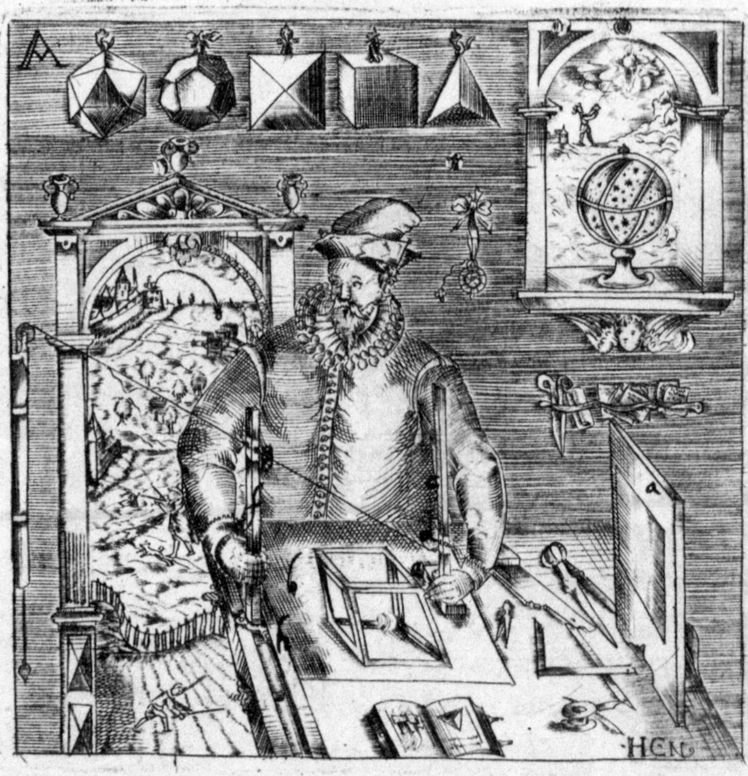

Cubus. N.B. wann der ligende Grund vff das ligende Täffelin gemacht wirdt / so kompt das Corpus vff den Tisch darunder / vñ hinwiderumb / ꝛc. Gebrauch zur Perspectiv.

Den vsum hab ich auff mancherley weiß erlangt / den gemeinesten modum will ich dir durch einen Cubum zu verstehen geben. Reiß den ligenden vnd auffrechten Grund deß Cubi auff Papier / das eine Papier mache mit Wachs auff ein auffrechtes Täffelein / das ander auff den Tisch fest. Darüber lege ein ander Täffelein / so an einer Spitzen vff den Tisch fein gangbar gehefftet werde / vnd kleybe auch ein Papier darauff / auff welches das Corpus kommen soll. Henck hernach die Seyten / welche den Augpuncten repræsentirt / nach Gelegenheit deines Wercks / wie in der Figur zu sehen / vnnd neme mit der hindern Perpendicular Regul / daran die Seyt gehefftet / durch den beweglichen Stefft die Höhe der vndern Lini / am auffrechten Grundt / vnd setz den vndern vnbeweglichen Stefft auff einen Puncten deß ligenden Grunds / vnd rück die vordern Regul mit jhrem Stefft an die Seyten / laß auch die Regul vnverrückt stehen / als dann

borrowing from Dürer's *Underweysung der Messung*, retransmitted Dürer's techniques for manipulating geometry back into Italian in his *La pratica della perspettiva*. What had begun as a joint Italian/ 30
German enterprise in the Ratdolt *Elements* and had given rise to the graphic revival and perspectival rebirth of the Platonic solids in Italy had regained its transnational mirroring and intermediality.

Whereas Euclid's instructions for the three-dimensional construction of the Platonic solids required deciphering several pages of highly technical Latin, within the visual realm, a realm newly essential to the pursuit of geometrical knowledge, significant contributions could be made by geometers working in and out of measured perspective. On the heels of the success of *De divina proportione*, physical models of Platonic solids were adopted as commonplace objects in the Renaissance *studiolo*, particularly for the training of abilities in geometrically modulated perspective, and were consequently reabsorbed into prints, painting, and drawing. Their casual ubiquity is preserved in various media and passing mention in texts, though few physical models or material specifications from the period remain. For instance, in the foreword to his *Geometriae practicae novae et auctae tractatus* (1641), the Nuremberg mathematician and professor of oriental languages Daniel Schwenter (1585–1636) posthumously confessed to having created Platonic solids (*der funff corporum regularium*) in his youth from paper, wood, and stone. The 1640 inventory in the 31
kurfürstlich-sächsische Kunstkammer in the Grünes Gewölbe, Dresden, is one of few instances to specify the type of paper used to construct polyhedral models, which are listed as being made "von türckischen papir," *turkisch* being a blanket adjective used to describe goods thought to originate from the Middle East. 32

More explicitly, thc Belgian author Samuel Quiccheberg (1529–1567) enumerated that polyhedra could be used as tools and produced as artworks themselves. *Inscriptiones vel tituli theatri amplissimi* (1565) (Inscriptions or titles of the most ample theater), a disputation on the organization of an ideal *Kunst-* and *Wunderkammer*, described a utopic space of making and learning that would house "mathematical instruments" together with "instruments for workshops and laboratories used by the more skilled artisans, such as the tools of sculptors, turners, goldsmiths, foundry workers, woodworkers, and indeed of all artisans whom this world supports in our age." 33

Here, he sited “regular solids of various shapes, beautifully constructed of transparent rods” within the same space as “mathematical instruments, such as astrolabes, spheres, cylinders, quadrants, clocks, geometric rods, and other objects to be used in measuring on land
34 and sea, in war and peace.”

Visual evidence suggests that in the first decades of the century, when the solids were still utilized primarily as teaching aids for Euclid, polyhedral models resonated in triangulated relation with the drawing surface and with *Elements*, or occasionally with another ancient text such as Ptolemy’s *Almagest* or Vitruvius’s *De architectura*. As the first of several examples, Vittore Carpaccio’s (1460–1520) sketch of a scholar at work displays several polyhedral objects hanging from strings above a workspace, possibly astrolabes, armillary spheres, or indeed models of the Platonic solids. The scholar holds a book with his left hand and a compass in his right, presumably gazing through the open window upon the celestial scene he is measuring. *Newe geometrische vnd perspektiuische Inuentiones* (1610) by Johannes Faulhaber (1580–1635), a mathematician from Ulm and associate of René Descartes, includes a similar image in which the Platonic solids are clearly depicted as tangible objects hanging from hooks above the door of a workshop where a man is pictured setting out a per-
35 spectival drawing of a cube. Faulhaber situates the solids as integral components of a network of measurement devices encompassing rulers, astrolabes, and compasses, suspending them adjacent to a Düreresque perspectival apparatus, drafting surface, and reference book opened to a page with a perspectival pyramid on one side and a generic architectural scene on the other.

The engraving of the philosopher Diogenes by Giovanni Jacopo Caraglio (ca. 1500/1505–1565) after a drawing by Parmigianino (1503–1540) takes the medial embeddedness one additional step. Diogenes is pictured pointing toward a reproduction of a dodecahedron from *De divina proportione* while consulting an open book whose pages are concealed from view. *De divina proportione* was the only available book in print that represented the Platonic solids side by side as *solidum* and *vacuum* in three dimensions (the corner of a skeletal dodecahedron appears just adjacent and out of the frame of the print), and the dodecahedron would have been recognized by knowledgeable contemporary audiences as a direct reference to Pacioli and indeed Plato.

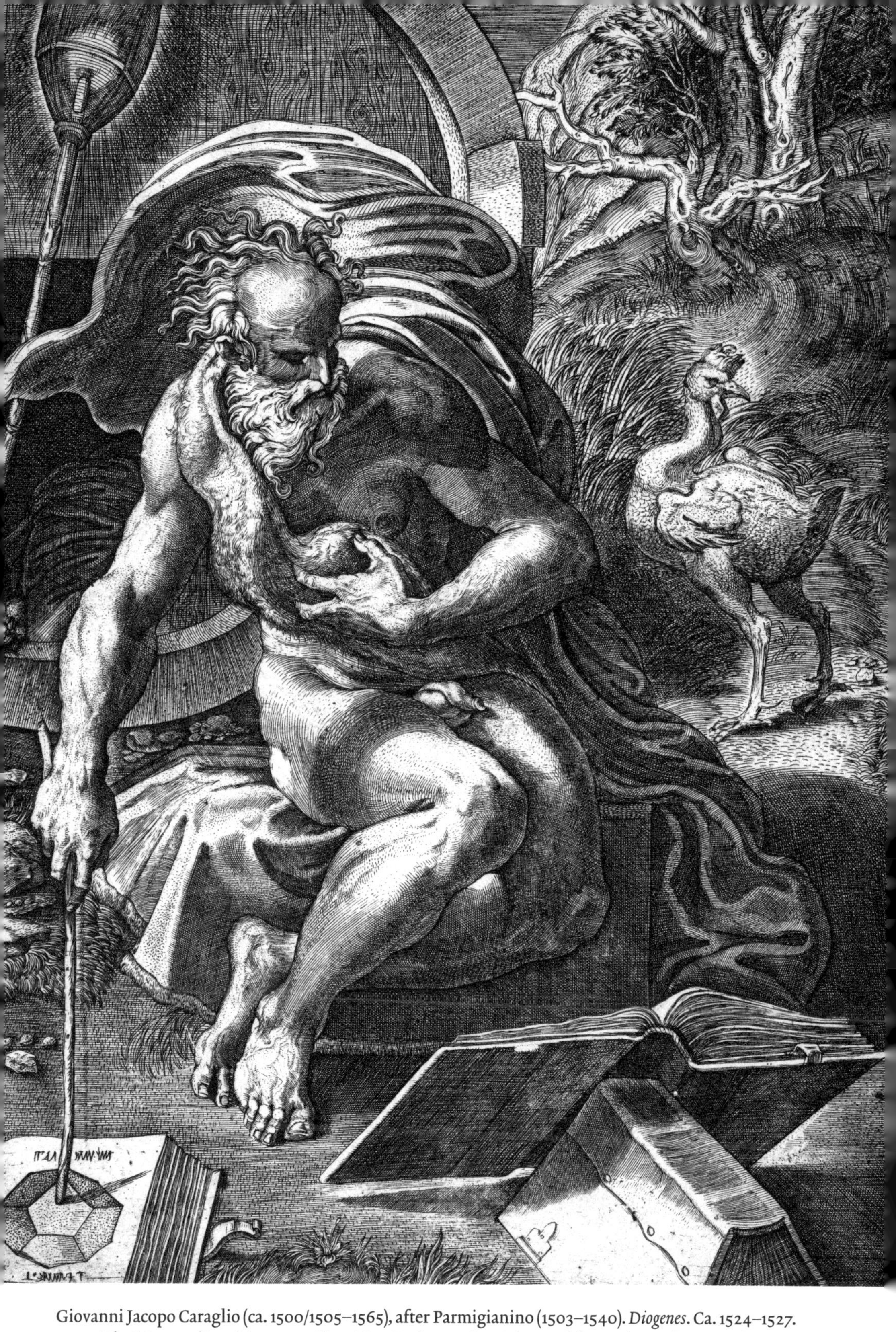

Giovanni Jacopo Caraglio (ca. 1500/1505–1565), after Parmigianino (1503–1540). *Diogenes*. Ca. 1524–1527. The Metropolitan Museum of Art, New York, Harris Brisbane Dick Fund, 1917 (17.3.3416).

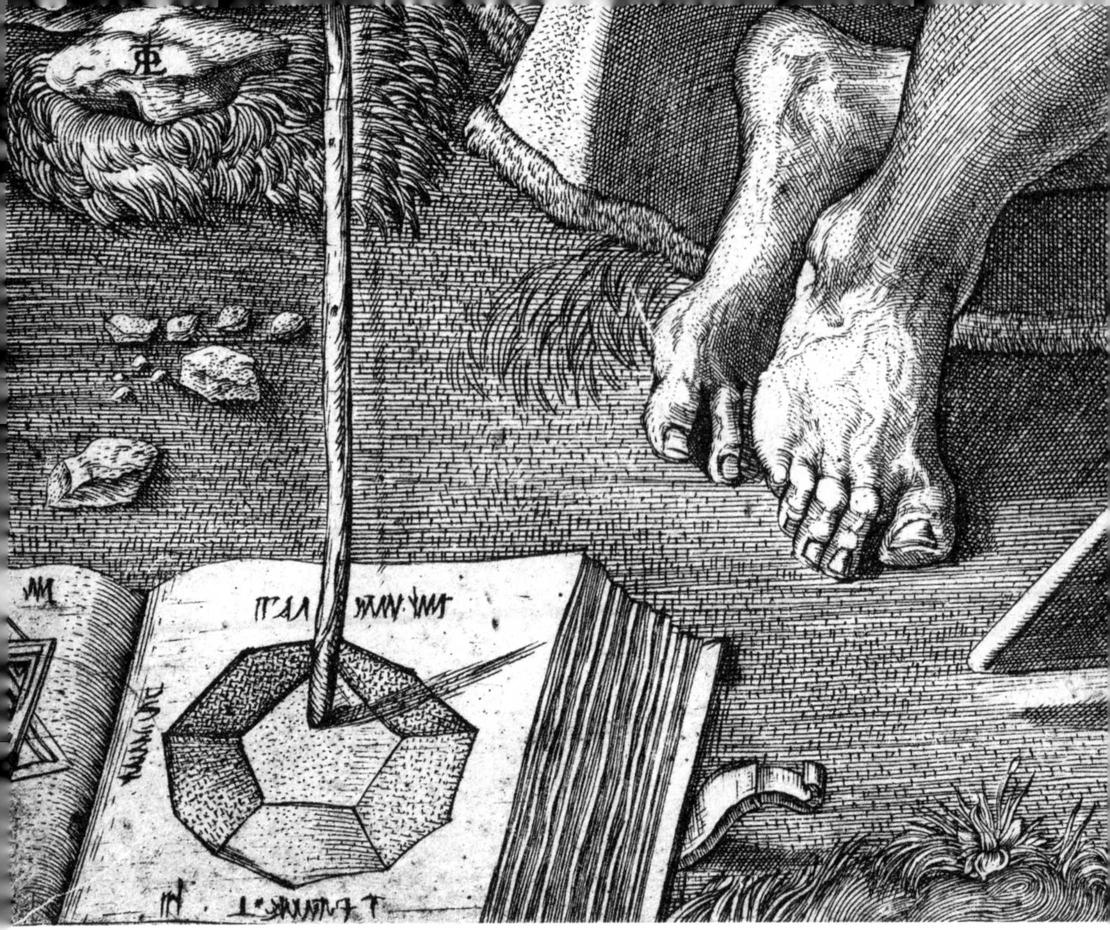

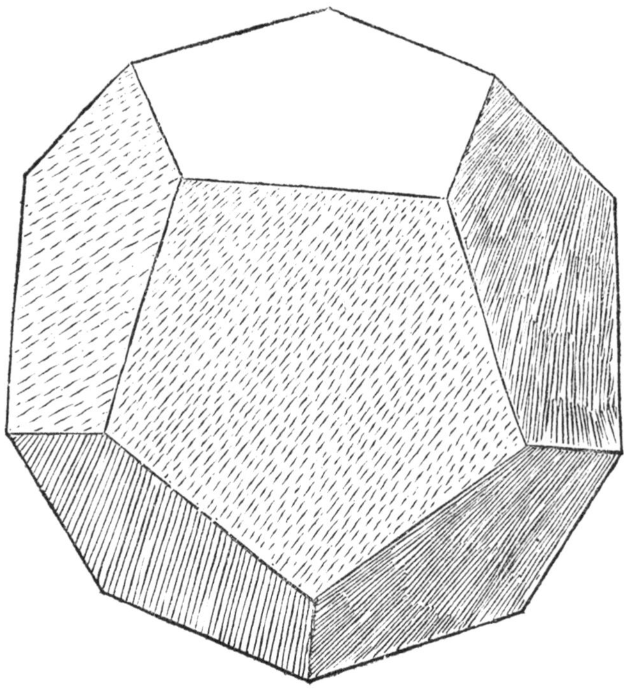

That Caraglio and Parmigianino referred to a real edition of *De divina proportione* is substantiated by the matching segments of diagrammatic lines that frame the dodecahedron. These correspond in location to the four sets of text on each of the Pacioli pages, specifically the dodecahedron entitled in Greek, Latin, and Italian and the page number on the upper righthand corner of the page. It is not unlikely that the second book consulted by Diogenes is a copy of Euclid's *Elements*. The relative sizes of the physical books available to Caraglio at the time seem to substantiate this theory. The 1482 edition of *Elements* published by Ratdolt was 32 cm (12.6 in) in height and the other editions popular in the 1520s, Pacioli's 1509 *Elements* and Zamberti's 1505 *Opera* by Euclid, were both an analogous 30 cm (11.8 in) high. By comparison, *De divina proportione* was a slightly smaller 27 cm (10.6 in), a size differential borne out by the relative scale of the books depicted in Caraglio's *Diogenes*.

Though Diogenes himself had a harsh opinion of mathematicians, he had written a biography of Plato in his *Lives and Opinions of Eminent Philosophers* that was commonly reproduced at the beginning of sixteenth-century editions of Ficino's *Opera*. It could well be that Caraglio's print references Diogenes as a biographer of Plato rather than any connection between Diogenes' philosophy and
mathematics or geometry. Still, Caraglio conscripts Diogenes into 36
a scene that mixes and matches antique references with the most up-to-date means of studying geometry. Diogenes himself gestures with a stick to the center of the dodecahedron from *De divina proportione* propped open in front of him, as if directly confirming the relationship between the text and its representation in perspective. Inadvertently perhaps, Caraglio's dramatization discloses the inadequacy of the crop of newly illustrated *Elements* populated with Ratdolt's diagrams, showing that more fully realized perspectival images, such as those by Pacioli, were critical for fully grasping the
reality of the solids. 37

Wielded, brandished, held limply or with vigor, the trope of the pointed stick (or compass, a "stick" with a double point) introduced didacticism into portrayals of mathematical learning. Both literal and metaphorical, the stick hints at the divide between culture and nature, the need for guidance, discipline, and punishment. In the spatiality it engenders, the viewer is conscripted into a

Caraglio's *Diogenes* (detail), and *Dodecaedron planum solidum*, from Luca Pacioli, *De divina proportione* (Venice, 1509), fol. xxvii. History of Science Colls., University of Oklahoma Libraries.

performance of knowledge, becoming witness to the act of transmission. Two stick-bearing masterpiece paintings, one in Nuremberg and the other in Naples, further explicate the pragmatics of how polyhedra were used as complementary participants in the acquisition of knowledge, though the precise nature of this knowledge, its shimmering relation to self and the textures of allegory, fluctuates between the two scenes. In de' Barbari's *Portrait of Luca Pacioli*, Pacioli, resplendent in the garb of the Franciscan order, stares intently at a suspended glass rhombicuboctahedron, half-filled with water to dramatize the volume of its container and de' Barbari's painterly skill in perspective. Each of Pacioli's hands unseeingly traces the contours of the same information, a chalk drawing of a tetrahedron inscribed in a circle on his right and the corresponding text from Ratdolt's *Elements* on his left. Pacioli is flanked by an aristocratic apprentice or student whose gaze is directed askance at the viewer, its vector pulling focus
38 to a wooden model of a dodecahedron resting on Pacioli's desk. In a similar vein, *Der Nürnberger Schreibmeister Johann Neudörffer mit einem Schüler* (1561) by Nicolas de Neufchâtel (ca. 1524–after 1567) depicts the "writing master" Neudörffer putting the finishing touches on a skeletal, wooden dodecahedron or training his attention upon one of its vertices. To his left, a student takes notes, presumably attempting to draw the dodecahedron in a notebook. Behind them both, a wooden cube is suspended with its vertices pointing up. On close inspection of the painting at the Germanisches Nationalmuseum in Nuremberg, the cube appears to have been hung on a painted nail protruding from the rear wall of Neudörffer's workshop. Neudörffer presumably will hang his new dodecahedron next to the cube once he finishes working on it.

Both scenes take place in the contiguous space of the *studiolo* or *Werkstatt*, artificially darkened by each painter to spotlight the actions of the main protagonists. Unlike Carpaccio or Faulhaber, or the many others who produced cluttered depictions of contemporary workshop activity, de' Barbari and Neufchâtel make use of a stripped-down and fathomless backdrop to ensure that their paintings will be read as allegories of pedagogy and not merely as depictions of more everyday mathematical practice. All extraneities have been eliminated in order to emphasize the essential components of study and teaching. Some of these, like the texts and measuring instruments, double as

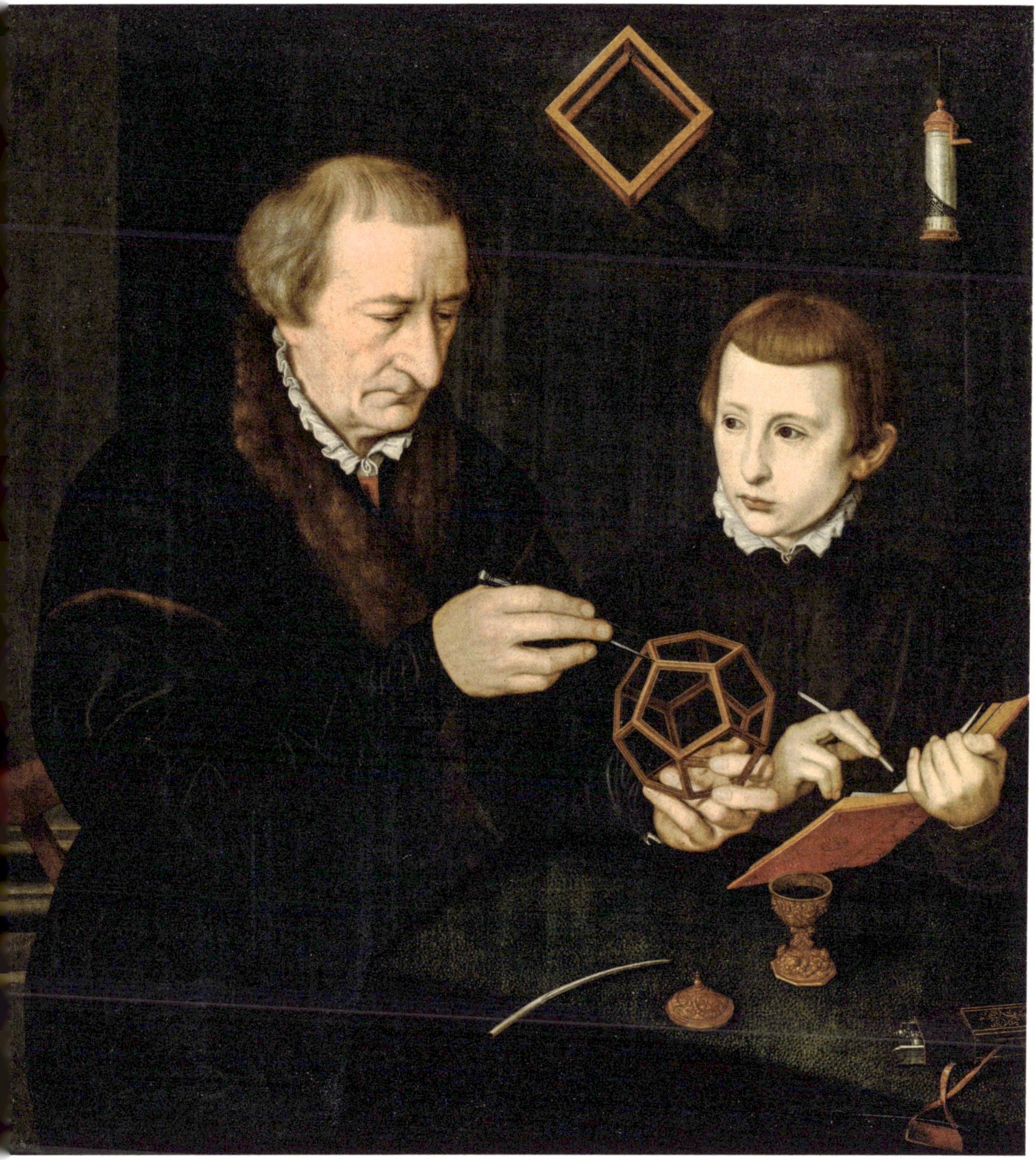

Nicolas de Neufchâtel (ca. 1524–ca. 1567/87). *The Nuremberg Writing Master Johann Neudörffer with a Student*. 1561. Germanisches Nationalmuseum, Nuremberg (Gm1836), on loan from the Bayerische Staatsgemäldesammlungen, Munich. Photo: J. Musolf.

props to precision, seriousness of purpose, and the robustness of the classical lineage. Other ephemeral qualities, such as the implicit distraction flickering in the eyes of Pacioli or Neudörffer's student or the relaxed grip upon their pointing sticks, indicates the softening of focus on the immediate surrounding required for intense mental labor and the social distancing needed to synthesize text, image, and model in the mind. At once real and unreal, the painterly polyhedra, much as for Ugo da Carpi and Plato, occupy a hybrid reality halfway between mental constructs and objects made physical.

As captured by Pacioli in the manuscript of *De divina proportione*, the representation of polyhedral models encouraged artistic transgression. Here the polyhedra are portrayed by (or after) Leonardo in states of suspension, if not suspense, gently hanging in space from string connected via rings drilled into each polyhedron's top facets or via elaborate knots tied to several linear segments for stability. The detailed attention to their fastening, striving for realism in the Milan copy, blatantly notional and represented by a single line in the Geneva copy, lends credence to the view that the artist drew from life, basing his drawings directly upon physical models. Nonetheless the appeal to physicality retains a productive ambiguity. The taut strings, tangled around each other and twisting in the breeze blowing through an open window beyond the limits of the page, appear held in palpable tension by the weight of the model, much as the length of string after the fastening knot frames each polyhedron's
39 nameplate in a physically impossible, decorative flourish. Elevated through the force of artistry and the mathematics of perspective to emblems of geometrical knowledge, polyhedra are also always still life and graphic folly.

Not so precious as to be stored out of view, but fragile enough that they are not kept on a low surface where the risk of being damaged would have been greater, polyhedral models could be put to use both as tools for helping to visualize ancient Greek geometry and as stereometric drawing aids for drawing precise representations of themselves. Models of the Platonic solids aided the understanding of Euclid's text, much in the same way that the impetus to invent and print diagrams in conjunction with Euclid's propositions was intended to help readers master the difficult material. Both de' Barbari and Caraglio overtly depict the act of learning to relate the visual properties of three-dimensional polyhedra as a method of decoding and working through their description in Euclid.

If polyhedra served the purpose of clarifying Euclid, they did so through their presence in printed editions of Euclid and in Pacioli's *De divina proportione*, often available for consultation in the study or workshop of an artist/mathematician. In turn, these printed images began to exert their own magnetic pull. As cases in point, Fra Giovanni da Verona's intarsia in the monastery of Monte Olivetto Maggiore near Siena and in the church of Santa Maria in Organo, Verona, both completed around 1520, display a seventy-two-sided sphere, an icosahedron, a truncated icosahedron, two elevated icosidodecahedra, a cuboctahedron, and a cube with equilateral pyramids affixed to each face, all derived from *De divina proportione.* And by 1525, the same year 40
as the publication of Dürer's *Underweysung der Messung*, Michelangelo had deigned to replace the customary sphere surmounted by a cross on the lantern commissioned by Pope Leo X for the New Sacristy in the Basilica of San Lorenzo in Florence with a gilded copper irregular solid or "elevated dodecahedron" (*Duodecedron elevates solidus*); aiming "to make it different from the others I have thought of making it faceted, which I think will make it look graceful." 41

But the vibrant color and technical mastery on display in these works and in *De divina proportione* obscure the more challenging impasses of polyhedral geometry in the first decades of the century. Not everyone agreed that the effort of producing *corpora* was worth the result. For Vasari, when the representation of geometry in perspective was the central purpose of an image and not subservient to the realization of a greater pictorial whole, the loss of labor and time of such intense and misspent concentration was wont to drive an artist to melancholy.

> For although these are ingenious and beautiful, yet if a man pursues them beyond measure he does nothing but waste his time, exhausts his powers, fills his mind with difficulties, and often transforms its fertility and readiness into sterility and constraint ...not to mention that very often he becomes solitary, eccentric, melancholy, and poor. 42

Perhaps such resignation weighs down Dürer's iconic angel in *Melencolia I* (1514), one of the best-known and yet elusive images in the history of art, whose posture hints that the oppositional thinking

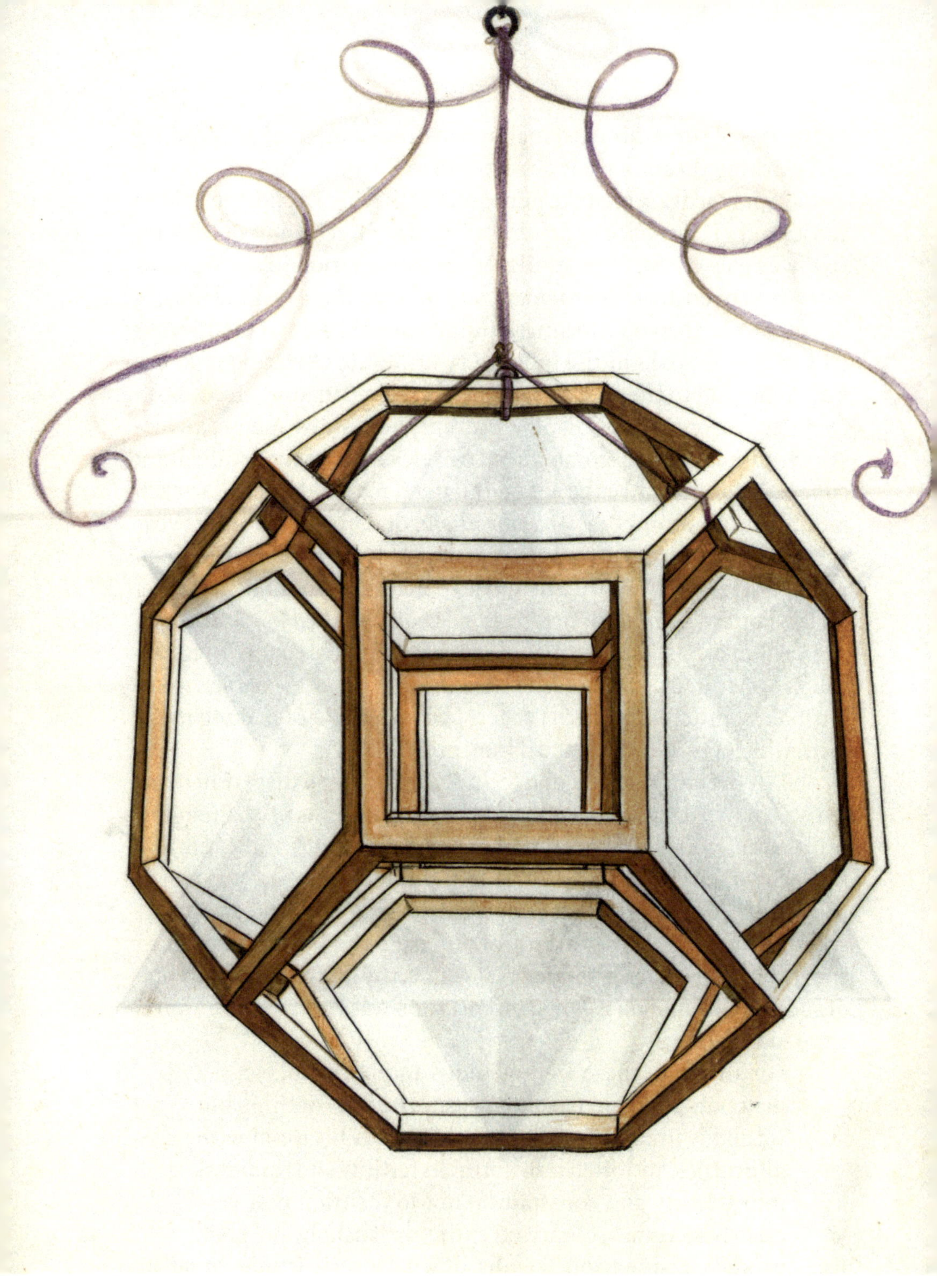

Stellated octahedron [*Octocedron abscisvs vacvvs*] and augmented rhombicuboctahedron [*Vigintisex basivm elevatvs vacvvs*], from Luca Pacioli, *De divina proportione* (manuscript completed in Milan, 1496–1498), fol. 100 recto, fig. XVIII, and fol. 110 recto, fig. XXXVIII. Veneranda Biblioteca Ambrosiana, Milan. Photo: ©Veneranda Biblioteca Ambrosiana/Mondadori Portfolio.

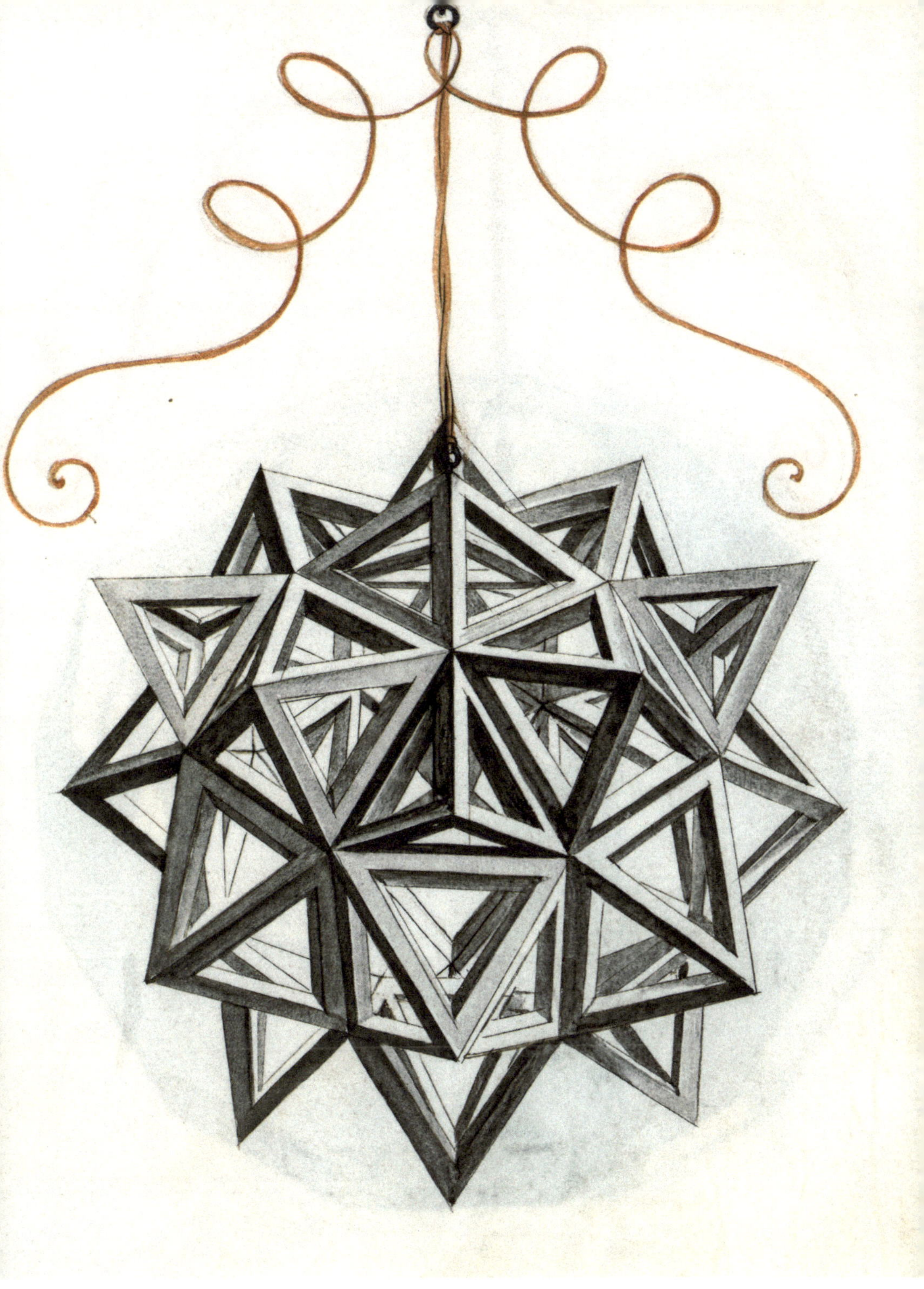

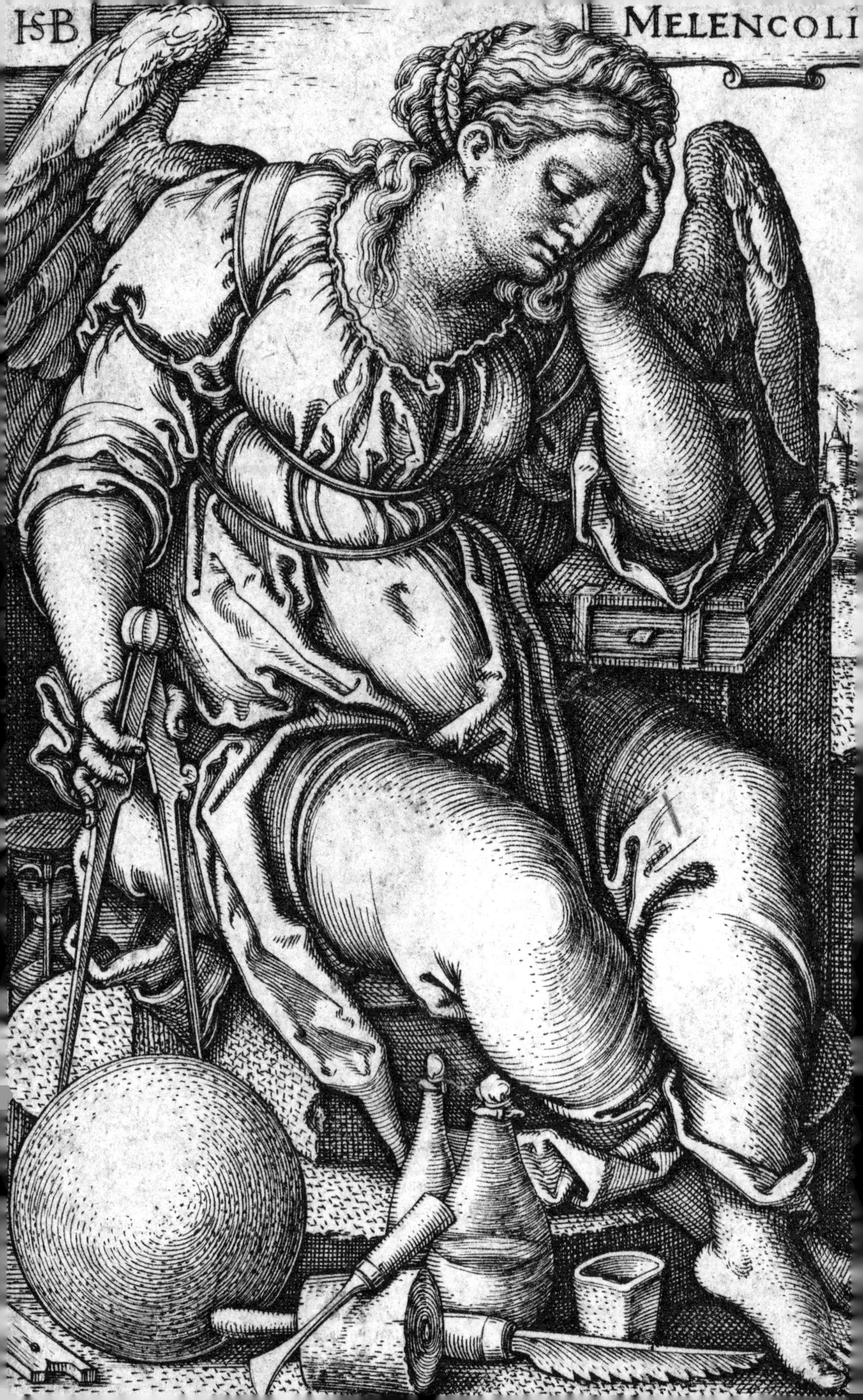

ISB
MELENCOLI

required to mediate the world of abstract geometrical knowledge with the concrete reality of objects, models, and visualizations was anything but a simple task. Eluding definitive analysis as it captures the hermetic spirit of an age in a claustrophobic panoply of symbolic artifacts—the magic square, the starving dog, the bat carrying the engraving's title, the scattered tools—*Melencolia I* remains unique by virtue of the sheer density and labyrinthine structure of its self-referential ambiguity. It manages to convey Dürer's complete mastery over the medium and somehow to represent access to his unconscious, as if, in its delirious precision and masterful array of intellectual conceits, the image had opened up a vista to a primal scene of memory and loss in excess of its own aspirations.

Melencolia I has garnered many readings, but central to most of them are the limits of geometry to describe both the world and artistic self-understanding. Surrounded by accouterments borrowed 43
from the traditional allegory of geometry as a woman engaged in acts of measurement, a truncated rhombohedron looms enigmatically in the middle distance, dividing the foreground from the background of a placid sea that vanishes into the horizon, ostensibly the type of natural territory the figure purports to measure. After Dürer, many 44
of the allegorical staples made famous by *Melencolia I* would remain in iconographic circulation in similar works by Virgil Solis, Abraham Bloemaert, Giovanni Benedetto Castiglione, and Hans Sebald Beham (1500–1550), whose own *Melancholia* (1539) is replete with a sphere, workshop tools, and the acquiescent expression on the face of the burly figure as she distractedly toys with a compass. Nevertheless, with the exception of the imitative copy of *Melencolia I* (1602) by the Antwerp-based artist Johannes Wierix (1549–1615), the *corpus irregulatum* from the 1514 original—a surrogate, in its absence, for all the three-dimensional geometry artists were newly struggling to invent, understand, and represent—was always conspicuously avoided. 45

The visual history of polyhedra is littered with false starts, poignant failures, and allegories unable to convey the weight of their subject matter. The work of Dürer or Paolo Uccello, the object of Vasari's scorn, is arguably not among them, though the disconnect between the budding enthusiasm for polyhedra and the actual capacity to visualize polyhedra in perspective palpably haunted lesser artists. *Geometry and Astronomy* (ca. sixteenth century) by the Dutch

Hans Sebald Beham (1500–1550). *Melancholia*. 1539. The Metropolitan Museum of Art, New York, The Elisha Whittelsey Collection, The Elisha Whittelsey Fund, 1966 (66.529.45).

artist Marin Bonnemer is one such image. Framed in ornamental
grotesquery, the title figures provocatively flaunt their measure-
ment tools in classical contrapposto. Below Geometry, an inscrip-
tion in French reads: “In order to wander & to measure the earth, to
conceive of all bodies [*corps*] & architecture, the man who wants to
46 acquire my knowledge of divine nature must follow me.” Below and
behind them, teams of surveyors and sailors valiantly demonstrate
the limitless possibility of measurement to conclusively represent,
and thus to know, the celestial and terrestrial spheres. The print is
layered with exotic vegetation, distant cities, portentous clouds, and
scientific instruments. But leaning against a rocky outcropping on
the ground to the left of Geometry, nearly camouflaged by their poor
quality and hesitancy, lie three sorry solids. Even in an image with
appreciably more intricate set pieces, to construct convincing poly-
hedra in perspective evidently tapped into deep structures of anxiety.
On the back of what might have been the simplest elements precisely
because they rely on the minimal use of a ruler and not on the whim-
sies of talent or inspiration, the epistemological edifice upon which
the allegory is assembled—the universal and divine certainty of
geometrical knowledge applied to the real world—buckles under
47 scrutiny and collapses in on itself.

Melancholy polyhedra, such as those of Bonnemer and the 1509 *De divina proportione*, hint at a subtext of artistic inadequacy premised on the inability to represent what were swiftly becoming the most iconic geometries in the early modern repertoire. While the world would be progressively mapped in geometrical terms and the certainty of the Euclidean proof increasingly supported by a new image culture, an overabundance of self-consciousness had created an undesired cap on perspectival realism. Polyhedra were singled out for special attention, as if they provided unique challenges that other representational constituents did not. Hence the vestiges of caution exercised when printing solids can still be traced at the level of the physical artifact. The ghostly outline around the edge of a well-tempered *corpus*, such as the rhombicuboctahedron in Ugo da Carpi’s *Archimedes*, indicates that it may have been designed on a separate block and imprinted upon the page as an autonomous element after the main figure had been finished. Equally, the existence of a rare copy of Caraglio’s *Diogenes* engraving in which Pacioli’s

dodecahedron has been omitted completely lends credence to the dodecahedron having been added after most copies of the print had been impressed. As there are no *Diogenes* with alternative subject 48
matter inserted onto the blank page consulted by the philosopher, it may be that Caraglio deemed the dodecahedron too difficult to risk incising into the primary copper plate. In the latter case, all of this precaution was for naught. Cursory inspection divulges the dodecahedron to have been engraved without a straightedge and its perspectival accuracy warped by the gesture of the human hand.

Polyhedra appeared in more or less proficient varieties, but artists overall became steadily acquainted with the solids without having to know the specifics of Euclid. That this familiarity would be beholden to technologies of their reproduction and thus to questions of representation signified a stark departure from eons of learning and thinking in the classical mold. Whereas Euclid trafficked in the absolute precision born from abstract and idealized quantities, the impetus to visualize mathematical concepts for general consumption had the counterintuitive impact of opening up geometrical-cum-theological reasoning to the peculiarities of aesthetics—to the condition of a line, to the quality of a print, and above all to the exercise of perspective and judgment. These tensions had always been present in the "speculative sciences," among which Thomas Aquinas had located mathematics as a mode of reasoning dedicated to speculative objects (lines, numbers, points). But the gap between geometry as text and as image had not heretofore been plumbed by artists, certainly not by artists of the caliber of Leonardo and Dürer, not to speak of de' Barbari and Neufchâtel. The strangeness of confronting the solids in vivid three dimensions, not as diagrams (themselves novel) but as fully realized objects, cannot be underestimated. Distilling a priori philosophical concepts to the level of tangible things consolidated models of thinking predicated upon the demand to see or witness what could no longer be trusted or taken for granted. Hereafter and despite the painterly pageantry at the turn of the century, the space of human perception would inevitably intervene in what had been the "divine nature" of geometry.

overleaf Marin Bonnemer (d. 1584). *Geometry and Astronomy*. Sixteenth century. Rijksmuseum, Amsterdam (RP-P-OB-33.909).

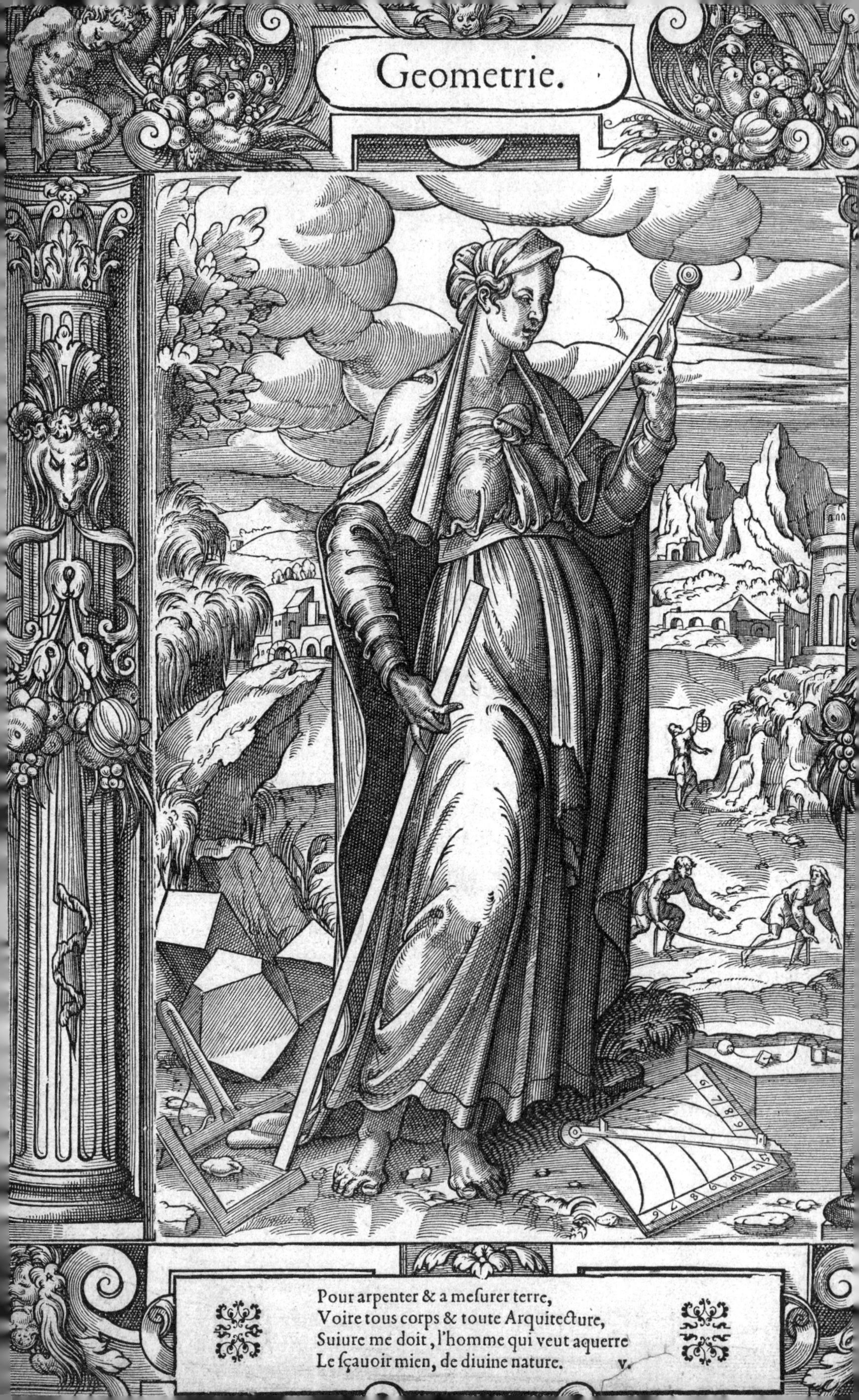
Geometrie.
Pour arpenter & a meſurer terre,
Voire tous corps & toute Arquitecture,
Suiure me doit, l'homme qui veut aquerre
Le ſçauoir mien, de diuine nature.
v.

Aſtronomie.

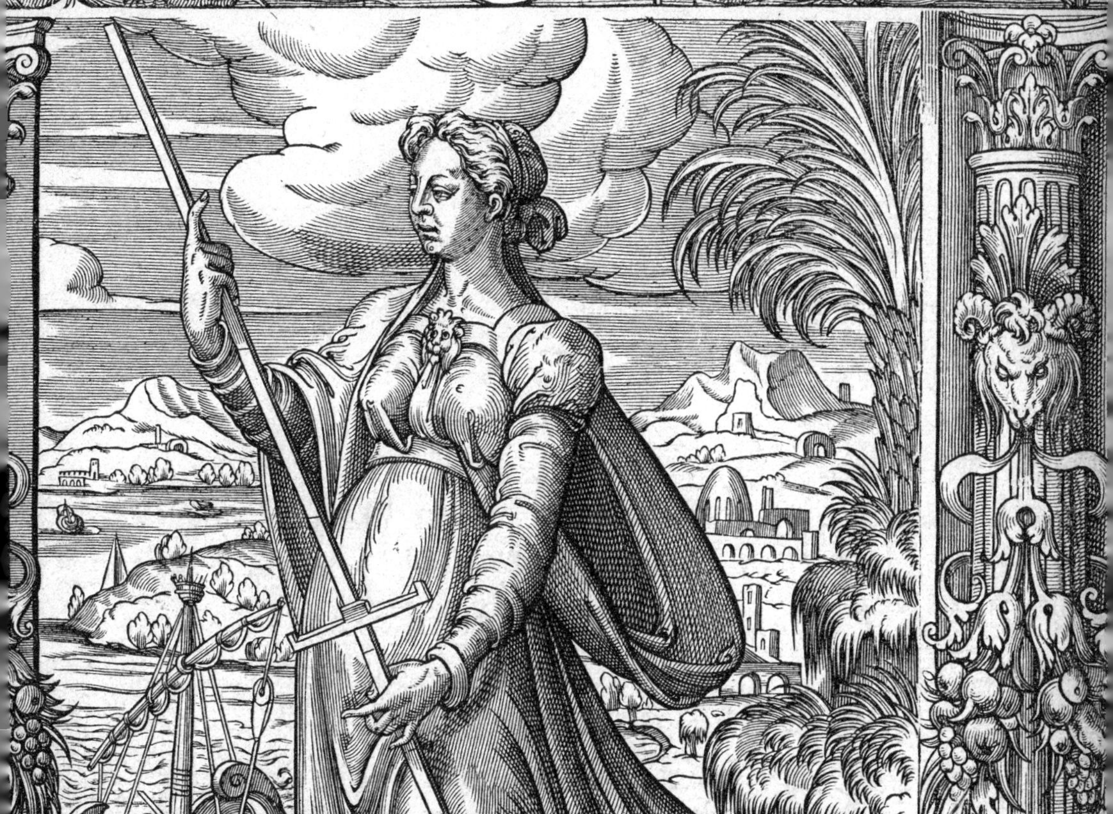

Qui veult ſçauoir le mouuement des cieux,
D'aſtronomye doit auoir congnoiſſance:
Car c'eſt vn art Diuin & precieux,
Par qui on voyt des Polles l'aſſeurance. vj.

I N S T R V C

M E A S V

TION IN

EMENT

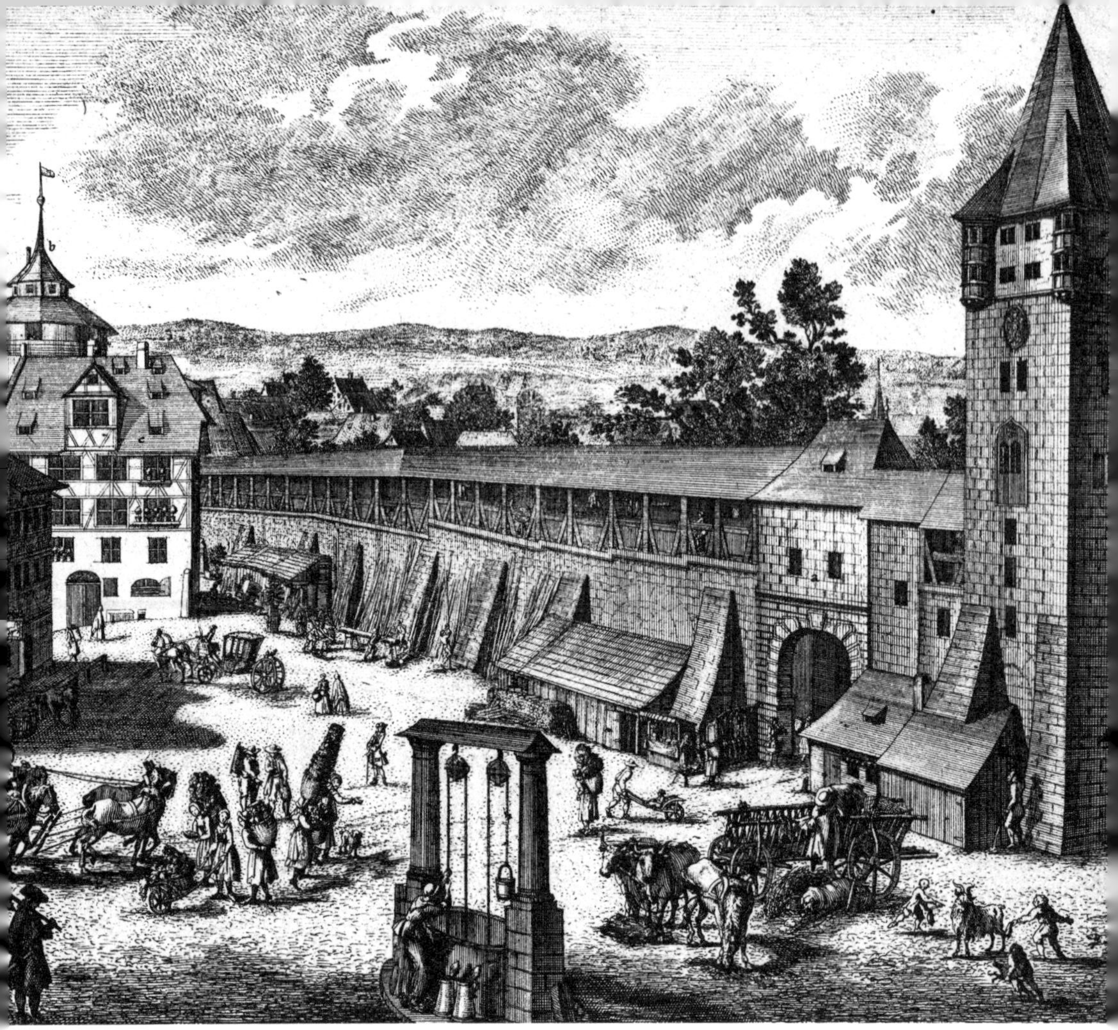

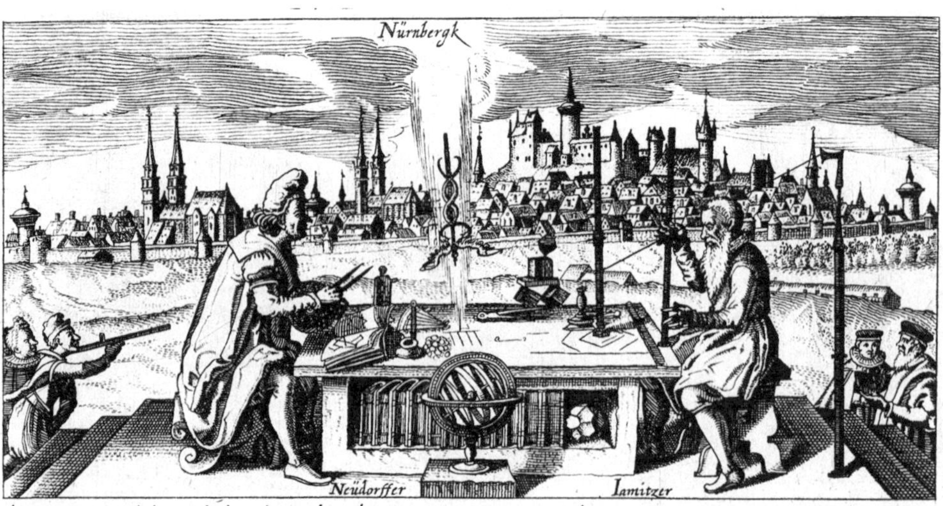

Arte nihil melius, nihil orbe salubrius arte; Est ea fida comes, comis amica, bonis.

In the northwest corner of the free imperial city of Nuremberg, one of the great cultural centers of the northern Renaissance, and close to the Tiergärtner gate, purportedly named because of the proximity to a former game enclosure beyond the city walls, rests a prominently situated, five-story building of robust if unremarkable volume, its bottom two floors composed of sandstone blocks and its upper levels in timber frame. It was here, in a city deemed "quasi centrum Europae" (more or less the center of Europe), that the great mathematician and astronomer Johannes Regiomontanus (1436–1476) spent the final years of his life and amassed his substantial library and collection of scientific instruments. That Regiomontanus would choose to settle in Nuremberg, after a peripatetic career which had taken him to Vienna, across northern Italy, and to the Hungarian court, had little to do with nostalgia for his upbringing in Königsberg some hundred kilometers north. With a level of wealth rivaling that of Florence, Nuremberg was the heart of German humanism, a bustling, high-end commercial hub strategically situated at the crossroads of trade routes leading south to Venice and northeast to Leipzig and onward to eastern Europe. Flush with celebrated think- 1
ers, artists, astronomers, instrument makers, printers, and a rapaciously wealthy and entrepreneurial merchant class, Nuremberg's prosperity and prestige made it an exceptional proving ground for the synthesis of science and art, distinguished by what amounted to an almost civic mandate to precision craftsmanship.

Casting an eye back on Nuremberg from the vistas of the early seventeenth century in his *Thesaurus Philo-Politicus* (1624), the poet Daniel Meisner (1585–1625) staged a didactic encounter between two of the city's greats: the master goldsmith and geometer Wenzel Jamnitzer (1508–1585) and Johann Neudörffer (1497–1563), historian, teacher of mathematics, "father of German calligraphy," and subject of the Neufchâtel portrait with wooden polyhedron. In an engraving that illustrates Meisner's work, under the heading "Nil melius arte" (Nothing betters art), Jamnitzer and Neudörffer face each other across a workshop table raised off the ground on a plinth and laden with the tools of their trades (including a dodecahedron tucked

Johann Adam Delsenbach (1687–1785). *View of the Tiergarten Gate*, with the Albrecht Dürer House in the background. 1714. Kunstsammlungen der Stadt Nürnberg (Gr.A. 12495).

Eberhard Kieser (1583–1631). *Nil melius arte* (Nothing betters art), portrait of Wenzel Jamnitzer and Johann Neudörffer. From Daniel Meisner, *Thesaurus Philo-Politicus* ... (Frankfurt am Main, 1624). Herzog August Bibliothek, Wolfenbüttel (VD17 23:285517W).

beneath the table at Jamnitzer's feet). Behind them Nuremberg looms in ever-present reminder of their urban ties. A caduceus staff, deployed as an antique symbol of commerce and fair exchange, descends from the heavens onto the protagonists' workspace, linking the act of measurement to the city as beneficiary and patron of their labor, and pointing upward toward knowledge of the divine structure of the world and nature. Fittingly, the German motto at the base of the engraving stresses the utility of art—not artistic genius, divine inspiration, or any other romanticizing of the creative process.

> Nothing on earth is better than art,
> Nothing more useful can be found than art:
> Art is a true friend,
> 2 Therefore all artists should be revered.

The engraving, by the German artist Eberhard Kieser (1583–1631), offers a baroque imagining of a uniquely northern Renaissance little resembling and easily differentiated from its older and more sensual Italian relative. Less beholden to patronage structures and with their livelihood dependent on market-driven economics, Nuremberg craftsmen were active and self-aware participants in the production of natural philosophical knowledge that in turn reflected favorably on their work and themselves. Thus, it is no surprise that there was much local interest in the Regiomontanus library, not just because of the significance of the great man himself but also because the library would have been eminently useful to the city's many artist-scholars. In due course after Regiomontanus's death, the house and its contents were purchased by Bernhard Walther (1430–1504), a merchant, proficient astronomer in his own right, and erstwhile collaborator with Regiomontanus on astronomical observations and in the establishment of a printing press devoted to producing astronomical trea-
3 tises revised with new observational data. Unsuccessful overtures were made to convince the new owner to part with Regiomontanus's books and possessions, most notably by King Matthias I of Hungary and Croatia who hoped to rehouse the collection in the royal library in Buda. But shortly before his own death in 1504, Walther decreed that the entire collection of books and instruments were only to be sold all together, and with the exception of several books which were sent to Krakow and Italy in 1512 and a selection of the brass instruments which were stolen in 1514, the council of Nuremberg lobbied

to keep the collection intact and under the city's auspices over the next fifteen years, though not likely *in situ*.

During this time, the "Regiomontan-Waltersche Bücherei" became an intellectual resource and reference library for a new generation of Nurembergers, such as Johannes Werner (1468–1522), who worked on spherical trigonometry and conic sections; Albrecht Dürer's close friend, the humanist Willibald Pirckheimer (1470–1530); Joachim Camerarius (1500–1574), the classical scholar and Dürer biographer; the influential globemaker and cosmographer Johannes Schöner (1477–1547), and many others. It was only in 1519, 4
after a failed attempt to sell the books to Elector Friedrich III of Saxony (1463–1525), that a portion was sold for the sum of 150 *gulden*, most of which went to Pirckheimer who recorded the purchase in his personal effects. 5

By Walther's death in 1504, Dürer had overseen his Nuremberg workshop for nearly ten years and was much celebrated for his mastery of print. He had published his popular and graphic rendition of the *Apocalypse* in a series of large woodcuts (1498), produced some of his greatest and most evocative early engravings including *The Sea Monster* (ca. 1498) and *Nemesis* (*The Great Fortune*) (1502), and completed the *Seven Sorrows Polypytch* (ca. 1500), an oil painting commissioned by Friedrich III, not to mention his epoch-defining *Self-Portrait* (ca. 1500). In 1504 Dürer also issued his masterpiece of engraving, *Adam and Eve*, an iconic synthesis of ideal proportion and humoral iconography rendered in heretofore unprecedented opulence.

As one of Germany's most celebrated artists, Dürer would have had ample reason to consult the Regiomontanus-Walther library, irrespective of his intimate friendship with Pirckheimer and the other Nuremberg scholars. Not only did Dürer fully participate in the circulation of knowledge among the city's elite, but his drawings reveal an artist grappling with quintessentially classical questions of human proportion. Though his early sketches evidence him using a compass to generate geometrical ratios of corporeal proportions, Dürer would move toward a Vitruvian system of ratios that balanced the dimensions of body parts to each other, sometimes surveying the ratios of the body's parts together in elaborate detail. In these later 6
proportional systems, Dürer used elements directly from *De architectura*, selectively modified with drawing instruments to suit his own

aesthetic sensibilities. To be sure, he mentions Vitruvius in a 1523 letter to Pirckheimer where he complains about how Jacobo de' Barbari, who had resided in Nuremberg in 1500–1501, had "showed me how to construct a man and a woman based on measurements. … But Jacobus, I noticed, did not wish to give me a clear explanation." In his retelling, Dürer took matters into his own hands and "read Vitruvius,
7 who has written a bit about human limb proportions."

It has become commonplace to append Dürer's deep engagement with Euclid and geometry generally to the story of his "discovery" of perspective during his first trip to Italy (1494–1495), a story reinforced by the illustrious German-Jewish art historian and arguably the greatest modern scholar of Dürer, Erwin Panofsky (1892–1968). And to be sure, Dürer did return from his second trip to Venice further acquainted with Italian geometrical-perspectival techniques and holding a copy of Zamberti's 1505 edition of *Elements*. In Bologna, Dürer may also have met Pacioli, who would have been working on his own annotated edition of Euclid at the time. Panofsky claims in addition that it was in Italy that Dürer first learned Piero della Francesca's
8 method of using foreshortening to create perspectival figures.

Then again, Dürer had long-standing access at home to the relevant literature on the subject that predated his engagement with Italian perspective. And given his direct connections to Walther, he may have had access to the library as a very young man and while Walther was still alive. His parents had known Walther personally and Walther's wife had been the godmother to Dürer's sister Christina, who was born in 1488. In 1509 he purchased the house from Walther's heirs and moved into it with his wife Agnes (1475–1539), living there until his death in 1528 when he was said to have been found with books from the house's original library scattered
9 near him. Maybe some of these books were those he had bought in 1523 from the Regiomontanus collection because of their "usefulness
10 to painters," when he had paid ten florins for a selection of ten.

It is hard not to think that Pirckheimer and his extended humanist circle would have been fully capable of answering any questions Dürer might have had on the mathematical issues that pertained to his work on geometry and measurement, prior to or irrespective of his debts to Italy. Indeed, *Underweysung der Messung / mit dem zirckel un richtscheyt / in linien ebnen unnd gantzen corporen* (1525)

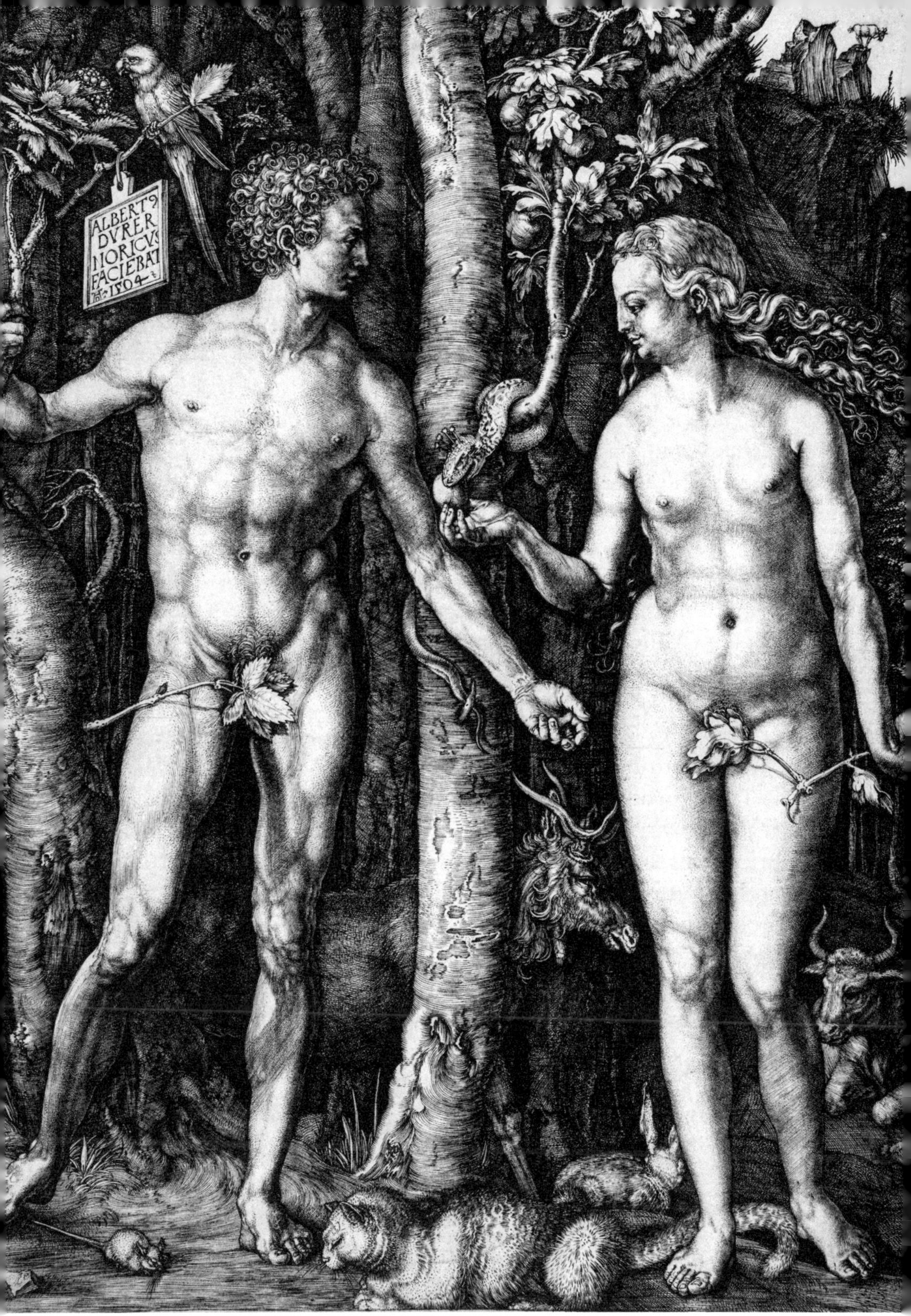

Albrecht Dürer (1471–1528). *Adam and Eve*. 1504. The Metropolitan Museum of Art, New York, Fletcher Fund, 1919 (19.73.1).

(Instruction on the measurement of lines, areas, and solids [whole bodies] with a compass and ruler), the first of Dürer's major late treatises on measurement, synthesized the internal structure, principles, and precision thinking of Euclid's *Elements* with various classical and contemporary geometrical texts readily available in Nuremberg, and with the knowledge of geometry Dürer had accumulated over a life-
11 time of art practice. Diagrams in the treatise are reproduced from several locally published books, namely the *Fialenbüchlein* (1486) by Hans Schmuttermayer, who was an acquaintance of Dürer's father; the *Püchlein von der filialen Gerechtigkeit* (Booklet concerning the correctness of pinnacles, 1486) by Mathes Roriczer; and the anonymous *Geometria deutsch, aus der geometry etliche nutzparliche stuck* (ca. 1472–
12 1484), possibly also attributable to Roriczer.

Moreover, Dürer was surrounded by the thriving book trade in Nuremberg and might well have seen books on pure and applied geometry from his prominent godfather Anton Koberger (ca. 1440/
13 1445–1513), the publisher responsible for the Nuremberg Chronicle. Due to the unique status of the Regiomontanus-Walther library as a substantial scientific and mathematical resource in Nuremberg, it is extremely unlikely that Dürer would not have taken advantage of the library to deepen his understanding of geometry and perspective and that he could have easily gained entry, either as a family friend or as an inquisitive artist, from very early on in his career, most certainly prior to his second extended trip to Italy in 1505–1507. In the comfort of the house that would become his home years later, he would have found himself surrounded by classics from antiquity and the Middle Ages on geometry, perspective, astronomy, and astrology (for example, Ptolemy, Sacrobosco, Ibn Al-haitan, and Archimedes) as well as contemporary works of relevance including calendars and astronom-
14 ical tables. From the two inventories of the Regiomontanus-Walther collection compiled by Pirckheimer in 1512 and 1522, the library is also known to have contained Ratdolt's illustrated Euclid from 1482, referred to as "Euclides. Impressus. (Geometria)," as well as a rare manuscript of the first translation of Euclid by Adelard of Bath (ca. 1080–ca. 1152). Of conspicuous note is the presence of a manuscript of *De pictura*, Leon Battista Alberti's definitive text on perspective, entitled "Liber de picture L. Baptiste de Albertis. (Geometria)" and referred to as "De picture babtis" in the library's inventory.

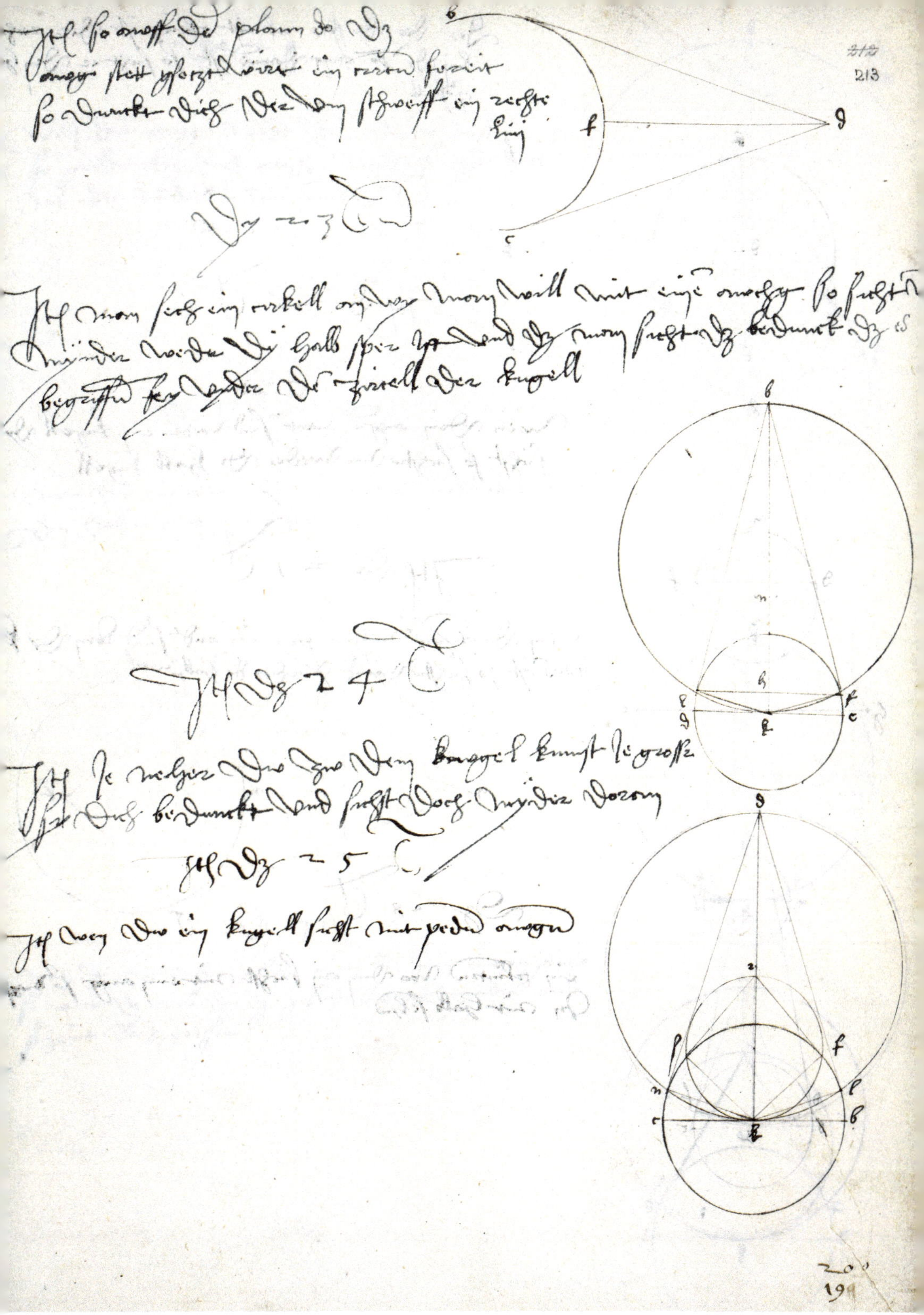

Albrecht Dürer (1471–1528). Translations and diagrams from Euclid's *Elements*. Early sixteenth century. The British Library, London (Ms. 5228, fol. 213 recto). Photo: ©The British Library Board.

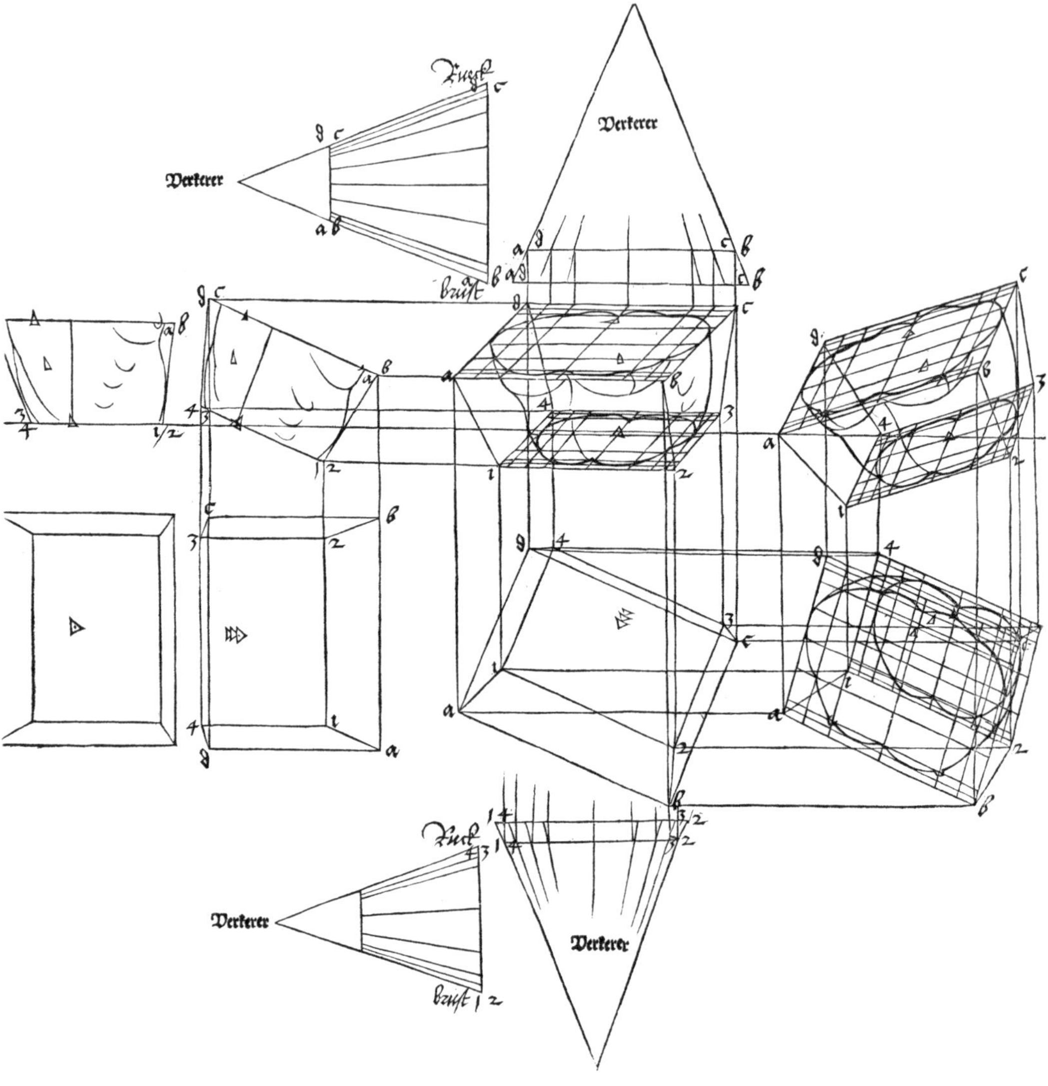

Graphic techniques for rotating a man's chest, from Albrecht Dürer, *Vier Bücher von menschlicher Proportion* (Nuremberg, 1528). Staatsbibliothek Bamberg (JH.Inc.typ.IV.340, fol. 125 verso). Photo: G. Raab.

Does it matter that Dürer may have been interested in Euclid and had access to texts on perspective before his firsthand exposure to Italian culture? This question bears on a range of historiographical issues pertaining to southern and northern European erudition and culture, as well as the romantic search for an origin that has so animated German culture and thought, art history included. But regardless of what he learned where and when, the last third of Dürer's prolific professional life would be indisputably intertwined with the production of several learned treatises on measurement and geometry, specifically *Underweysung der Messung* and the more well-known and posthumously published *Vier Bücher von menschlicher Proportion* (1528) (Four books on human proportion), which Dürer considered to be a more advanced treatise appropriate for study only after mastering the "measurements [of] lines, surfaces, solid bodies" in
Underweysung. In *Vier Bücher*, Dürer deployed geometry as a stereomet- 15
ric aid in the representation of the human body and its mutual relations to other bodies. In order to be reformed into workable parts, the body is disassembled into more easily manageable sections that confine and define it volumetrically. These geometries, which are simpler to manipulate in three dimensions than irregular body parts, are then rotated to reveal multiple angles on the graphic space of the page, such as in the figure depicting the upper part of a man's torso.

As convincing as the Italianate orientation may be in certain iconological respects relevant to the evolution of Dürer's oeuvre, it cannot account for his innovations in the field of applied or practical
geometry, which have no artistic precedent of equivalent intensity. 16
The representation of the human body as geometrical volumes was an original graphic strategy essential to the progression of Dürer's system of teaching drawing and proportion. His stereometric studies and sketches for *Vier Bücher* as well as the final printed renditions depict bodies as unnervingly semiabstract figures composed partially, or entirely, of geometric solids in various poses and states of animation. Not a Renaissance of passive contemplation, these bodies are plastic, mutable entities made from geometrically composed shapes that provide little resistance to visual dissection and may be manipulated, sliced, or stretched to achieve the desired aesthetic effect. In promoting the efficacy of his stereometric method, Dürer claimed that it "may be useful for sculptors beginning to learn this

craft who intend to cut a figure from wood or stone. In order to copy a figure exactly they can chop away from the square surfaces what is
17 necessary, without cutting off too much or leaving too much." Two heavily faceted, stereometric heads from Dürer's *Skizzenbuch*, dated 1519, may illustrate this intermediary step in the production of life-like sculptural figures, though Dürer's comment may not have been entirely genuine. Why would a sculptor practice a graphic technique that mimicked, in the most superficial of ways, an imagined and intermediary step in sculpting, somewhere between the material of the stone and final figure? Having stumbled upon an arena whose implications he may not yet have fully understood, perhaps Dürer had no other way to justify to himself what he had uncovered than to treat the stereometric as a work in progress. In seeming anticipation of modern regimes of technological determinism and social alienation mapped onto the form of the body, the artists Erhard Schön (ca. 1491–1542), Luca Cambiaso (1527–1585), and others would go on to develop this proto-digital aesthetic into a visual genre that masked a distressed and incompliant geometrical humanoid behind the mollifying palliative of the "unfinished."

Dürer's incursions into geometry predate any "two cultures" debate or interdisciplinary phobia and comprise a landscape of experimentation and testing with its own peaks and slopes, both gradual and vertiginous yet parallel to the development of his artistic oeuvre. And so while he might very well have bought Alberti's *De pictura* as one of his ten "artistically useful" purchases from the Regiomontanus-Walther library in 1523, it may not astonish to know that the only book from the collection identified as having definitively been in Dürer's possession was Regiomo ntanus's own man-
18 uscript copy of Euclid's *Elements*. We know Euclid captivated Dürer and that he collected multiple editions of *Elements* in addition to the copy he bought in Venice. Dürer's letters rank among the first translations into German of eleven suppositions and forty theorems from Euclid's *Perspectiva naturalis*, some of which are in Pirckheimer's hand, alongside numerous hardline geometrical sketches and illustrations
19 copied from the printed diagrams in Zamberti's Euclid. Though he owned at least two copies of *Elements* by 1523 and had continual access through the Regiomontanus-Walther library to the precious Adelard of Bath manuscript as well as Ratdolt's *Elements*, Dürer remained

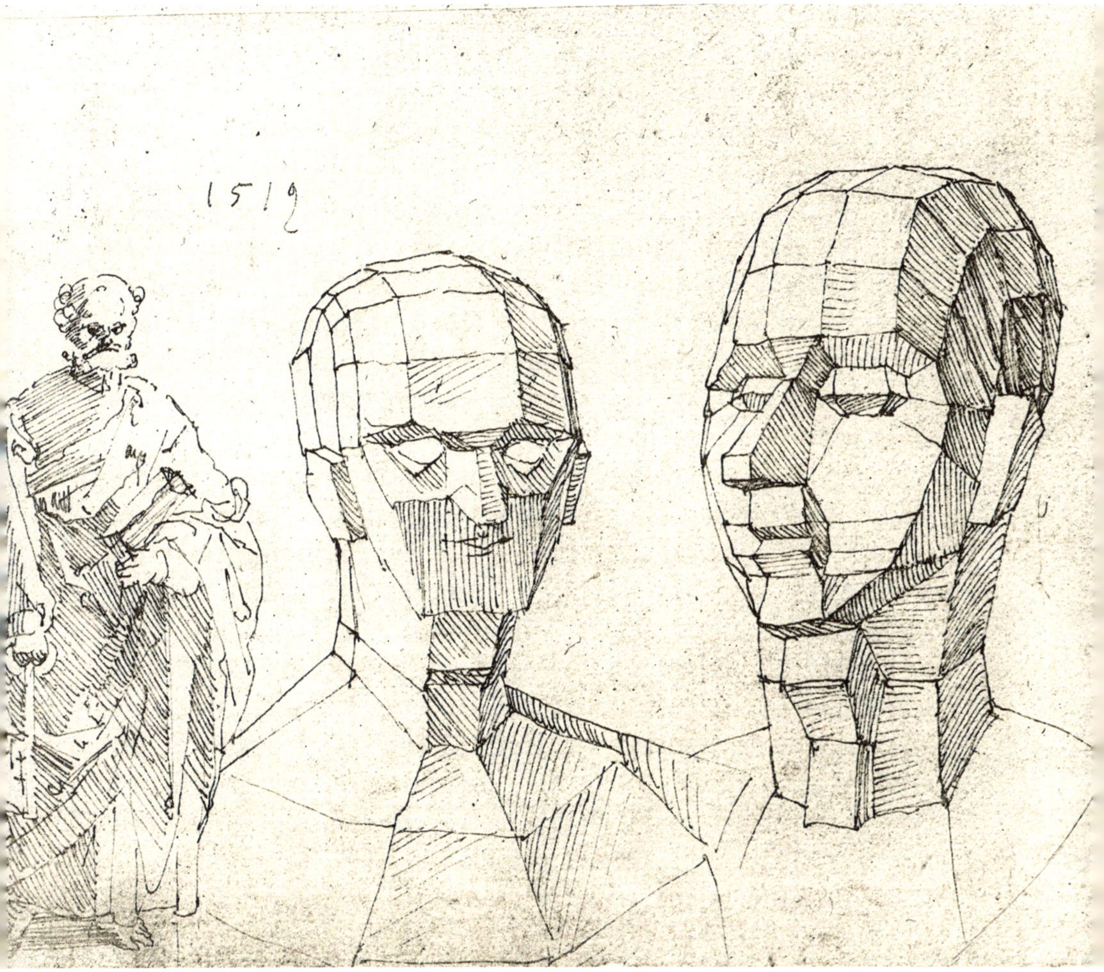

Albrecht Dürer (1471–1528). Sketch of Saint Peter and two stereometric heads. 1519. From the *Dresden Skizzenbuch*. Sächsische Landesbibliothek—Staats- und Universitätsbibliothek, Handschriftensammlung, Dresden (Msc. Dresd. R147 f. 101b). Photo: SLUB Dresden/Digital Collections/S. B.6023.

interested in the most up-to-date translations. In a letter dated
December 5, 1524, to the German mathematician Nikolaus Kratzer,
who was living in London at the time, Dürer inquires anxiously how
long it would take until Kratzer's planned new German translation
20 of Euclid would be ready.

Despite having ready access to this mathematical literature and
being familiar with Latin, Dürer might have needed help reading the
ancient texts, including the Zamberti translation that he had
21 acquired in Venice for one ducat. His contemporary biographer
Camerarius claims that although Dürer had acquired skills in "the
natural and mathematical sciences" (*naturalium et mathematicarum
rerum scientiae*), he had not undertaken literary studies, which meant
in this context "the analysis and also the productive imitation of clas-
sical Latin writers." Camerarius does admit, however, that the natu-
ral and mathematical sciences were predicated upon ancient texts
which Dürer had "almost learned" (*fere didicerat*), and that Dürer's
geometrical treatises evidenced his understanding of scientific prin-
22 ciples and his ability to set them down into words. Pirckheimer
would have been an inimitable asset here. There are numerous
instances of Pirckheimer's writing in Dürer's capacious literary
remains that confirm their close collaboration on mathematical
issues and translations, including corrections made by Pirckheimer
23 to drafts of Dürer's *Underweysung*. Pirckheimer had studied law in
Italy from 1489 to 1495, was fluent in Greek and Latin, and routinely
acted as an intermediary between Dürer and Emperor Maximilian I,
going so far as to act as an advisor on antique iconography for some
of Dürer's highest-profile commissions including the *Triumphal Arch*
or *Ehrenpforte Maximilians I* (first edition printed 1517–1518)—a mon-
umental composite image printed on thirty-six large sheets of paper
from 195 individual wood blocks and saturated with transhistorical
24 references. Pirckheimer also possessed his own well-appointed
library which, he claimed in a 1503 letter, held a copy of almost every
text printed in Italy, most of which were presumably in Latin and
25 Greek. Dürer surely had access to this library as well, including to the
volumes on geometry and perspective, such as Luca Pacioli's *Somma
di aritmetica, geometrica, proportioni e proportionalita* (1494), and is known
26 to have personally illuminated a selection of his friend's books. The
solution to several classical mathematical problems that concerned

Dürer and would later appear in *Underweysung* can likewise be traced to texts residing in Pirckheimer's library. 27

Underweysung is divided into four books covering a range of topics in the visualization and invention of various geometrical figures, emphasizing the construction of complex shapes from basic elements. Beginning with Euclidean principles and spreading into a diverse set of mixed-mathematical applications such as column design, sundials, calligraphy, perspective, and the modeling of polyhedra, *Underweysung* systematically organizes a wealth of accumulated knowledge and was intended, as Dürer declared in the work's introduction, to be "not only for painters, but also for goldsmiths, sculptors, stonemasons, carpenters, and all those for whom using measurement is useful." Book I is devoted to the definition and con- 28
struction of lines, ranging from parallel lines to multiple ways of constructing spirals, spiral projections, hyperbolic lines, conic sections, and parabolas. Book II discusses plane surfaces, the construction of polygons, tile patterns, and the Pythagorean theorem. Book III covers a variety of topics, including column and monument design and construction, sundials, and the geometrical construction of the letters of the alphabet. Book IV returns to the Platonic and Archimedean solids and concludes with the Delic problem, studies of proportional lines, and various perspectival theories and apparatuses. 29

This structured approach contrasts markedly with Dürer's more ruminative *Vier Bücher*, in which he mixes hard truths about the limits of training artistic ability with meditations on the elusive nature of beauty. In his "Aesthetic Excursus," appended to the end of Book III, Dürer casts the mind of the master artist as a bottomless resource "full of images they might conceivably make" and yet also beholden to an untrainable talent, a "marvelous gift" granted by God so that "it follows that one man may draw something with his pen in one day on a half-sheet of paper, or incise something with his graving tool in a little woodblock, which turns out better in its art than the great work of another who has spent a whole year's endeavor on it." If *Vier Bücher* preserved the spread of Dürer's explorative negoti- 30
ations of the proportional rules governing nature vis-à-vis the variability of observable phenomena, the authorial voice he presents in *Underweysung* is unburdened by uncertainty, staging a progressive complexification of geometrical rules centered on idealized

graphic constructs existing primarily in an imaginary paper space.
Thus Book I begins with his evocative description of a point (*Punck*)
as "such a thing with neither size, length, width, or thickness. And
still it is the beginning and end of all corporeal things we may want
31 to construct, or which we may conceive in our minds." He contin-
ues by designating points as the building blocks of the geometrical
imagination, "to the extent that no point takes up any space, for it
is indivisible and can in our senses or thinking [*synnen oder gedanken*]
be placed anywhere. I may use my mind to throw a point high up
in the air, or drop it into the depths where I cannot reach it with my
32 body." Given that points for Dürer are thought constructs, to con-
nect them to each other mentally requires an unrepresentable and
impalpable (*unsichtig*) line. Drawing becomes necessary to visualiz-
ing geometry so as to make legible the connections between points,
and therefore, he says with characteristic awareness of his audience,
"I want all matters that I have described in this treatise by way of a
sketch, drawn in a very succinct way described therein, so that the
young, through the help of the imagination, will be able to see with
33 their own eyes and grasp what has been described here."

In this extraordinary introduction, Dürer makes his most
direct and compelling case for including the visual realm as essen-
tial to the learning of geometry, and geometry as essential to the uni-
versal learning of art. His use of the word *Einbildung* (*ein* = inward,
bildung = image/imagination) points toward contemporary under-
standings of sensation and the faculty of the imagination—one of
the five "internal" senses together with cognition, memory, fantasy,
and common sense—namely, to the closely related meaning of the
34 image of an object created and imaginatively manipulated. Already
inching toward Kant's *Einbildungskraft*, Dürer's *Einbildung* flags the
importance of images (*Bilde*) in the production of a fully formed and
educated person (*Bildung* and *Ausbildung*). To draw geometry with
attention and dedication, in other words to use lines to connect
precisely plotted if imaginary points, served to yield a new capacity
for vision, strengthening the interdependence between the tactile
manipulation of drawing instruments to create exact representations
of ideal things (columns, polyhedra, spirals) in a graphic space, and
the imagining of these things in a mental space. To follow Dürer's
logic was to imagine, conceivably for the first time for many German

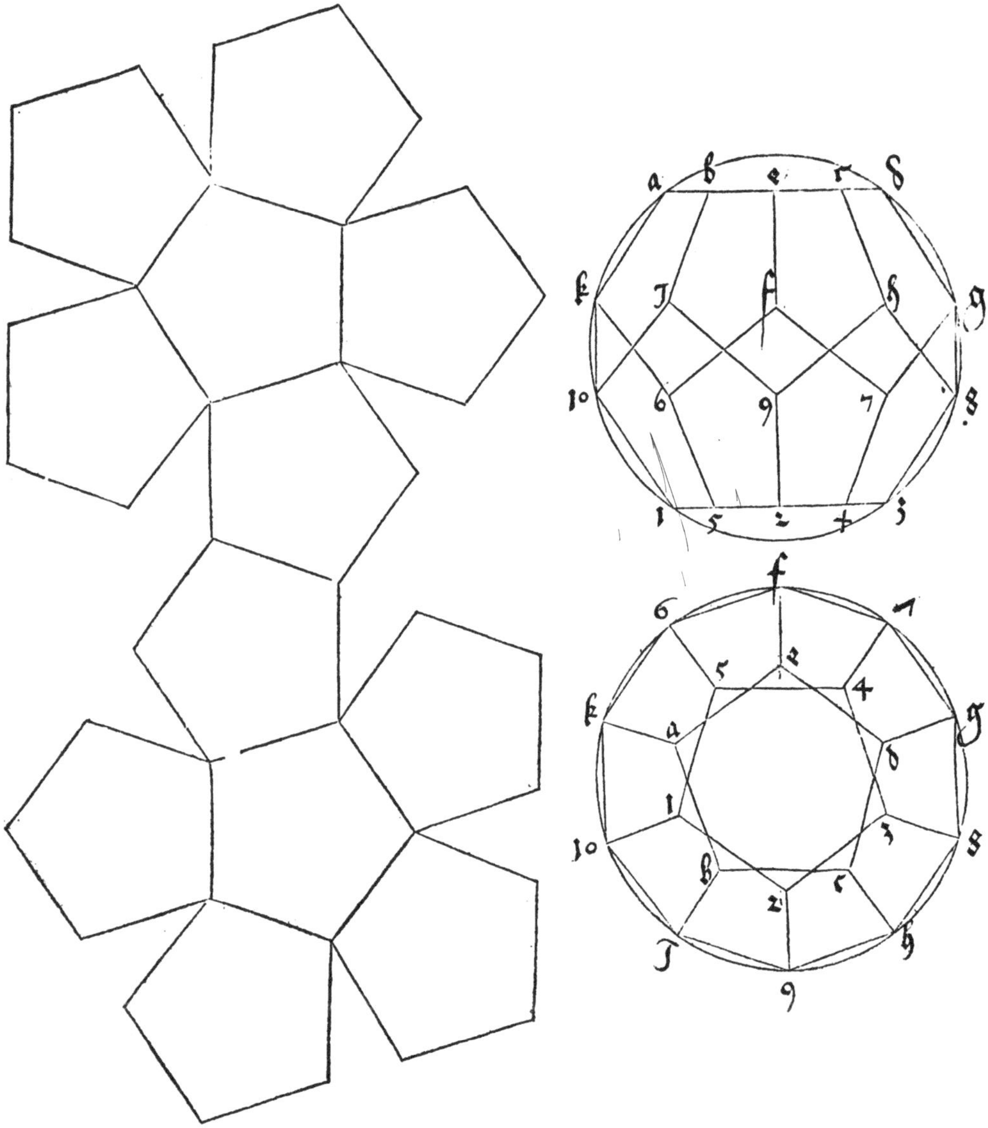

Page with net of a dodecahedron, from Albrecht Dürer, *Underweysung der Messung* (Nuremberg, 1525). Universitätsbibliothek Erlangen-Nürnberg (CIM.I 6, fol. 71 verso).

craftsmen, that virtual images and designs, either plucked from the *Synnen* or completely fabricated without visual referent, ought to be translated to the page with measured fidelity. The centrality of visualization to learning geometry acts as the structural motif tying together the representation of all subjects, whether an unfolded dodecahedron or convex columns. *Underweysung* was Dürer's manifesto on the proper and exact use of geometry *made visual.*

In accord with the Euclidean tradition, Dürer dedicates the concluding section of *Underweysung* to the Platonic solids and the question of polyhedra more generally. But rather than rely upon perspective, a science he had undoubtedly long ago mastered, Dürer takes the decisive step of rendering the Platonic solids as "unfolded" polyhedral surfaces, a decision that would have catastrophic, if unforeseen, effects on the stability of geometric knowledge. In the example of the dodecahedron (the technique is the same for each regular and irregular solid he describes), the polyhedron is splayed into twelve equal five-sided polygons (*zwelf blatte funf eckede feld*), "laid out in plan" (*in grund gelegt*) and both outlined and unfolded (*aufgerissen*), as per the double German meaning, into a completely open (*ganz offen*) and developable surface. There is no three-dimensional solid in sight, nor any painterly attempt to reduce the edifice of Platonism to five geometrical emblems. That this material follows directly after an extended, typographical analysis of the alphabet à la Pacioli's *De divina proportione,* in which Dürer demonstrates the utility of geometry for teaching how to form correct letters (*recht bustaben*) since "builders, painters, and others are required to make
35 inscriptions on high walls," reveals that the decision was purposeful. Retrieving the polyhedral gauntlet decisively thrown down by the Italian artists, Dürer had chosen to proceed in a manner fundamentally different from any of the treatises of the preceding decades.

Dürer was far from the only artist in Germany taking advantage of print technology to create developable surfaces with real-world applications, and there are surely ways to read Dürer's folds in *Underweysung* within complementary traditions of paper instrument experimentation. Martin Waldseemüller (ca. 1470–1520) and Johannes Schöner unrolled a globe into twelve paper gores that could be cut out and reassembled onto a wooden substructure to form a representation of the earth or celestial sphere. Further afield,

Gothic traditions of cutting paper to create architectural elements
such as vault designs were still commonplace, as evidenced by the
sketchbook and pattern book (1560–1572) of Master WG in Frankfurt
am Main. Particularly germane is the work of the Nuremberg-based 36
engineer and instrument maker Georg Hartmann (1489–1564).
Hartmann, who likely collaborated with Dürer on the illustration
of a block sundial in *Underweysung*, printed interactive paper sundials
in polyhedral or even cruciform shape that could be carried around
and unfolded as necessary. 37

But printing unfolded polyhedral nets was a relatively new rep-
resentational technique with only the most scattered of precursors.
Medieval manuscripts of Calcidius's *Commentary* on the *Timaeus*, such
as a tenth-century copy in Brussels, instruct how to reduce cubes to
a single plane face and are accompanied by directives guiding their
mental reconstruction in three dimensions. Toward the end of his 38
De geometricis corporibus (1510), the French mathematician, and Dürer's
contemporary, Charles de Bovelles (ca. 1475–after 1566) made use of
small, rudimentary representations of several unfolded solids, both
regular and irregular, to illustrate his synthesis of Christian theology
with the polyhedral physics of the *Timaeus*. For Bovelles, directing 39
attention to the sacred geometry of the regular solids provided access
to the divine in a perfect synchronicity of Christianity and Platonism.
Not unlike Pacioli if further developed and articulated, the geome-
try of sensible things served to guide the mind upward toward the
contemplation of celestial hierarchies, though there is no indication
that Bovelles's diagrams were intended to be physically reproduced
as models. If his text is suffused with visual metaphor, it addresses
neither tactile construction nor materiality, nor are measured draw-
ings or translations into French provided. The emphasis is rather on 40
how the angles of plane figures can be used to construct the three-
dimensional solids and further in extracting divine trinities from
sets of geometrical relations—namely that "every regular solid angle
[angle of the regular solid] is procreated [*procreari*] and brought forth
by either the power of one in three, or of three in one." 41

While Pacioli, unlike Bovelles and more expertly, relies on per-
spective to illustrate all five Platonic solids and three of the thirteen
Archimedean solids, Dürer chooses, conspicuously, to render the sol-
ids as large, sheet-size unfolded nets, in addition to several *corpora* of

overleaf Georg Hartmann (1489–1564). Cruciform sundial. 1529. Germanisches Nationalmuseum, Nuremberg (H 665). Photo: M. Runge.

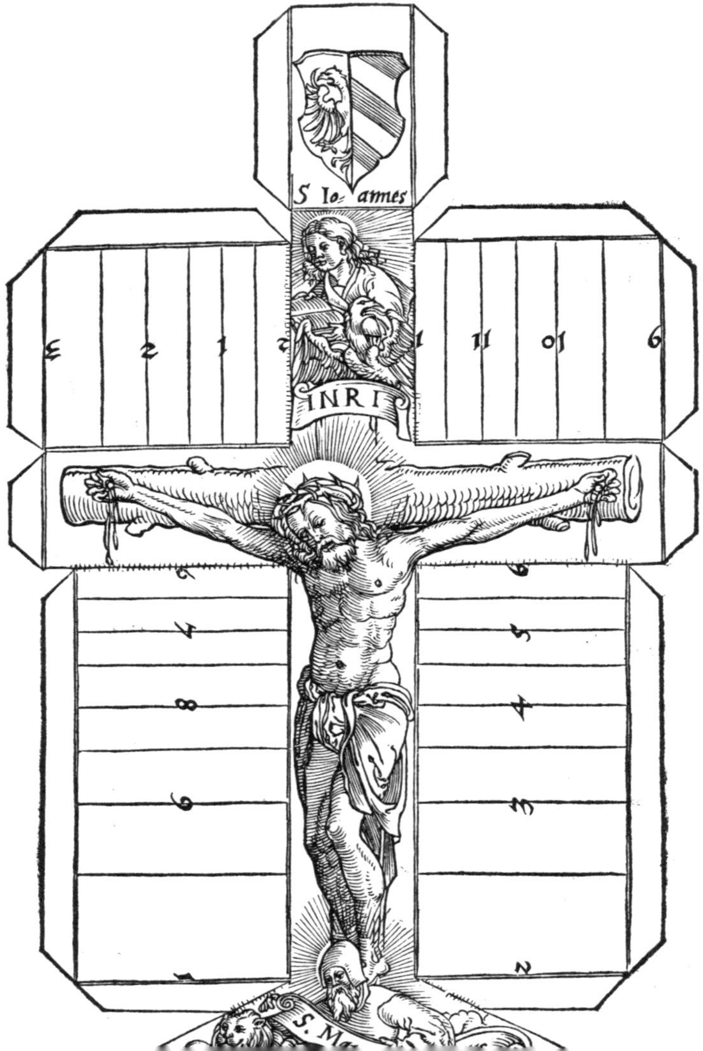
S Io. annes
INRI
S. Ma

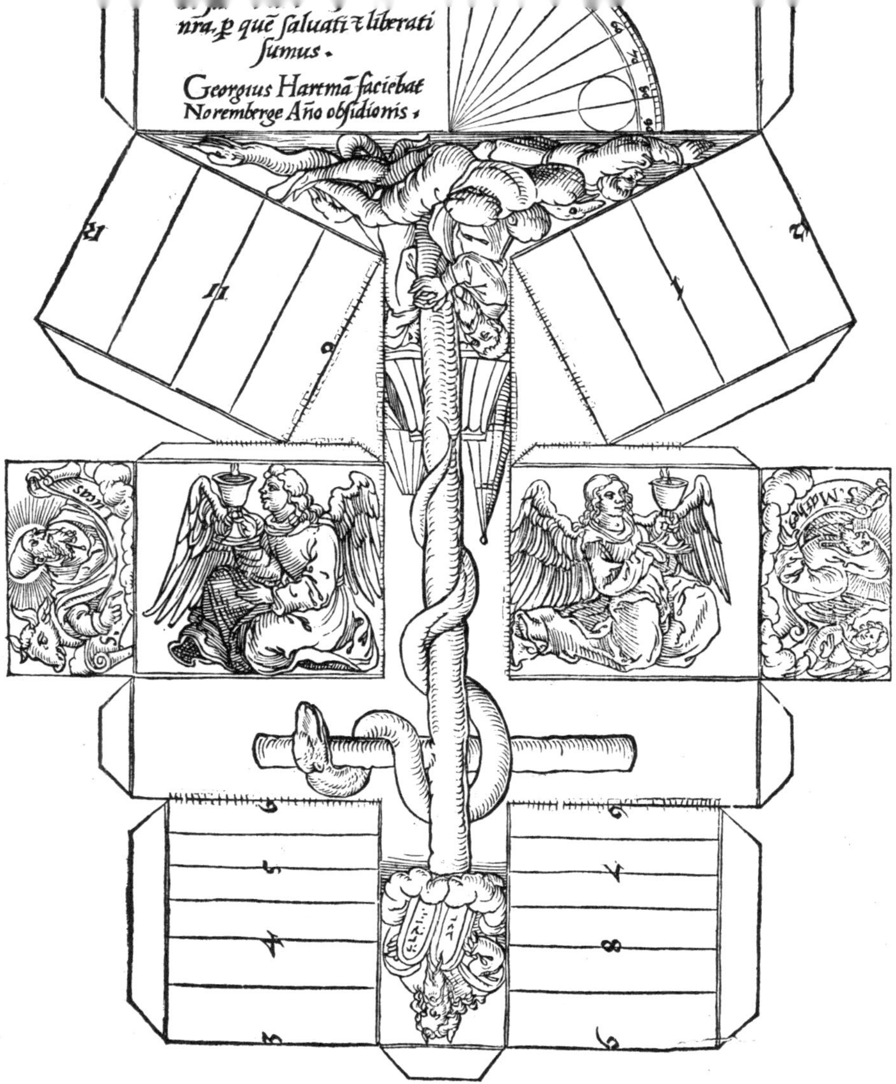
nra. ꝑ quē saluati ꝛ liberati
sumus.
Georgius Hartmā faciebat
Noremberge Año obsidionis.

his own invention, namely one body composed of eight dodecagons, one from twenty-four isosceles triangles, and one from eight equi-
42 lateral triangles. At once practical and phenomenological, Dürer is unambiguous in enumerating his logic in providing solids *ganz offen* rather than adhering to the more traditional Euclidean diagrams or
43 the perspectival stylings of *De divina proportione.*

> Should one wish to create even more beautiful bodies, which touch a hollow sphere with all of their corners but have uneven [*ungleyche*] surfaces, I will lay them out in their parts, totally accessible in the following section so that each of their surfaces touches the other. Whoever wants to replicate these surfaces should trace the larger [i.e., unfolded] figure on a duplicate piece of paper attached to the original and cut this paper along the figure with a sharp knife so that all the lines on the original paper remain intact. Then place the body together along the lines of the drawing. One should pay attention to this technique because it will be useful for
> 44 the following figures.

Simply put, shifting the visual definition of *corpora* to their developable surface on paper increased the ease of reproduction and facilitated the potential for limitless geometrical invention.

The lack of emphasis on the perspective of complex objects and the absence of perspectival elaboration suggests that Dürer aimed for *Underweysung* to serve a different purpose than *De divina proportione* or later Italian geometry books such as Barbaro's *La practica della perspettiva*, both of which relied on polyhedra as epitomes for
45 the demonstration and study of perspective. By not including perspectival polyhedra, Dürer intended the polyhedral nets to ensure that readers would be required to construct the solids out of paper in order to study them, or, indeed, to learn how to imagine them or other *ungeregulierten corporen* in three dimensions solely on the basis of their nets, much as the medieval readers of Calcidius's *Commentary* had been prompted to do with much simpler forms. Dürer, it seems, separated his interest in measurement and geometry from the teaching of perspective. While perspectivally constructed solids were finite, laborious to produce, and restricted analysis to a view predetermined by an artist, the most essential qualities of paper polyhedra

were the number and length of their surfaces and the way their surfaces fit together. The only solid he addresses is the cube, and that rather curtly, leaving the science of perspective up to the reader's imagination. "Through this [I will] show how one may also treat all
of the other bodies." For all intents and purposes, the epochal inven- 46
tion defining the Italian Renaissance in a revolutionizing of pictorial space had been reduced to barely half a page.

As Panofsky famously noted, Dürer was striking out into new linguistic and mathematical territory—territory that tested the limits of the German language while exploiting its capacity to convey meaning through description. German did not yet have words for many of Dürer's geometrical constructions, requiring him to invent a new scientific prose of descriptive-graphical terms. The use of words like "'Fischblase' ('fish's bladder') and 'der neue Mondschein' ('crescent') for the figures resulting from the intersection of two circles," as well as new terms like "'Gabellinie' ('fork line') for hyperbola ... 'Schneckenlinie' ('snail line') for spiral" meant that the typesetters, who would have given a first treatise by Germany's legendary artist their undivided attention, may also not always have been able to follow the challenging logic of the text/image relations. 47

Whether due to the difficulty of interpolating diagram and text or for another unknown reason, many of the pages in Book IV juxtapose the unfolded polyhedral net of a solid (regular, semiregular, or irregular) with the definition of the next solid in Dürer's sequence. For example, the first and most simple unfolded solid to be presented—diagram 29, the pyramid (*dryanglich corpus*)—is flanked by two definitions. Its own definition comes in the paragraph just above the diagram, following an introductory paragraph, while the description of the octahedron, the next solid in the sequence, has been printed beneath the image of the pyramid. The doubling of definitions on this first page creates an intermittent staggering of visual and textual classifications that confounds the desire to read each page as neatly captioned images. The description of the cube is set below the image pertaining to the icosahedron, the description of the dodecahedron is beneath the image of the cube, the description of the sphere lies beneath the image of the dodecahedron, and so on. Given the close attention paid by Dürer to editing and revising drafts of *Underweysung*, and the reverence accorded these edits by

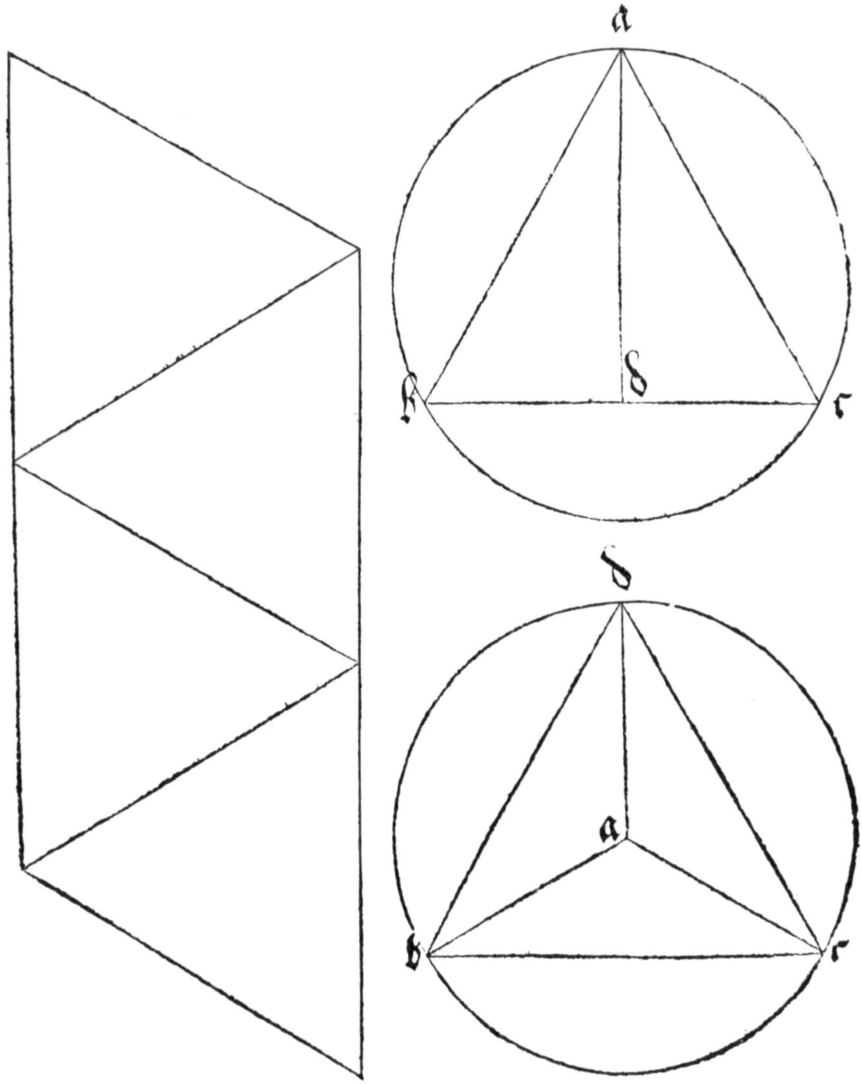

Page with net of a pyramid, from Albrecht Dürer, *Underweysung der Messung* (Nuremberg, 1525). Universitätsbibliothek Erlangen-Nürnberg (CIM.L 6, fol. 69 verso).

the workshop of Hieronymus Andreae (died 1556), the book's printer,
when the text posthumously went to print again thirteen years later,
the fact that Dürer did not address or correct the misleading prox-
imities of polyhedral diagrams and definitions likely implies that
he did not find the arrangement to have posed a problem of legi-
bility. He seems to have conceived of his unfolded polyhedral solids 48
as a continuous scroll of geometrical information unbound by the
printed page: the reader was simply to read the definition and glance
at the image following it, regardless of whether that image rested
on the same page or on the next page, which would have required
the challenging skill of determining whether each polyhedral net
matched the definition printed adjacent to it. 49

Dürer's polyhedral nets would have been produced from wood-
blocks analogous to the outstanding collection at the Albertina
in Vienna, and probably originated from the Andreae workshop. 50
Andreae himself was one of the finest woodblock cutters in Nurem-
berg and a close associate of Dürer's who had garnered a reputa-
tion for carrying out demanding assignments efficaciously, such as
the Herculean task of printing Dürer's life-size *Triumphal Arch* for
Emperor Maximilian. The blocks in the Albertina collection would
have been fabricated from traced drawings framed tight to the image
as they would have been composed for use with type set on a page; the
little ink preserved on the surfaces of some but not all of these blocks
suggests less than extensive use and is in all probability evidence of
only occasional test prints having been made. Though several of the
blocks appear to be copies from the *Underweysung*, most of the nets
do not match the orientation of Dürer's layouts, as strips of geomet-
rical faces radiate outward from an arbitrary central face, giving the
overall nets a starlike impression. Many of the polyhedral patterns 51
carved into the woodblocks sport concentric polygons absent from
Dürer's prints. A few of the blocks use this geometrical ornament to
demarcate the relations between the regular solids and the semireg-
ular solids (as well as prisms and antiprisms), in essence preserving
the potential act of translation by indicating the truncation lines
required to transform the Platonic solids into the Archimedean. 52

Dürer's innovative strategy of seeing solids not just as graphic
objects but as unfolded paper surfaces served his contention that a
knowledge of and willingness to engage with precision geometry

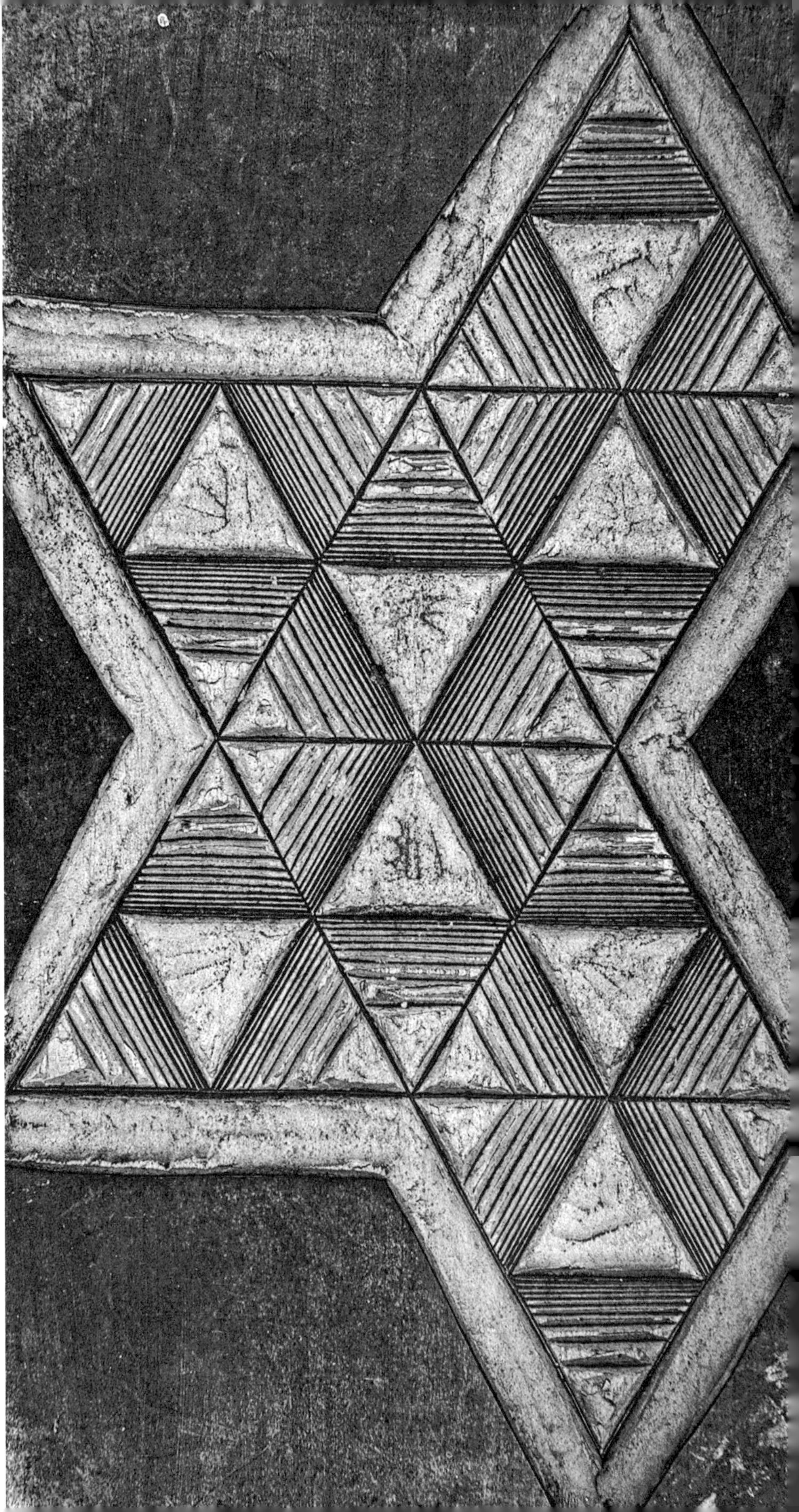

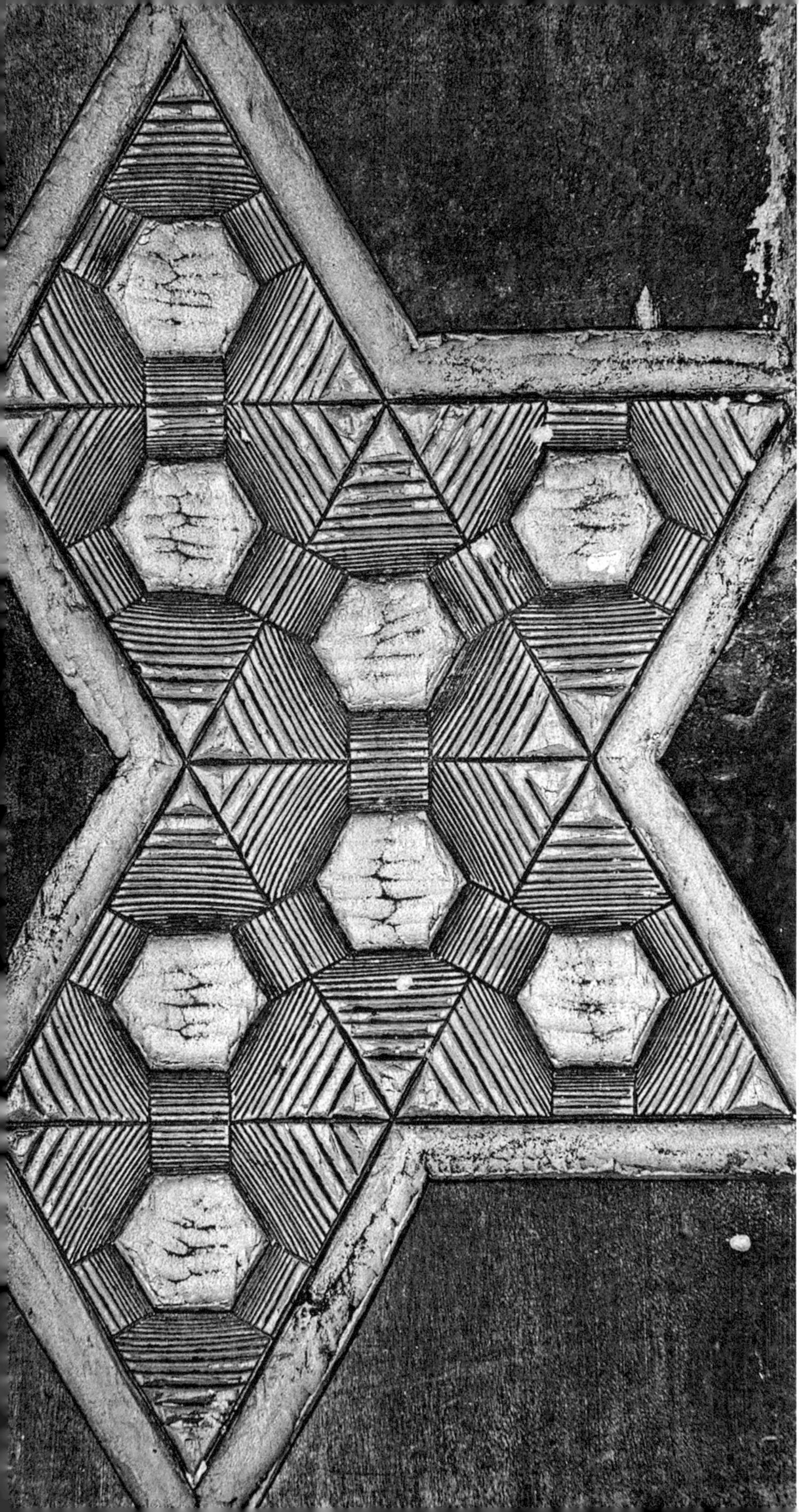

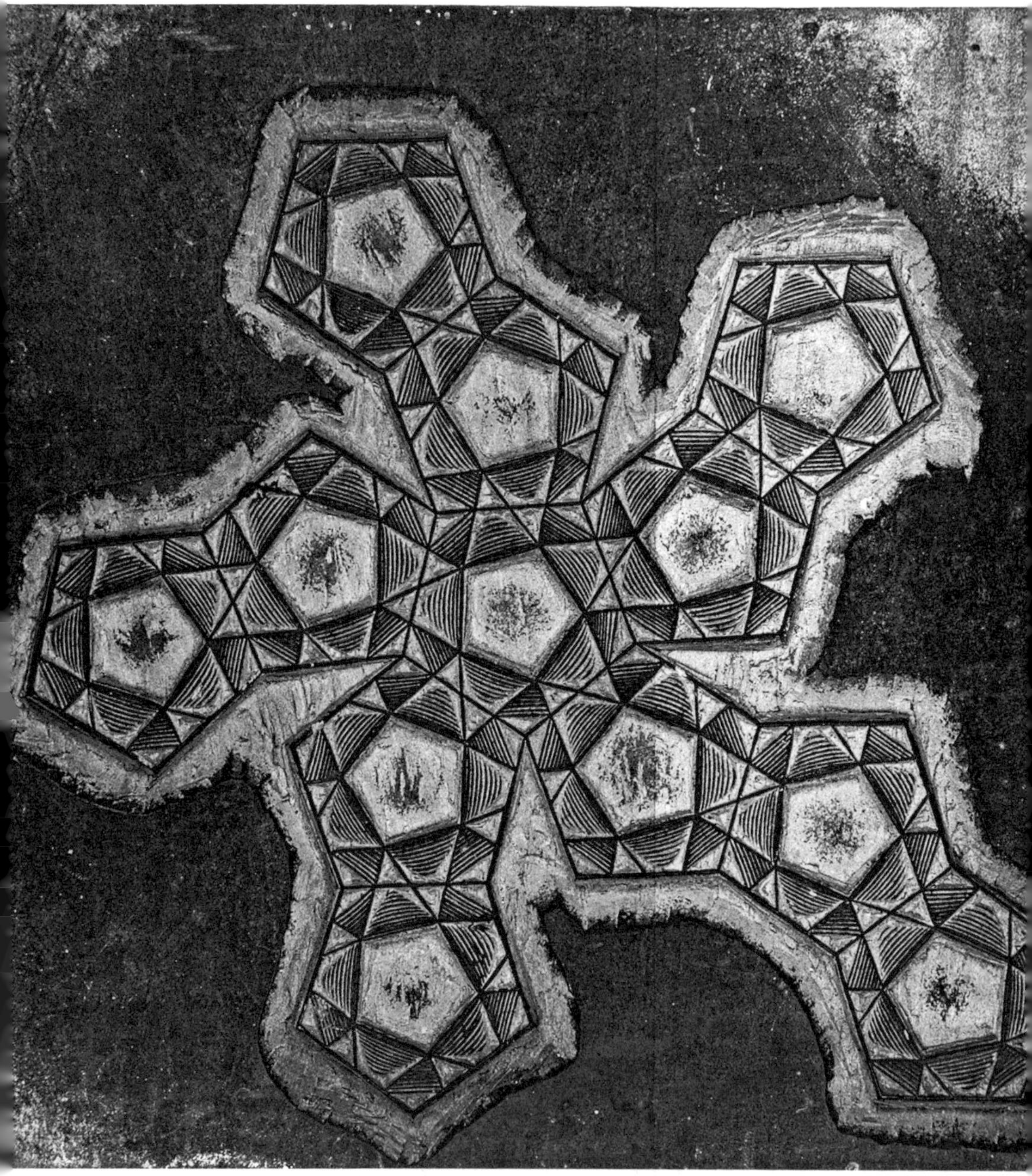

above and preceding spread Woodblocks by Hieronymus Andreae (d. 1556), after drawings by an unknown artist. Net of an octahedron, with incisions indicating a cubocatahedron; net of an octahedron, with incisions indicating a truncated cuboctahedron; net of an dodecahedron, with incisions indicating a smaller dodecahedron. After 1538. The Albertina Museum, Vienna (H02006/706, H02006/70, H02006/723)

was a skill very much in demand. His dedication to Pirckheimer in *Underweysung*'s introduction decries the state of German art as a milieu in which “the sole reason that painters haven't learned the art of measurement is because they have taken pleasure in their errors, and thus they cannot and will not ever be true artisans.” To be a “true 53
artisan” (*rechter werckman*) was to be impervious to seduction by the *appearance* of correctness and accuracy, and to steadfastly pursue the full embodiment of an artistic vision down to the microscopic level.

In contradistinction, “the taking of pleasure in error” (*wolgefallen in iren yrthumben*) generated a panoply of “dishonest mistakes”—dimensions or components of a work born from accidents in the creative process, although these might be imperceptible to anyone other than the artist, or might possess unintentionally pleasurable qualities. These uninvited guests, which could take up residence in a work and seek to convince the artist that their place was rightful, were pernicious entities for Dürer, signifying a general infatuation with impressions, a malaise of imprecision in German art that could only be expunged through a commitment to structuring reciprocities between thought/vision and content/image. Dürer's fastidious attention to language mirrors his faith in the foundational veracity of “the art of measurement” (*die Kunst der Messung*), an art whose business it was to align calculable quantities with graphic qualities and in which *Underweysung* sought to provide useful instruction.

The requirement of cutting out and folding perfect polyhedra cannily inculcated habits attuned to exact observation and measurement. If the net of a polyhedron was not perfectly measured and drawn on paper and then just as precisely cut out, the facets of its geometry would not exactly meet when folded up. Unlike in perspectival drawing, which then as now could fool the eye into believing that an image is the exact and realistic rendering of an object from a specific angle, the precision required from the unfolded polyhedron meant that the slightest inaccuracy would be noticed as soon as the net was folded. Folding, here, was the standard against which the precision drawing (or copying) of a polyhedron could be used to test the accuracy of its baseline drawing and assembly. Unless the edges of the surface were connected seamlessly to each other, the illusion of solidity, and indeed the very identity of the polyhedron, made up as it is by identical polygonal faces, would be destroyed.

Despite its influence, and much like the enigmatic Albertina woodblocks, editions of *Underweysung* do not tend to bear the scars of reproduction, the pricking and tracing that characterize the most quotidian acts of copying. It could be that the pristine page and the absence of cutouts indicates reverence accorded to a master at the height of his fame and the status of the treatise itself as one of the last masterworks to be completed before his untimely death three years later. Devotees may well have used tracing paper or constructed measured arrangements on separate pieces of paper—a relatively straightforward task given that each face of the solid is identical to the other and could thus be arranged in a number of equally plausible configurations. But some of the readers of *Underweysung*, those who did not feel called to enact the instructions as Dürer had prescribed them, may have encountered the unfolded solids not as physical tasks but as imagined possibilities of physical tasks. Here the approachability of the solids, made more so through the removal of Euclid's lengthy descriptions, would have led to envisioning transformation through physical analogue.

Disdaining three-dimensional images of polyhedra for illustrating techniques of *making* polyhedra, Dürer had reconceptualized geometry as something disposable that could be traced, cut out, and folded together. By proxy the nature of geometrical knowledge had become intimate and approachable, capable of being touched, handled, manipulated, and experimented with. In this sense, removing polyhedra from Euclid and reframing them as easily reproducible surfaces had begun the process of making them radically more accessible. Geometrical knowledge would become more portable, easier to traffic, circulate, and copy on paper and, in its portability, less mysterious or theologically inflected; more associated with making and practice, less abstract, more ephemeral and useful for testing, graphic experimentation, and geometrical play. As Dürer claimed in Book IV of *Underweysung*,

> And one is capable of drawing these bodies intersecting one another in a larger duplicate [image], so that one corner sticks into the other plane [surface]. ...Also one can place a point above or below each plane surface to which the surface can connect should one want to create new
> 54 sides [angles] for the surface.

Dürer details specifically how through intersecting two regular bod-
ies and then adopting the outer edge condition of this intersection the
outline of a new geometrical body might be formed, one that could
also be further modified by stretching or depressing its vertex points.
At the basic level, by interlacing the regular polyhedra and accommo-
dating for points of variable altitudes, Dürer had defined a class of
polyhedra of a potentially greater size than Pacioli's, who had limited
himself to adding regular and equilateral pyramids to the faces of the
Platonic and Archimedean solids. But since the Platonic solids had 55
always presupposed a categorical logic in which divinity was to be
found through the divine proportion, it would not be irresponsible
to assume that Dürer's geometrical fecundity had dared, at least on
paper, to broach the profane if not the downright sacrilegious.

The burgeoning instability of the Platonic solids during the
Renaissance was a dimension overlooked by Panofsky's mentor,
Aby Warburg (1866–1929), in his iconological composite of Western
art, the *Mnemosyne Atlas*. Here in the Library for Cultural Studies 56
(Kulturwissenschaftliche Biblothek Warburg) he had founded in
Hamburg, a hub for German intellectual life much like the librar-
ies of Regiomontanus, Walther, and Pirckheimer some four hundred
years earlier, Warburg showcased an unfolded dodecahedron from
La plaisant jeu du dodechédron de fortune ... (1560 [1556]), a text errone-
ously ascribed by its Renaissance publishers to the medieval author
Jean de Meung. Originating from *Underweysung* and pinned along- 57
side excerpts from a variety of astrological pamphlets and images,
Warburg draped the dodecahedron in the garb of the sacred, a fig-
ure "[that] had been estimated by the ancients as the most excellent
and perfect of all geometry, so much so that they had compared it to
the great machine of heaven." Warburg appears not to have recog- 58
nized, or to have been aware of, the epistemological implications of
the newly unfolded polyhedra. Pushing at the uncertain boundary
between geometry as an abstract body of textually described knowl-
edge and the demonstrations of this knowledge on paper, the lines
that make up Dürer's polyhedral nets are no emblems of divine order
but agents of their own erasure; edge conditions prescribing future
actions of cutting and folding; temporary directives extracted from
the morass of wood guiding imaginary cuts between each *Punck* of
the polyhedral net. Unlike Pacioli, who would have fit Warburg's

more harmonic, celestial vision of geometry, Dürer had naturalized *corpora* as strange and wondrous products of the human hand.

By introducing media into the study of polyhedral geometry, forms that do not readily exist in observable nature, Dürer had irrevocably thematized the possibility of making abstract concepts palpable as matter (through the coordination of paper, printed lines, and scissors). The images of polyhedral nets are not "final," nor do they serve as visual substitutions for textual definitions. Rather, the real/physical world of drawing, printing, and making acts as a limiting condition into which the infinite possibilities of geometrical knowledge are made concrete as objects. *Underweysung*, a treatise perpetuating the conditions under which the knowledge that comes from studying objects may be developed, generates paper objects that may be observed, studied, and represented as if one were drawing live from nature or while viewing the exotic collection of a *Wunderkammer*. No longer merely static, ideal forms, the Platonic solids had become open for interpretation, available for use and reuse, unfolding and folding again.

While Pacioli's perspectival solids were watershed examples of new and innovative attempts to use the power of the image to augment Euclid's textual descriptions, Dürer went one step further, transforming the Platonic solids from abstract philosophical concepts, forbiddingly rendered in *De divina proportione*, into cheap, easily copied and manipulable artifacts that might be unendingly modified and adorned. *Underweysung* distinguished itself as the first Renaissance treatise to strip the solids of their purported solidity if not also their metaphysical and Christological symbolism—a solidity that Pacioli, Caraglio, Ugo da Carpi, and others had struggled so assiduously to achieve in perspective. Envisioned more as an instructional provocation than a public dissection of his creative process, *Underweysung* would go on to form the theoretical base on which the next generation of German artists would acquire proficiencies in geometricality. In further adopting the physicality and tactility of Dürer's geometry, they would contribute to the altering of the nature of geometrical knowledge from the abstract and theoretical discussions of Euclid into an intimate and approachable field of investigation with broad popular appeal. Although Dürer saw the measurement and construction of the solids as an intermediary step in building

Unfolded dodecahedron, from Jean de Meung, attrib., *La plaisant jeu du Dodechédron de fortune...*, rev. ed. (Paris, 1560). Warburg Institute, London (FMH 3210, fol. Aii recto).

ENtre tous les ieus & passetemps de fortune dont i'ay eu cognoissance, cestuy que iadis cõposa feu maistre Iean de Meun, excellent poëte François, du temps du Roy Charles le Quint, est à mon iugement le plus subtil, & artificiel. Car il y procede selon les reigles, & demonstrations de l'astrologie iudiciaire, obseruant ses effectz & proprietez assignées aux douze maisons du ciel. Faisant ses demandes & responses tant bien à propos suyuant icelles, que rien n'y est à desirer. D'auantage il prent pour l'instrument de son ieu vne figure geometrale appellée Dodechedron, fort propre & conuenable à son dessein: car elle est semblablement composée de douze faces pentagones egalles, & contient vingt angles, ou carnes solides, & soixante plains, pource que trois plains y font vn angle solide. Ceste figure a esté estimée par les anciens la plus excellente & parfaicte de toute la geometrie, tellemẽt qu'ilz l'ont accomparée à ceste grande machine du ciel, la ciconferen-ce duquel contient douze signes, & douze maisons, & est diuisée en 360 degrez, aussi le Dodechedron contient douze faces pentagones, comme demonstre la presente figure.

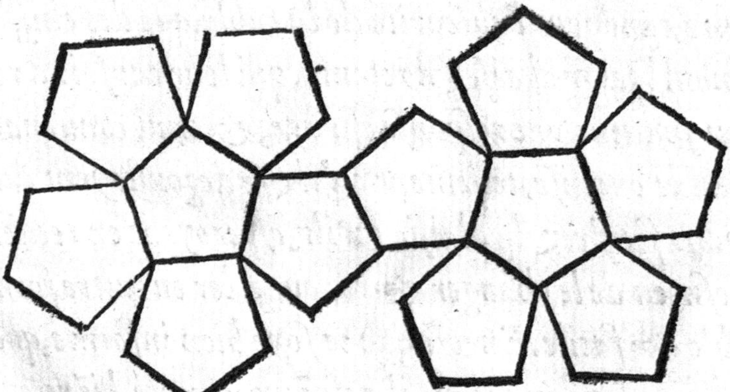

proficiency with geometry that would aid the depiction of nongeometrical bodies, modeling irregular variations of the solids became a convention in its own right—first by diffusing into southern German artistic circles via widely distributed *Lehrbücher*, and then by engendering new geometrical objects intended to be touched, handled, and manipulated, such as the many three-dimensional and irregular polyhedral solids that appear in German print culture and decorative arts. In *Underweysung*, as in Schöner's stretching, flattening, and rolling up of the maps of the earth and stars, geometry became fluid, inclusive of implied conceptual movement and manipulation in which its graphic form served as a temporary resting point—the fixing into a singular representation of a potentially endless process of reinvention, of folding, of transforming solids to surfaces and back again. These were not antique models of Euclidean knowledge, inherited from Greek times and carefully transmitted from one generation to the next. After the publication of *Underweysung*, it would become impossible to imagine the Platonic solids—maybe *geometria* in general—as a transcendently stable, antique entity. A new strand of graphical infinity had emerged through artistic experimentation with geometry.

The epistemological implications of unfolding the Platonic solids were many and various. Now mutable and even graphically mutilated, the solid was no longer an enduring guarantor of solidity. The seeming reality of the perspectival image had been traded for the flexibility and indeed the fragility of a paper model. Though it would take until the maturation of topology for mathematics to exploit the assumption of a theoretical object's hollowness, in their unfolding, as both event and material technique, Dürer's polyhedra emptied out their allegorical and theological content—the trinities and divine sensibilities alongside the crabs, tridents, guns, winds, compasses, and all the other items associated with the five elements—and reduced the solids to their formal properties. This was an act of radical violence disguised beneath the inoffensive manual operations of opening and closing paper bodies. The Platonic solids had been ripped apart in order to reveal a space of possibility and artistic experimentation, substantively free from the elemental and cosmological significations associated with geometry since Plato, free from the weight of antiquity and its Christian synthesis.

The great reveal, however, was not an elusive interiority hidden inside the solids and which, having eluded the hegemony of the perspectival gaze, had been released by the transmutation of geometry from solid to surface. Dürer's desolidification did not uncover secret spaces or hidden depths, but rather showed that, inasmuch as depth had been invented in the case of perspective, so too had the Platonic solids' solidity. Geometry was changing from laborious image to disposable and covetable thing—easily substituted, easily reinvented. Dürer had discovered a new kind of infinity, if not divine provenance, bound not to the finitude of divine proportion but to cycles of perpetual destruction and resurrection. Even as the freedom to invent and distort, to rotate and handle, to crush and dispose that so many later German artists, artisans, and mixed-mathematicians made use of in their geometrical investigations signified the disintegration of the stability of inherited geometrical knowledge, Dürer refused to put things in perspective. In adjoining the misshapen and the rogue alongside the so-called harmonious, the irregular had become the new regular.

GEOM

AND

LEHR

ETRIA

THE

BVCH

The "Renaissance man" was not an exceptional category. Beyond what it meant to be an artist after Dürer, to be a professional in the sixteenth century was to operate in diverse fields defined by opportunities that spanned from academia to private practice and the court, where geometrical knowledge often overlapped with military pursuits (specifically ballistics and fortifications), precision instrument design, and mechanics as much as it did with more mundane duties such as the casting of horoscopes. The astronomers Johannes Kepler and Galileo were expected to do as much for their patrons, the Holy Roman Emperor Rudolf II (1552–1612) and Cosimo II de'
1 Medici (1590–1621) respectively. One was never solely an architect or a goldsmith, a surveyor or an astronomer. All of these professional identities existed in productive tension with one another, animated by the networks of patronage, institutional support, and personal associations that bound natural philosophers and artisans to practical mathematics (sometimes all within the same person).

The capacity to conceive of and accurately construct geometry in three dimensions joined the micro- to the macrocosm in a harmonic universe of vast spatial orders that covered everything from the precision plotting of alphabetical letters in Christoph Stimmer's *Alphabet* (1549) to the determination of the position of celestial bodies as the basis for constructing linear perspective, as in Cesare Cesariano's edi-
2 tion of Vitruvius's *De architectura* (1521). Textbooks invariably began with common Euclidean exercises and progressed toward the elaboration of bodies of specialized knowledge bound together by geometry's expanding relevance. In student notebooks, mechanical parts and ornamental motifs could be imaged on pages preceding calculations for fortification designs; astronomical diagramming could be prefaced by perspectival constructions. Alongside obligatory editions of *Elements*, the collection of "Geometrie" at the famed Herzog August Bibliothek in Wolfenbüttel catalogues numerous books on embroidery, architecture, goldsmithing, weaponry, optics, astronomy, cartography, surveying, and machine design. To learn and work with geometry was to study the basic tools necessary for all forms of mixed-mathematical practice. Diverse functions remained loosely grouped, united by geometrical principles. As the French pedagogue Petrus Ramus (1515–1572) put it, "the description and measuring of the Starres, Countries, Lands, Engines, Seas, Buildings, Pictures, and

Statues or Images ... [need] the help of no other art but of Geometry." 3
Distinguished by the breadth of its use in the definition, prediction, and representation of real-world phenomena and the overlapping similarity of the shape-making techniques it proffered on paper, geometry had decisively emerged from the shadows of astronomy,
music, and arithmetic, its fellow *quadrivium* subjects. 4

In particular, southern Germany, known for its unprecedented concentration of manufacturing talent, strict standards of quality control, and intense competition for commissions, emerged as a singularly rich site of geometrical innovation. Practice was seen to generate geometrical knowledge that was used to burnish the credentials of author-makers, despite the rhetoric of utility not neces-
sarily indicating a reliable link to the practicalities of craftsmanship. 5
If, in universities, the study of geometry was just one level in a pyramid of knowledge, for craftsmen and practitioners, many of whom deployed geometry in service of other more recognizable titles, geometry *was* the central theoretical focus of their profession. The verve of the practical textbook genre or *Lehrbuch*, sometimes referred to as *Kunstbüchlein* and primarily published by German artisans outside of the university system, is emblematic of this centrality. In contrast to the cloistered spaces of the German university—the hallowed hall, the codified text, the scholarly disputation, the learned chains of command—*Lehrbücher* filled a nonhierarchical gap in the acquisition of geometrical knowledge deemed increasingly critical to the development of skills in measurement and representation. Self-contained, open-ended, and in their unmediated relation to the reader antithetical to Reformation-era educational philosophy and traditional courses of university study in mathematics, *Lehrbücher* served as novel graphical spaces of scientific experimentation in which a haptic and tactile form of geometrical knowledge was practiced, contested, manipulated, and invented.

The contributions of the workshop as a site of geometrical innovation unfolded within the broader place of geometry within the Reformation-era development of the German higher educational system. Flare-ups of violence and disobedience by Protestants in Swabia, Thuringia, and the Black Forest threatened the stability of the German states, while the brutal defeat of the revolutionaries in the German Peasants' War (1524–1525), Europe's largest popular

uprising prior to the French Revolution, resulted in casualties num-
6 bering in the hundreds of thousands. Under these conditions, the German university could not afford to stay neutral. Founded in 1502 by Friedrich III, Elector of Saxony, the relatively young University of Wittenberg converted into a hotbed of Protestantism under the influence of Martin Luther (1483–1546), who had been appointed chair of theology in 1512. Philip I, Landgrave of Hesse (1504–1567), inaugurated the University of Marburg in 1527 as the first Protestant university. Even the storied University of Heidelberg, established in 1386 and Germany's oldest university, was profoundly affected by what would come to be called Luther's Heidelberg Disputation, held at a university lecture hall in 1518, and would oscillate between Lutheran and Calvinist allegiances for much of the sixteenth century.

That the Reformation would find purchase within the ideological heart of the ivory tower made the university a command center responsive to the travails of current events in real time. Thus when Philip Melanchthon (1497–1560), the influential German educational reformer and Luther's colleague at Wittenberg, came to believe that Luther's messages were being transformed by evangelical Lutherans into a broad and unintended antiestablishmentarianism that surpassed and overshadowed Luther's grievances with
7 the Catholic Church, the university was primed to act. Together Melanchthon and Luther turned toward a reestimation of reform in which initiative would come not from the populace but from the
8 ruling princes and governments. During one such political crisis, no less than the Augsburg Confession was drawn up by Melanchthon with the hope of reaching a compromise with the Catholic princes. And although future geopolitics would test the theories of nonresistance upheld by the leaders of the Reformation, the lawfulness of opposing the emperor (and state) would remain a fraught philosophical-religious issue for the Lutherans, requiring serious debate and
9 differing forms of justification.

Pedagogy played a key role in Melanchthon's method of instantiating respect for order, civility, and the law. His new humanist curriculum, which was based on interpretations and selective emphases of Aristotelian natural philosophy, actively sought to inculcate a worldview in which civil disobedience was understood to be antithetical
10 to the divine order of nature and society. To study was to recognize

the balance and proportion underlying the natural and divine order
of all things. From man's moral behavior to the proper regulation of
cities, the values of just societies and the apprehension of celestial
movements, the rule of law and authority was intentionally natural-
ized—a pedagogical agenda to flatten out dissent and revolt while
reinforcing respect and deference for institutional establishments.

The Protestant university incentivized deference by doling out
discretionary stipends and maintaining close vocational ties both
to the swelling ranks of the reformed clergy and to the staffing of
municipal Latin school systems. The student body, many of whom 11
came from middle-class families and aspired to acquire bachelor of
arts degrees, were funneled into teaching jobs at the discretion of
the university rector who had the authority to fill regional vacan-
cies. In turn, graduates spread the Reformation and Melanchthonian 12
ideology through preaching, positions in local government, and as
teachers in Latin preparatory schools. Together this new generation
of an educated elite worked to consolidate the changes instituted
by Luther and Melanchthon in regard to educational administra-
tion, the new church, and the territorial state system, creating a self-
perpetuating feedback loop geared toward the development of a new
class of state-oriented political and social leaders. 13

The pedagogical and curricular reforms of the German school
and university systems developed by Melanchthon were the result of
direct consultation with the universities of Tübingen, Leipzig, and
Heidelberg, and radiated out across Europe from his new arts curric-
ulum at the upstart University of Wittenberg. The common syllabus 14
at the storied and timeworn universities of Oxford and Cambridge,
in existence since the eleventh and thirteenth centuries, was recon-
ceptualized in a Melanchthonian manner in 1549 to make the math-
ematical sciences a much more prominent component of English
education; these were now to be taught four days a week, for one
hour from 12 pm, and were based on Euclid and Ptolemy. Of geom- 15
etry, the English administrator and royal advisor Cuthbert Tunstall
(1474–1559) echoed Melanchthon's own sentiments, proclaiming
that "God, architect of all things gave their form to the fabric of the
world and every created thing in it so that all would reveal symmetry
among themselves. ... The power of proportions ... witnesses that God
has arranged all things." 16

Melanchthon's educational philosophy was aided by the legacy of medieval Scholasticism, which prided itself on transmitting a well-trodden and tiered system of knowledge. His reforms built upon the classical liberal arts model of the *trivium* and *quadrivium*, maintaining the placement of the mathematical sciences within a hierarchy of subjects topped by theology and metaphysics. In contrast to Italian universities, which concentrated on terminal degrees in law and medicine, theology was the most prestigious graduate study in German universities, with the majority of students pursuing bachelor's degrees in the arts, a category that encompassed the humanities,
17 logic, and philosophy. In the Carthusian prior Gregor Reisch's (ca.
1467–1525) *Margarita philosophica* (Pearl of wisdom) (1503), one of the first printed encyclopedias, education is rendered as a symbolic tower of learning populated by historical figures representing the progressive levels of universal knowledge—Cicero for rhetoric, Ptolemy for astronomy, Aristotle for logic, and others. The study of geometry is associated with Euclid, who is identified on the tower's second tier. Depicted as Nicostrata, the Roman goddess of childbirth and inventor of the alphabet, Grammar beckons an eager pupil to follow her out from the nearby urban surroundings, a chaotic panoply of buildings in the middle distance. Education will serve as a form of intellectual rebirth through engagement with the heroes of Western culture, who will in turn afford the student access to the eternal knowledge of the ancients. While the design for the building is sound and everlasting, it is only through ascending the tower of learning that critical perspective can be gained on the vagaries of modern life.

The discovery of the New World lent a particular urgency to the development of expertise in mixed-mathematical fields such as navigation and cartography, resulting in the establishment of specialized trade schools disconnected from universities in seafaring
18 countries like Spain. But in central, landlocked Europe, university lib-
eral arts requirements for the study of mathematics reflected neither the demand for increased training in geometry nor the mixed-mathematical skills that characterized southern Germany's thriving precision industries. The heavy reliance of the humanist curriculum upon the classics—Ptolemy's *Almagest* and Pliny's *Natural History*, alongside time-honored medieval staples such as Sacrobosco's *De sphaera*—meant that university courses remained committed to the

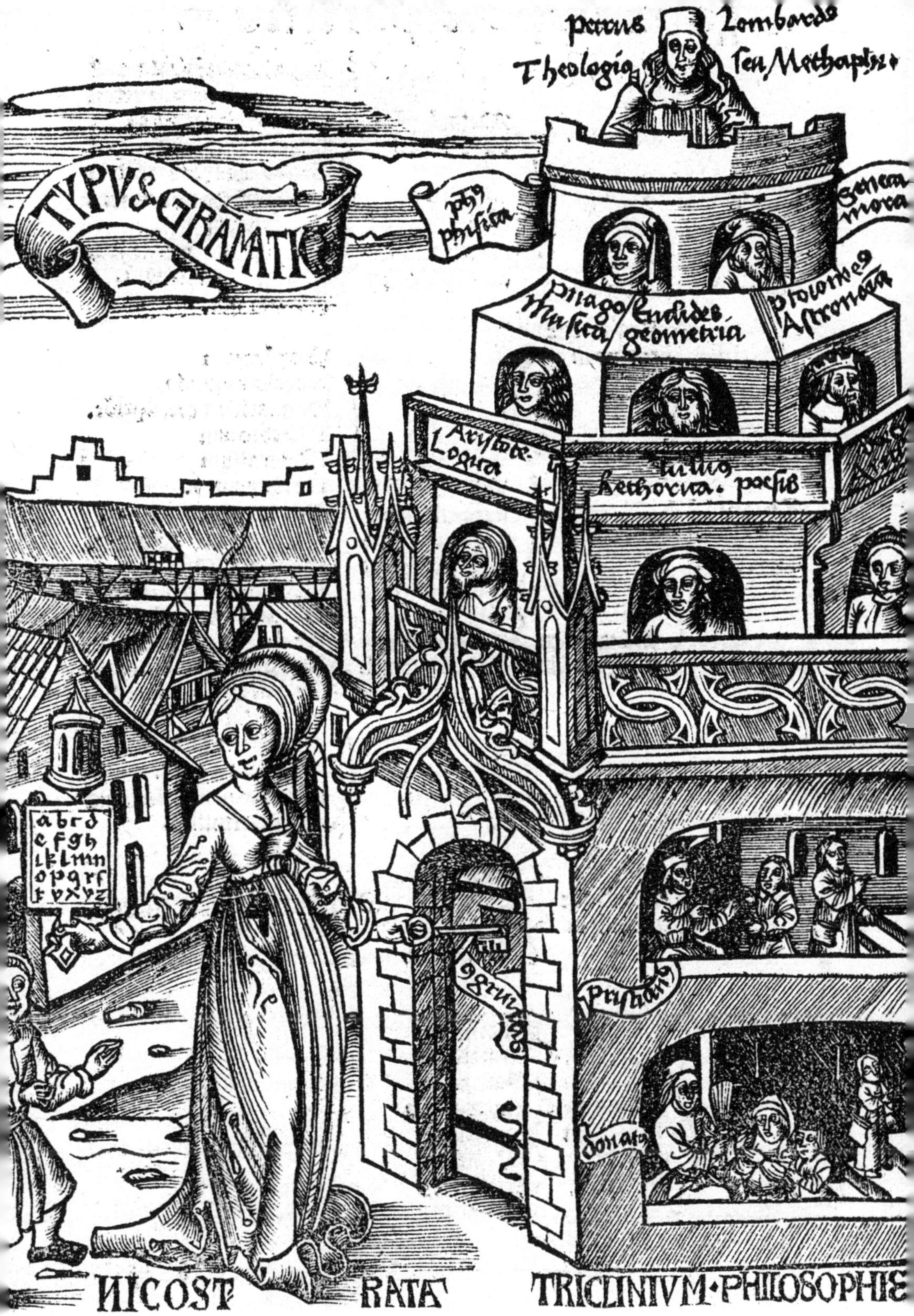

Allegory of Knowledge, from Gregor Reisch, *Margarita philosophica* (Basel, 1517).
Bayerische Staatsbibliothek, Munich (Res/4 ph.u 121, p. 1).

teaching of geometry through the theoretical exercises and proofs in
19 Euclid's *Elements*. The most notable changes to occur in mathematical
instruction had less to do with the content of the curriculum than
with the arrival of new textbooks, such as a modern translation of
20 *Elements* from the purified Greek text. Furthermore, though math-
ematical instruction since the Middle Ages had bridged into optics,
music theory, and astrology, most universities only demanded a min-
imal competence with the first few chapters from *Elements*, creating a
vast gap between the average student's academic knowledge of Euclid
and all the possible real-world and professional scenarios to which
21 geometry might be usefully applied. As an example, the statutes of
the faculty of arts at the University of Bologna required that only the
first three books of Euclid were to be taught, one per year, in the first
three years, while the 1389 statutes from the University of Vienna
22 stated that only the first book of Euclid was to be taught.

By contrast, craft knowledge was a highly guarded commod-
ity in the sixteenth century, particularly in artisan-dense cities like
Nuremberg. Apprentices did not expect a codified course of study
and had to learn skills through observation and assistance in the
23 workshop. In most cases, they were required to spend anywhere from
two to five years in training and could range in age from twelve to
eighteen years old. As most were illiterate, their mastery of a trade
was independent of reading and writing skills, though literacy
and knowledge of contract law were considered useful for success-
24 fully navigating through the guilds' citizen-run administrations.
Practical instructions for the obligatory apprenticeship required by
master craftsmen were usually not written down, though standards
of conduct and the quality of workmanship were enforced by the
ordinances of each city's guilds or council bodies. Few if any journey-
men had formal or theoretical training in any form of mathematics,
25 art, or mechanics prior to commencing their apprenticeship.

Though not substituting for the mentorship of a master, image-rich *Lehrbücher* published in the vernacular conveyed ideas that could be apprehended visually, teaching sets of conceptual and theoretical skills beyond what might be experienced firsthand in the workshop. For elite craftsmen or journeymen with aspirations to improve their grasp of geometry and measurement, university textbooks in the mathematical sciences were of little help, as they

were invariably published in Latin for learned audiences and thus excluded the majority of object makers. For most craftsmen and artists in the early years of the Reformation era, the majority of whom necessarily worked outside of the university, the only way to better understand mathematics was by directly observing their peers or through studying Dürer's large-format and highly sought-after treatises. Here the new genre of geometrical *Lehrbücher* published locally in the south were essential to the communication and distribution of geometrical knowledge. Circulated in tandem with hands-on practice and in place of an official university system or institutional training, these *Lehrbücher* contributed to an informal network of training in geometry and its manipulation in perspective targeted to German object makers. Given Dürer's notoriety, *Lehrbücher* were often also pitched as explications of his writings, which were considered difficult or opaque, and consequently enjoyed a great deal of popularity throughout Germany. While in surveying and measuring dis- 26
ciplines innovation occurred in the manufacture of tools, in the trades that produced ornate objects innovative graphic strategies circulating within the numerous *Lehrbücher* offered conceptual bases for learning how to design in the abstract—apart from material, cost, and any other practical considerations that might later impact a design during development. Indeed, these books surged to fill a gap in artisanal education by purporting to specifically target and develop the ability of object makers to think about and to design complex form in three dimensions.

Dürer's pioneering use of geometry in *Underweysung der Messung* was immensely influential, introducing media to the Renaissance category of *geometria*, particularly through his opening up of the *corpora regulata* as unfolded surfaces. Central to the text's philosophy was the transformation of the solids—the most advanced geometries described by Euclid—into paper polyhedral "nets" that could be printed, cut out, and glued together to create quick mockups for perspectival drawing. Shedding their mystique of unapproachability, these copies enabled a tactile and playful relationship to geometry that expanded beyond the initial subject matter, heralding a new era of abstract experimentation by later German artists and artisans such as Augustin Hirschvogel (1503–1553) and Heinrich Lautensack (1522–1590), as well as Wenzel Jamnitzer, Johannes Lencker (active

1551–1585), Hieronymus Rodler (d. 1539), Erhard Schön, Lorentz Stöer
27 (1537–1621), and others. In turn, these artists' exploration of geometrical form making completed the transformation of polyhedra from neoclassical stereometric drawing aids in the Italian *studiolo* to the premier objects of northern European geometrical curiosity.

The precedent of Dürer's unique engagement with the Platonic solids created a substantial research track in southern Germany, further developed in the *Lehrbücher.* These practice-oriented texts diverged from the priorities of Melanchthon's humanist curriculum and suffused the study, measurement, and production of geometry with a spirit of mathematical inventiveness, newness, and erudition. Moreover, the improvement and development of techniques for constructing new geometrical forms sharply diverged from a mathematics predicated upon learning the basics of *Elements*. Though they were not mutually exclusive, at stake was the identity of geometrical knowledge. Geometry both described a preexisting classical order, eternal, timeless, and distinct from the messiness of everyday life, and defined a site for the exploration of new shapes, forms, and utilities which had never before been seen.

Still, there was a degree of porosity between practice-based mathematicians and academic life. Makers could be asked to participate in universities or collaborate on projects with local sovereigns. Protestant universities, often emulating Melanchthon's Wittenberg model, provided necessary expertise in mixed mathematics on request, particularly surveying and practical geography, when required by local princes. The University of Helmstedt offered students geographical instruction in *mathematicus inferior*, although the instructional texts were the standard *Arithmeticae practicae methodus facilis* (1576) by Gemma Frisius, Johannes Honter's *Rudimenta cos-*
28 *mographica* (1546), Pliny's *Historia naturalis*, and Ptolemy's *Geographia*. Germany's first *Gymnasium*—St. Egidien in Nuremberg, also guided by Melanchthon—hired Johannes Schöner, one of the most prominent mapmakers and builders of globes in Nuremberg, as professor
29 of mathematics. Nevertheless Schöner's monolithic *Solidi ac sphaerici corporis* (1517) is substantially bereft of explanatory graphics and makes no reference to or use of the geometrical techniques inherent in the author's mapmaking and globe-building practice. And while some German universities like Helmstedt employed practical

mathematicians who taught an "inferior" strain of mathematics, the pedagogical trajectory still led to "mathematicus superior," which remained firmly based on Euclid and Ptolemy.

Melanchthon too promoted a mathematical pedagogy that delved deeper into text as opposed to developing skills in the graphical manipulation of geometry. Far from taking advantage of the blossoming knowledge and techniques for visualizing geometry in the workshops of the southern German cities, academic textbooks shied away from physical practice and making. The politics of keeping geometry restricted within a newly Protestant framework, separate from pressing geopolitical concerns and removed from practical application, suggests a schism between the mathematical study of classical texts occurring within the university/secondary school system and that occurring within environments predicated upon the production and selling of goods and services, for which the theorems of Euclid were less useful than the practice of learning how to create and manipulate geometry. At the same time as Melanchthon's educational reforms encouraged a critical investment in the mathematical sciences which was not necessarily common across other humanist-inspired school systems, neither the higher nor lower mathematical curricula adopted the mixed-mathematical *Lehrbücher* widely circulating around Germany at the time. 30

The newest developments in the swiftly accumulating knowledge of the visual possibilities of geometry, the *Lehrbücher* for teaching geometry and perspective, were neither destined for nor products of university education, but emerged in tandem with the priorities and conceptual innovations of mixed-mathematical practices. Seizing on the potential of geometry to train the visual imagination, the coming century's *Lehrbücher* turned the invention and execution of new graphic geometries into a fundamental tenet of artisanal education, beginning a trend of learning geometry through visualization that would continue through the following centuries, as can be seen in Jean François Niceron's *La perspective cvrieuse* (1652) or the geometrical sculptures populating the pastoral landscapes of Christoph Andreas Nilson's *Anleitung zur Linearperspective* (Instructions on linear perspective) (1812). Removed from the drafting board and installed in a pleasantly wooded glade, Nilson's stacked architectonic examples (*architektonischen Beyspielen*) illustrate the unaffected movement from

Euclidean geometry into the rendering of stereometric forms in perspective as "being the most natural of steps, as geometry and perspec-
31 tive are very closely connected with each other." Though there was no overarching directive, no singular institution that sponsored all the transformations of geometry, these investigations preserve the methodological approaches to the development of design intelligence by early modern German artisans, and in accumulation constituted the origins of a groundswell of research into the morphological and topological possibilities of geometry.

Arising out of the vibrant artisanal culture of southern Germany, *Ein aigentliche vnd grundtliche anweysung in die Geometria* (1543) by Augustin Hirschvogel and *Des Circkels vnnd Richtscheyts* (1564) by Heinrich Lautensack were *Lehrbücher* that taught shorthand methods of conceptualizing and inventing geometry on and with paper, no knowledge of Euclid required. Both had little explanatory text, in contradistinction to academic texts on geometry, and were filled with exuberant prints marketed to their generation of German artisans/mixed-mathematicians for whom knowledge of geometry and the ability to precisely manipulate and measure three-dimensional form on paper, in fulfillment of Dürer's directive, had become essential. *Geometria* and *Des Circkels vnnd Richtscheyts*, like other *Lehrbücher*, were understood as records of prior experimentation with geometry by their authors as much as they were also experimental spaces encouraging explorative, graphic form making. As such, unlike *Elements*, these texts were concerned with a unique type of utility or practicality—what *geometria* was and could be in and of itself—aside from geometry's fundamental role in measurement and description, and apart from the repeatable theorems of Euclid. At once pragmatic and abstract, Hirschvogel and Lautensack demonstrate some of the unique contributions made by artisans to the way that geometry was understood, taught, and used in Germany, and testify to how the invention and manipulation on paper of three-dimensional geometry had become integral to artisanal training.

A famous artisan in his time, Augustin Hirschvogel is described by the eighteenth-century mathematician Johann Gabriel Doppelmayr in his *Historische Nachricht von den nürnbergischen Mathematicis und Künstlern ...* (1730) as a glass painter with great skill in design, etching, painting, and enamel work who, later in life,

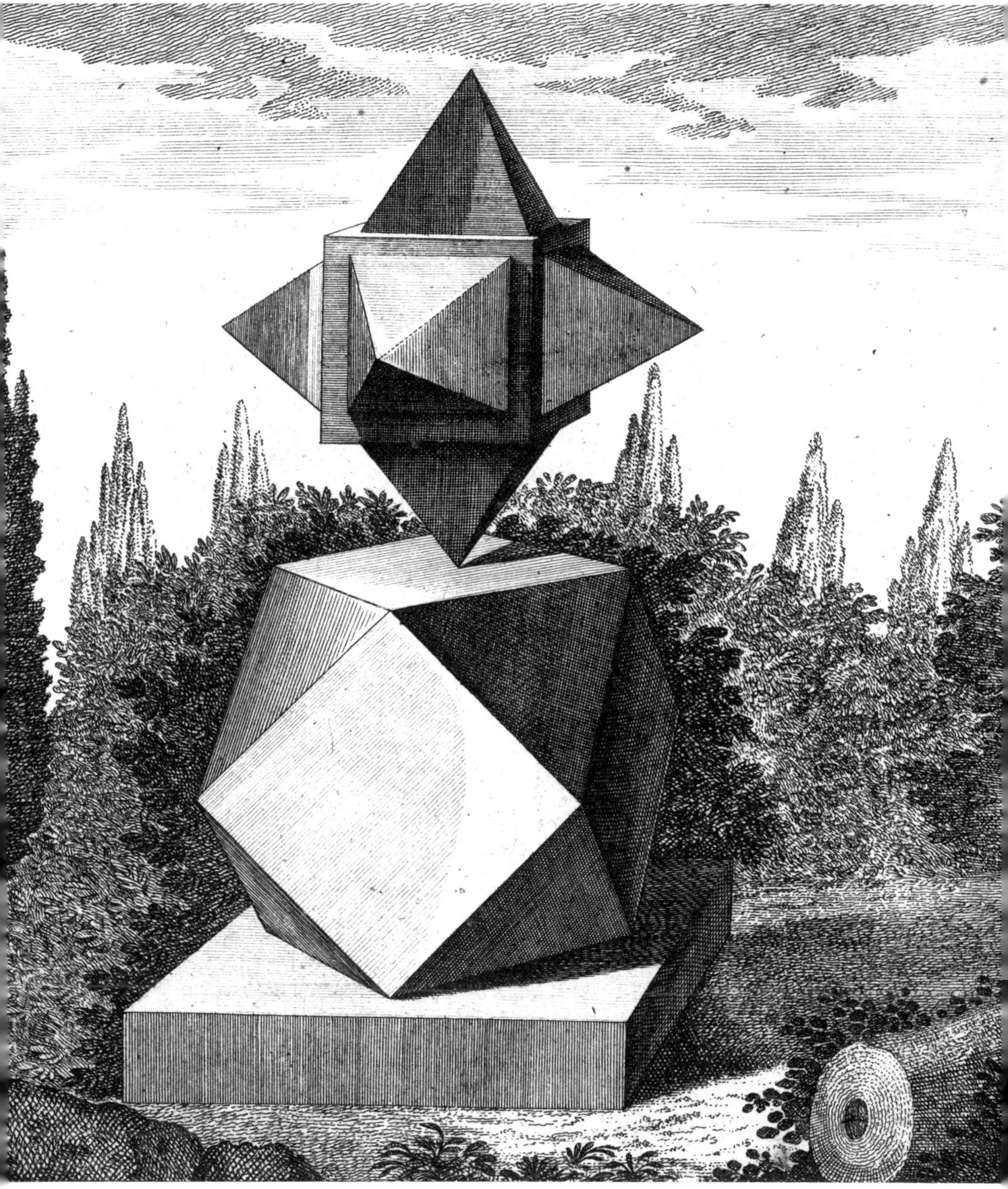

Stacked stereometric figures, from Christoph Andreas Nilson, *Anleitung zur Linearperspective* (Augsburg, 1812). The Huntington Library Collection, San Marino, California (706131, fol. A).

32 also engaged in stonecutting and mathematics. By mathematics Doppelmayr was referring to Hirschvogel's *Geometria* (1543), a slim, leatherbound *Lehrbuch* in two volumes locally published shortly before Hirschvogel set off to survey Europe and design maps of Moscow
33 and Austria dedicated to Ferdinand I, the Holy Roman Emperor. Hirschvogel produced over three hundred etchings of gold vessels, ornament for goldsmiths, book illustrations, religious, historical, and mythological scenes, coats of arms, and representations of hunting and animals, though he is only known to have produced one
34 *Lehrbuch*. Despite being limited to a single edition, *Geometria* was appreciated for its visual clarity and found a broad audience. Besides appearing in many contemporary libraries, including that of August the Younger, Duke of Brunswick-Lüneburg (1604–1666) and founder of what became the Herzog August Bibliothek, *Geometria* was extensively referenced in the Spanish artist and anatomist Juan de Arfe's (1535–1603) *De varia commensuracion para la escultura y arquitectura* (1585) and was still in use at least one hundred years after its publication, as evidenced by the foreword to Daniel Schwenter's *Geometriae practicae novae et auctae tractatus* in which he explicitly acknowledged his debt to Hirschvogel. "Hereupon I was lent by a good acquaintance of mine the well-known Geometria of Augustin Hirschvogel. I have studied it with great zeal, and because it is easy, not complicated, and well-
35 argued, also with great pleasure."

Hirschvogel's text is split into twin volumes in which the corresponding expositions and visualizations are cannily separated—an organizational strategy likely intended to maximize its use among apprentices and craftsmen. The tractate's full title indicates the extent of Hirschvogel's ambition: *Ein aigentliche vnd grundtliche anweysung, in die Geometria, sonderlich aber, wie alle Regulierte, vnd Vnregulierte Corpora, in den grundt gelegt, vnd in das Perspecktiff gebracht, auch mit jren Linien auffzogen sollen werden* (An authentic, thorough instruction in geometry, especially how all the regular and irregular bodies are to be inscribed in the ground, and brought into perspective, also illustrated with their lines). In tacit allusion to the *Vogel* in the name Hirschvogel, the frontispiece to *Geometria*, the second book of images, sports an owl, symbol of Athena, the goddess of wisdom, jauntily perched atop an Archimedean solid, specifically the small rhombicosidodecahedron, outlined with the Latin phrase *Spero Fortvne Regressvm*

Title page, from Augustin Hirschvogel, *Geometria* (Nuremberg, 1543). The Metropolitan Museum of Art, New York, Harris Brisbane Dick Fund, 1932 (32.18[1–2]).

GEOMETRIA

DAS PVCH GEOMETRIA IST MEIN NAMEN ·
AL FREYE KIST AVS MIR ZVM ERSTEN KAMEN ·
ICH PRING ARCHITECTVRA VNDT PERSPECTIVA ZVSA-
MEN ·

15 AF 43

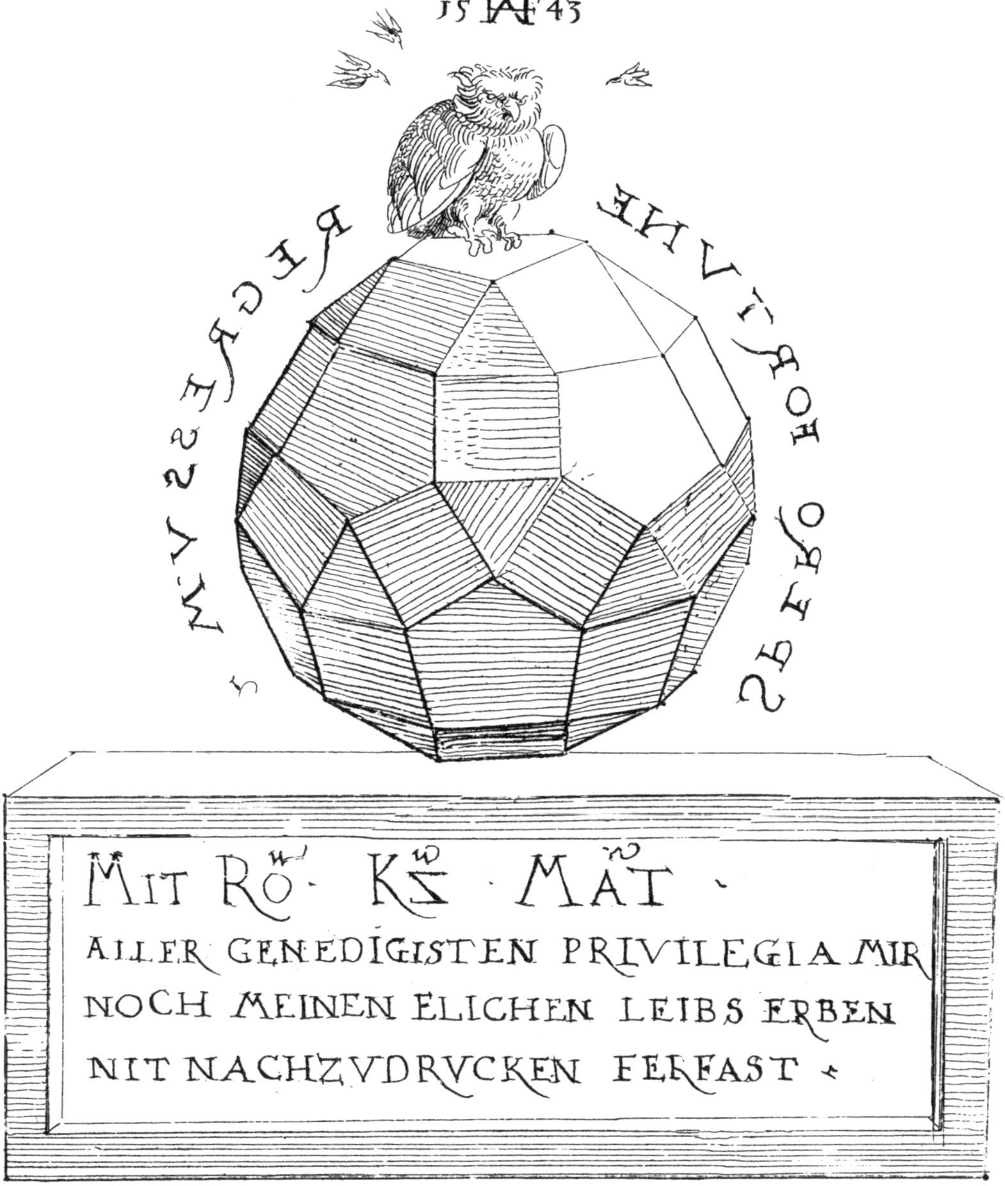

(I hope for a return of fortune) with reference to *Aeneid* 11.413: *neque*
36 *habet Fortuna regressum* (Fortune cannot reverse her course). Inscribed underneath the title is a light-hearted rhyming phrase, “Das Buch Geometria ist mein Namen. All freye Kvnst avs mir zvm ersten kamen. Ich bring Architectvra vnd Perspectiva zvsamen” (The book Geometria is my name. All free art came first from me. I bring architecture and perspective together). In the style of Diderot’s *Encyclopédie*, if less encyclopedic in scope and two hundred years earlier, the main text is apportioned into numbered points directing the reader to corresponding images in the companion volume. The length of text is unusual for a German *Lehrbuch* and points to Hirschvogel’s desire to give additional theoretical instruction on perspective drawing, which, as he states in the introduction, is “a noble and useful art of measurement (known in Latin as Perspectiva) which has been kept hidden in the German language, and which for the common man is hard to acquire and learn, given that the better part of it was written
37 in Latin and Greek.” The transference of geometrical knowledge from text and image to eye and mind required no professor, no rector, no course of classical studies nor the satisfactory completion of an arts degree, nothing more than turning the pages of a book.

Following a Euclidean-style progression from the simplest geometrical elements to the most complex, Hirschvogel begins with an exposition on lines and circles before turning his attention to the main subject of *Geometria*, namely “Der Ander teyl de corporibus” (The other part concerning the Platonic solids): double-page spreads of perspectival polyhedra floating alongside their planimetric and unfolded views illustrating a pattern for building each solid out of paper as well as a scaled reference model of the solid once assembled. Much as Dürer had outlined in *Underweysung*, Hirschvogel prescribes a method of construction, placing emphasis on the necessity of using scissors and the practice of deliberate folding or bending (*biegen*). “So you should cut and fold thick paper, glued and held
38 together, and thereafter you will have a material body.” The polyhedral exemplars are depicted consistently both as the nets (*Rete*) of adjacent, equal polygons making up each solid’s unfolded surface and as their corresponding perspectival assemblage demarcated within a concentric sphere. Even when Hirschvogel describes further articulated polyhedra covered in increasingly more facets than

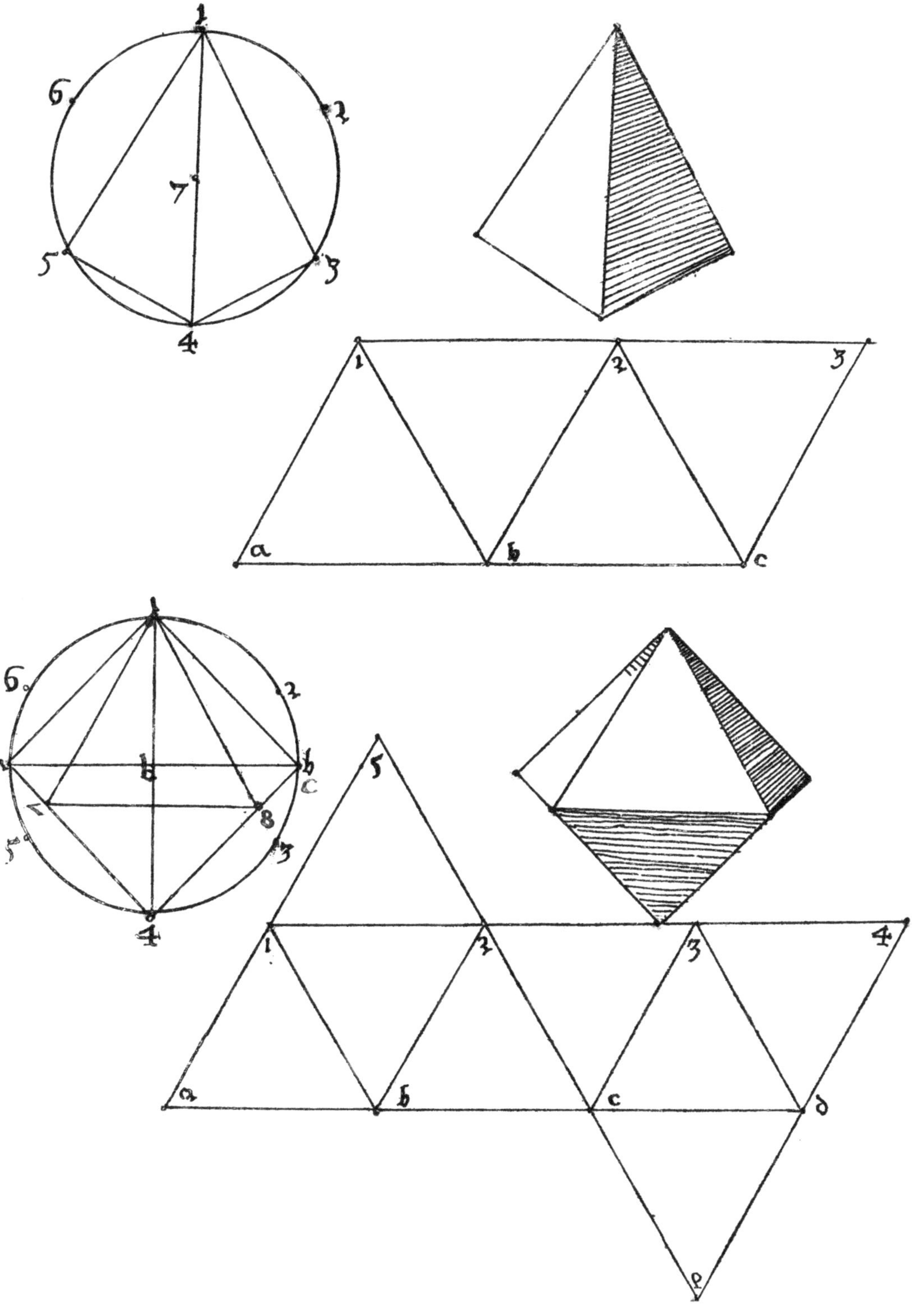

above and overleaf Pages with geometrical models, from Augustin Hirschvogel, *Geometria* (Nuremberg, 1543). The Metropolitan Museum of Art, New York, Harris Brisbane Dick Fund, 1932 (32.18[1–2], fols. B I and A VII).

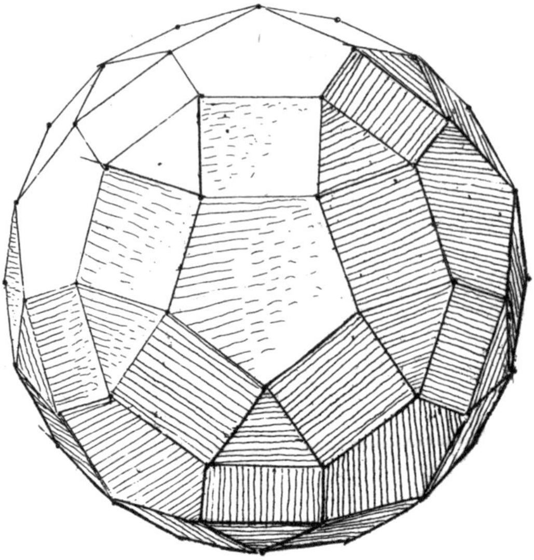

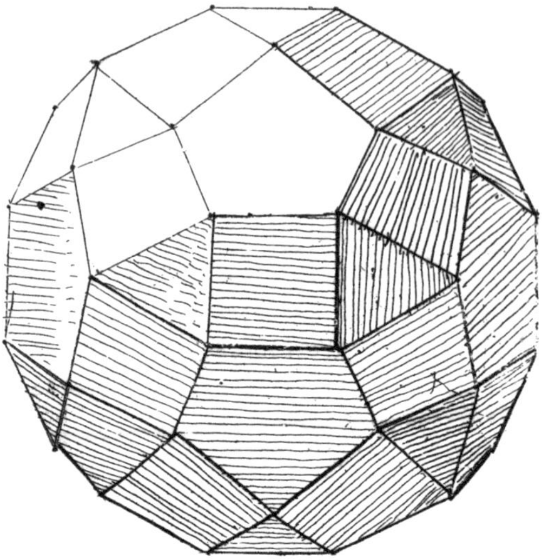

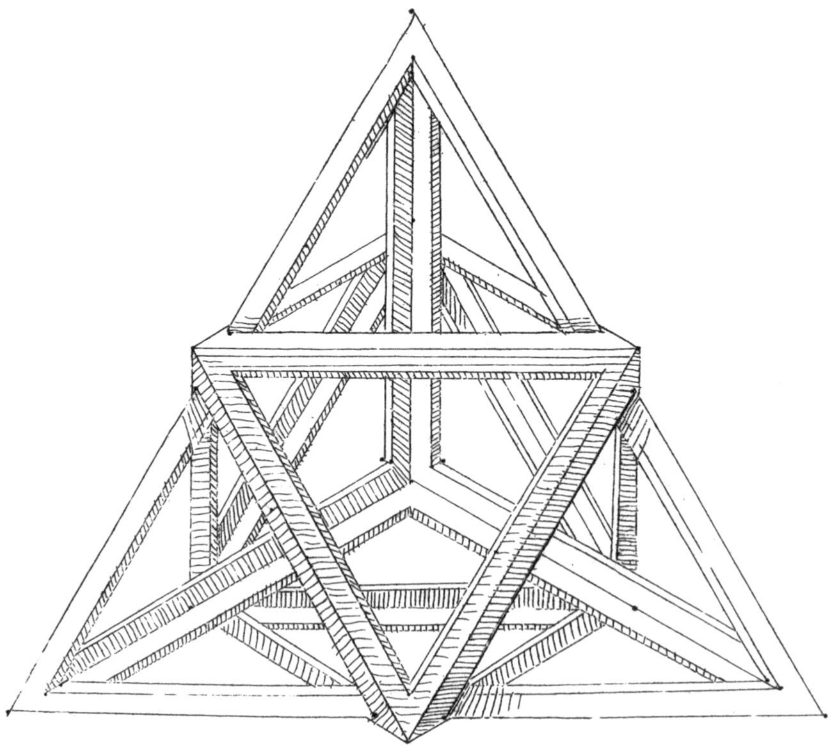

the *corpora regulata*, these geometries are rendered within a sphere as well. Moving through all the regular solids in *Geometria* and beyond the realm of the Platonic and into Archimedean territory, it becomes implicitly understood that geometrical evolution for Hirschvogel is equated with a further and further segmentation of surface, which, at the edge of an infinite process of subdivision, would begin to approach their limit condition as spheres.

While the aim of a *Lehrbuch* was often to explicate principles by which objects could be *in das Perspecktiff gebracht* (brought into perspective), these principles varied from author to author, as the graphical techniques for constructing and inventing objects in perspective depended on personal or stylistic preferences. Hirschvogel, for his part, speaks about perspective almost as a new state of being into which geometries and other objects could be carried. The word *gebracht* from *in das Perspecktiff gebracht* (in the title of *Geometria*) implies both a directional vector whereby a quantity is moved from one place to the next and the stationary state of a thing rendered in perspective. The phrase *in den grundt gelegt*—meaning to situate/locate a geometrical body in/on the ground (or in plan)—indicates the act of "setting down" a polyhedron as an unfolded graphic net to prepare it for translation into perspective, first through building it as a paper model and then by rendering the model perspectivally back on paper. Perspective, then, was a virtual space, albeit a graphic one, where geometry—made physical as a haptic object—could be transported onto the page through close observation, precise measurement, and sustained practice. The phrase *mit jren Linien auffzogen* (with their lines illustrated) refers to the unfolded sides of his solids laid flat against the surface of the page. Similar to Dürer's earlier directives, Hirschvogel's instructions make it explicit that these solids were to be constructed from the paper-as-ground. The woodblock-printed lines double as tracks guiding the path of scissors, the net itself most probably having been transferred to another page by pricking the edges of the figure with a needle through an additional loose leaf of paper, which could then be cut up to form each solid without ruining the original book. The logic guiding Hirschvogel's geometry is the development of a more expanded repertoire of shapes predicated specifically upon their constructability from paper. Defined by the technologies he used to draw, the terms of Hirschvogel's inventiveness were wholly dependent upon these specific techniques of making.

As represented in the Augsburg stone engraver Peter Halt's (1575–after 1634) *Perspektivische Reiß Kunst* (1625)—a printed compendium of geometrical models on ornamental bases—once a paper model had been cut out and glued together, the next step was to capture its outline as a drawing. While this could have been accomplished by sight alone, workshops used a variety of tools to help direct the eye to the correct point on the object to be brought into perspective. For instance, Halt shows an octahedron placed upon a simple measuring device with a rotating hinge, which enables it to be traced
39 by a pencil held vertically by an adjustable bracket. The octahedron would accordingly have been transferred into a perspective drawing and then graphically operated upon, its surfaces chiseled, protuberances added, and embellished with further geometrical subdivisions to create irregular variations on the regular solids. This is to say that *Geometria*, in concert with graphical and measurement tools, was not just a static thing to be admired nor a paper instrument to be made once, but an agent setting into motion uncompromising phenomenologies of its own deconstruction and assembly. Collapsed onto the page of a book, the world was reconstituted in two dimensions merely to be excised, folded together and brought back to life in three dimensions, only to be captured once again in perspective, translated into a planar net, and made anew again. These cyclical regimes of transformation and intermediality allude to the various editions of Caraglio's *Diogenes* or the tripartite correspondences between text, model, and drawing in de' Barbari's *Portrait of Luca Pacioli*, though without privileging the erudition of the print connoisseur or the Neoplatonic philosopher. Geometry had been pulled out from under Euclid and transplanted to a terrain more hospitable to artists, architects, designers, and all those for whom the cultivation of the geometrical imagination was a practical and material necessity.

In his *Ein schöner kurtzer Extract der Geometriae vnnd Perspectiuae* (1599), Paul Pfinzing von Henfenfeld (1554–1599), cartographer and chronicler of perspectival technology, elaborates on the working spaces of Dürer and Wenzel Jamnitzer as well as the perspectival apparatuses they and the Nuremberg artists Heinrich Lautensack, Hans Lencker, and the instrument maker Hans Haiden (1536–1613)

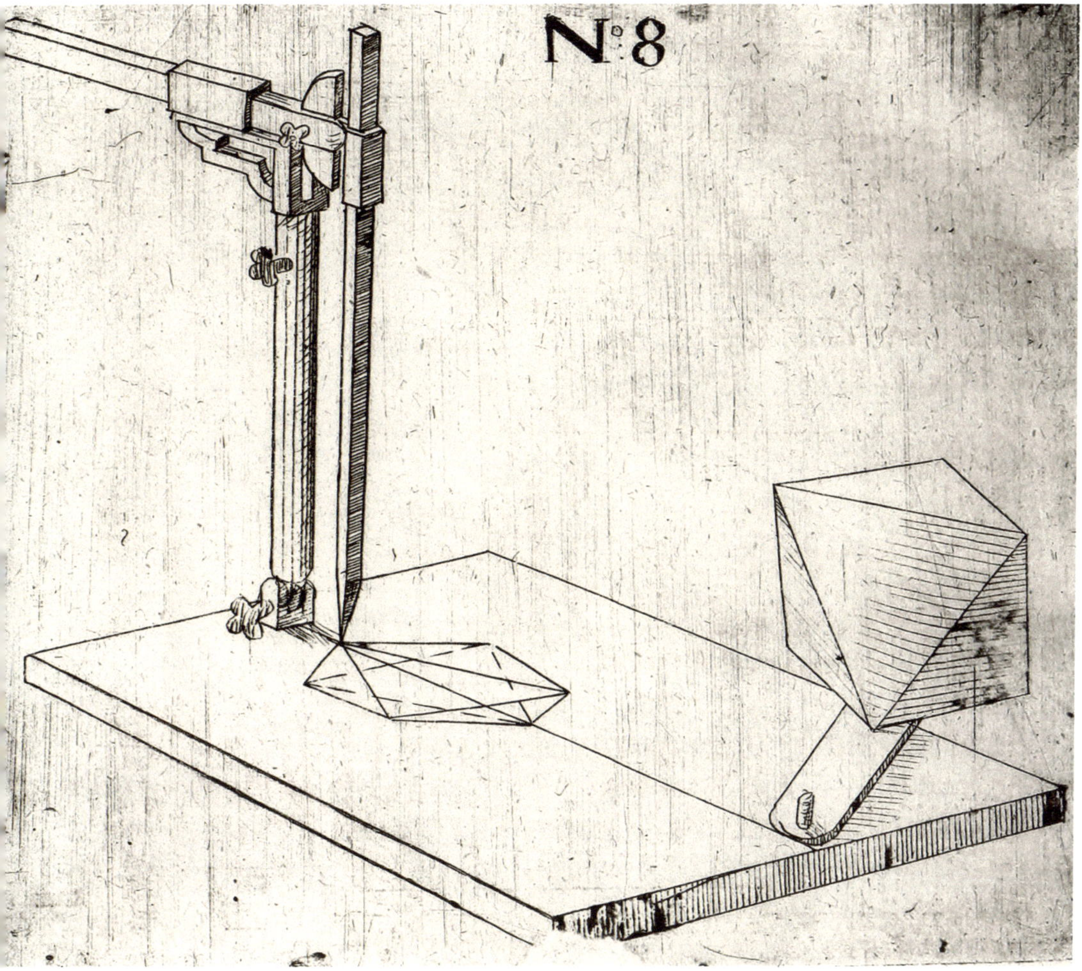

Image of a drawing device (no. 8), from Peter Halt, *Perspectivische Reiß Kunst* (Augsburg, 1625).
Getty Research Institute, Los Angeles (89-B24866).

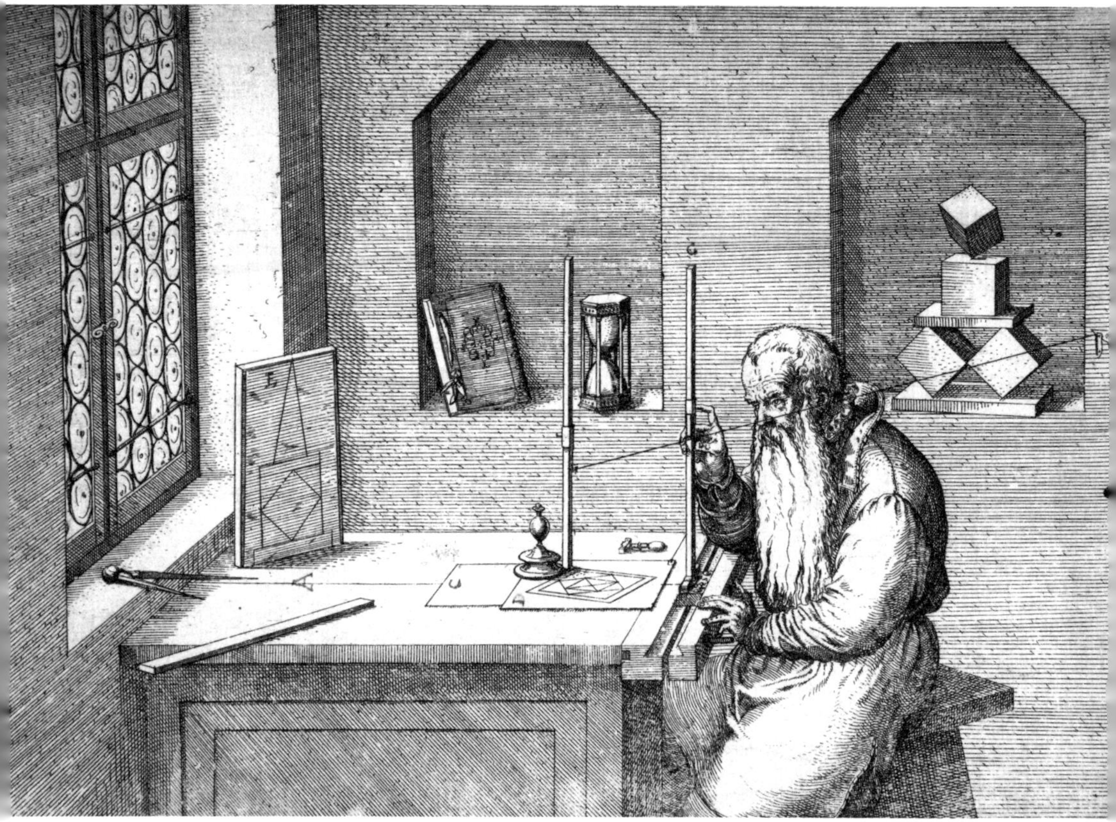

Jost Amman (before 1539–1591). *Wenzel Jamnitzer*. Late sixteenth century. The Metropolitan Museum of Art, New York, The Elisha Whittelsey Collection, The Elisha Whittelsey Fund, 1956 (56.510.2).

used to execute their graphical-geometrical models. In particular, Pfinzing claims that Jamnitzer, who lived at Zisselgasse 17 (since renamed Albrecht Dürer Straße), less than one hundred meters from the Dürer-Haus, made use of a *Perspectiu Tisch* (perspective table) in composing *Perspectiva corporum regularium*, his treatise on the *funff Regulierten Corporibus* (five regular bodies), though as much can be gleaned from Jost Amman's etched portrait of Jamnitzer. 40

> He then let a book begin at the five regular bodies, and the infinite number of bodies that originate from it, which he then brought out through this way of perspective. He had his perspective desk in a special room in his house, screwed in so that it could not move; standing so that he could attach to it strings from a screw on the room's wall. And thus in such a place of his house, his perspective had been realized in his works. 41

The colored images accompanying Pfinzing's description of Jamnitzer's *Perspectiu Tisch* show an experimental apparatus integrated into Jamnitzer's domestic interior via the attachment of a counterweighted string to the living room wall for stability, transforming one room into a giant "perspective machine" for making drawings. According to Pfinzing, Jamnitzer's perspectival armature represented an evolution from Dürer's originary setup in *Underweysung*, one that would be further developed by Haiden. Speaking to the importance placed upon technology in generating geometrical form, the page following the imagining of Jamnitzer's workspace depicts a pyramidal model made from slices of wood, each rotated and of decreasing size, copied directly after a model from Jamnitzer's *Perspectiva*. Balanced atop the copied model like 42
Hirschvogel's owl is Pfinzing's reconstruction of the perspective apparatus Jamnitzer used to draw the model in the first place, a rhetorical conceit he repeats on a similar page dedicated to the geometrical work of Haiden. The foregrounding of the supportive tools and apparatus typically rendered invisible by a final image highlights the critical importance of technology to the production of perspective. Though the end result is uncannily self-referential and devoid of human presence—the Renaissance equivalent of presenting an image of a CAD model supporting an image of the computer used to design and render it—Pfinzing's entangled technogenesis condenses

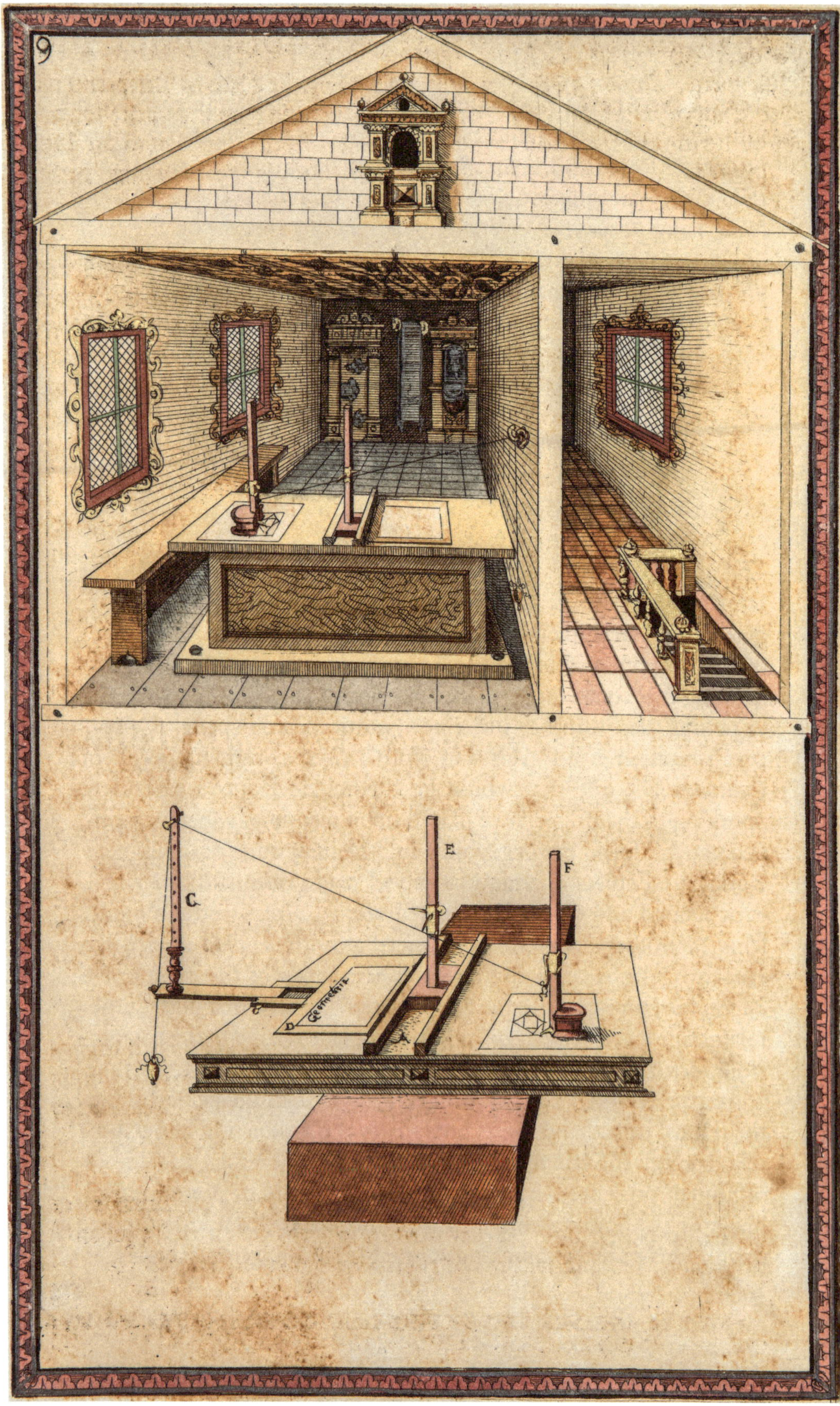
9
C
E
F
D
Geometria

the sophisticated enmeshing of graphic and machinic expertise that enlivened the study and mastery of geometry.

Geometria is evidence that by the mid-sixteenth century, geometry had become a self-conscious topic for inventive mixed-mathematical research by German artists, and that this research was informed by practice, making, the physicality of materials and technologies, and the work of Albrecht Dürer, rather than by the time-honored tradition of repeating Euclidean propositions, as in Melanchthon's university curriculum. While Dürer's new investigations into the constructability of the Platonic solids from unfolded paper surfaces had introduced the geometrical net to the broader artistic community, it was the *Lehrbücher* that were responsible for disseminating what might otherwise have been a novel graphic technique to a wider public. Within a matter of decades, the unfolded net, signifier of geometrical constructability and initially confined to cadres of creatives, had become a constituent property of mathematical textbooks, eventually finding its way back into the academy. Petrus Ramus, the French humanist and educational reformer, moved in the direction of integrating "pure" geometrical knowledge (*mathematicae purae*) with the knowledge gained from the mixed-mathematical experiments with geometry. His *Arithmeticae libri duo* (1569) collected visual cues of the existing mathematical knowledge on the Platonic solids, displaying shaded, skeletal, and unfolded *corpora*—further evidence
that unfolding techniques had become standardized. The grand *Livre* 43
de perspective (1560) of Jean Cousin (1490–1560/61) deals explicitly with
producing "les cinq corps Reguliers de Geometrie" in perspective
and from a surface. The section "De solidorum constructione" in 44
Christian Wolff's (1679–1754) wide-ranging *Elementa matheseos universae* (1743) still describes how to construct the Platonic solids out of nets two hundred years later, picking up on Hirschvogel's term "rete": "Rete describere, ex quo cubus construi possit" (to describe the net, from which the cube can be constructed), "Rete pro prisimate
describere" (to describe the net for a prism). 45

Colored image of Wenzel Jamnitzer's drawing apparatus, from Paul Pfinzing von Henfenfeld, *Ein schöner kurtzer Extract der Geometriae vnnd Perspectiuae* (Nuremberg, 1599). Staatsbibliothek Bamberg (JH.Ma.f.5, pl. 9). Photo: G. Raab.

following spread Colored images after Wenzel Jamnitzer's *Perspectiva*, to which has been added a reconstruction of Jamnitzer's drawing apparatus balanced on top; and Hans Haiden's drawing apparatus balanced atop geometrical forms. From Paul Pfinzing von Henfenfeld, *Ein schöner kurtzer Extract der Geometriae vnnd Perspectiuae* (Nuremberg, 1599). Staatsbibliothek Bamberg (JH.Ma.f.5, pl. 10 and 11). Photos: G. Raab.

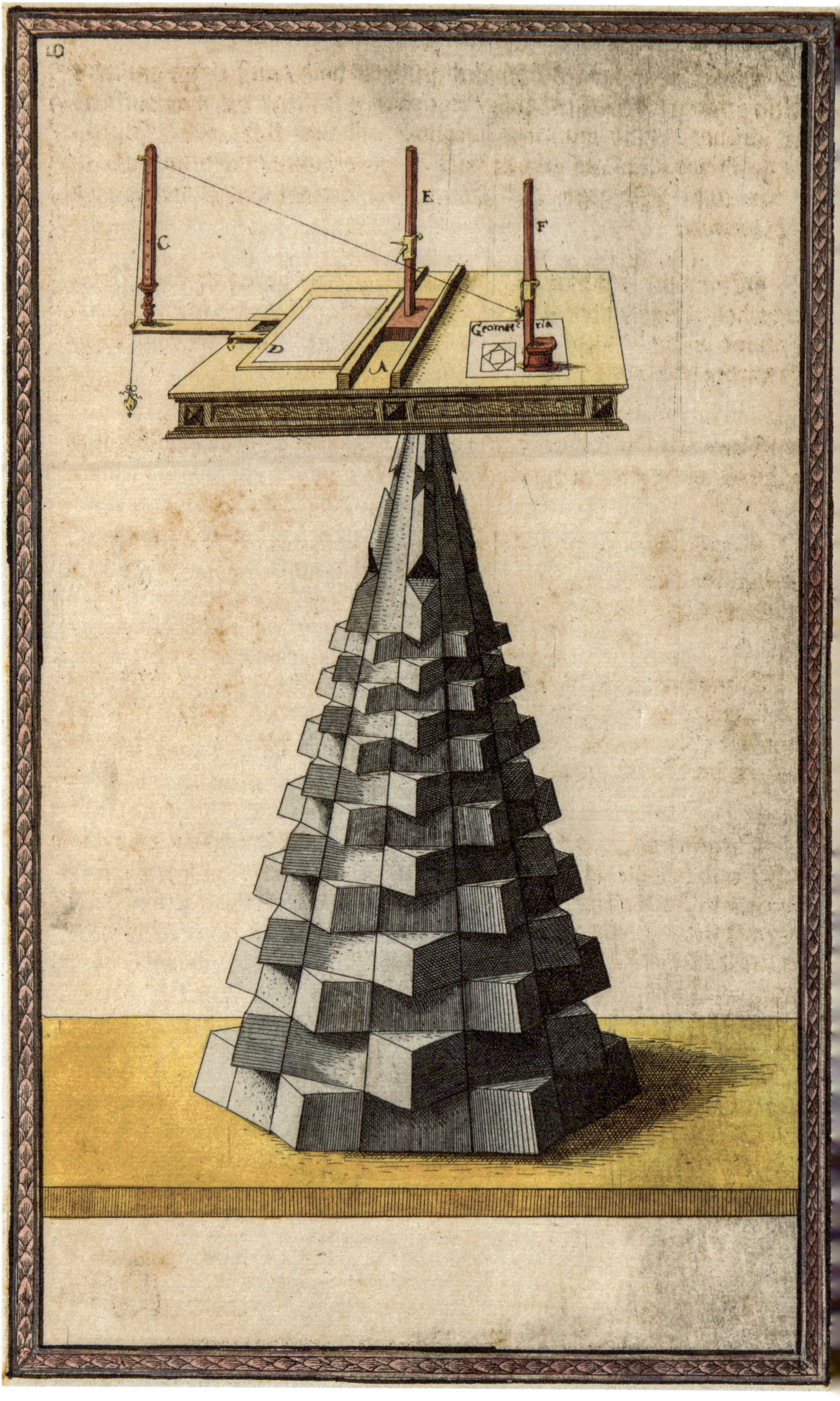

10
C
E
F
D
A
Geomet
ria

II.

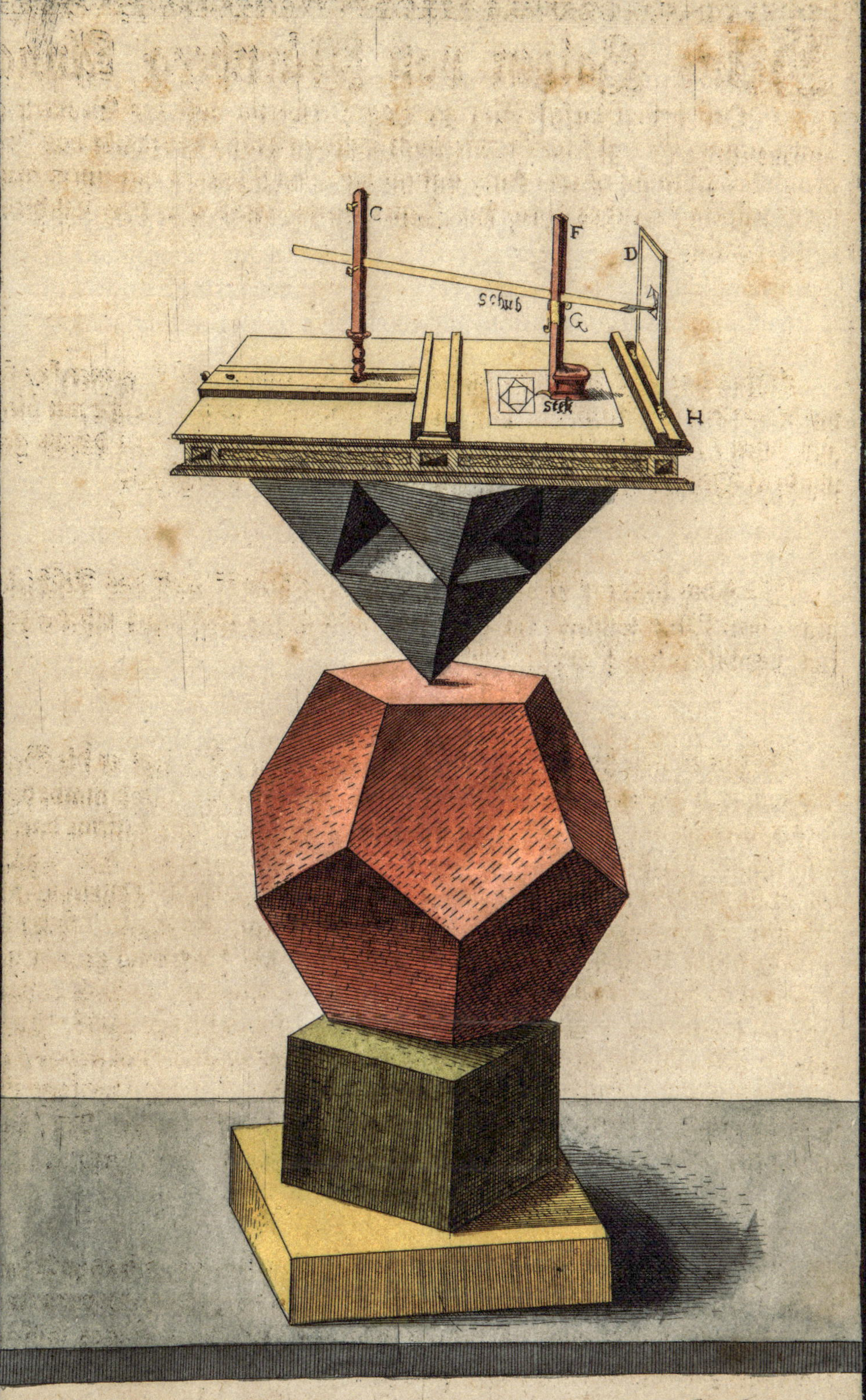

The widespread popularity of *Lehrbücher* confirms the utility that the construction and representation of geometry were seen to have. Economical to make and experiment with, paper models could serve as objects to be drawn in perspective or as the base forms for more complicated geometrical confections. Hirschvogel's use of the Platonic solids for teaching geometry and perspective highlights their status as the pedagogical objects of choice for experiments in the new science of perspective. Geometry was increasingly an area of invention, not regurgitation, and the *corpora regulata* were the most common trial subjects to be operated upon.

Much like Hirschvogel, Heinrich Lautensack, the author of *Des Circkels vnnd Richtscheyts* and an "enthusiast of geometry and perspec-
46 tive," belonged to a prominent family of Nuremberg artists. His father, Paul Lautensack, a German painter and organist from Bamberg, had moved to Nuremberg in 1525 after embracing the Reformation, and Heinrich Lautensack himself later settled in Frankfurt where
47 he made a living as a goldsmith, painter, and engraver. As with Hirschvogel's *Geometria, Des Circkels vnnd Richtscheyts* is a *Lehrbuch* self-avowedly modeled on Dürer's treatises. It is similarly pitched to goldsmiths, painters, stonecutters, carpenters, and builders and covers a spread of topics running from basic geometry through the construction of the polyhedral solids in plan and perspective. In its latter sections, *Des Circkels vnnd Richtscheyts* also expands its scope to cover the drawing of architecture and the human body in perspective as well as human proportion, much of which is copied directly from Dürer's *Vier Bücher von menschlicher Proportion.*

Assembled from multiple discrete objects balancing precariously one on top of the other, Lautensack's most complex geometrical composites, which were known to Pfinzing and almost certainly to Nilson as well, do not get a diagnostic treatment. But Lautensack does provide sets of compact instructions on how to draw an array of original geometrical components in a planimetric view, as in his rendition of a hexagonal polyhedron shown in plan and as a skewed plan in perspective. The construction of the hexagonal polyhedron, labeled as no. 25 in the accompanying figure, is described as follows:

> If you want to make the tetragon into a hexagon, you need to discard from it a seventh part on one side. ...

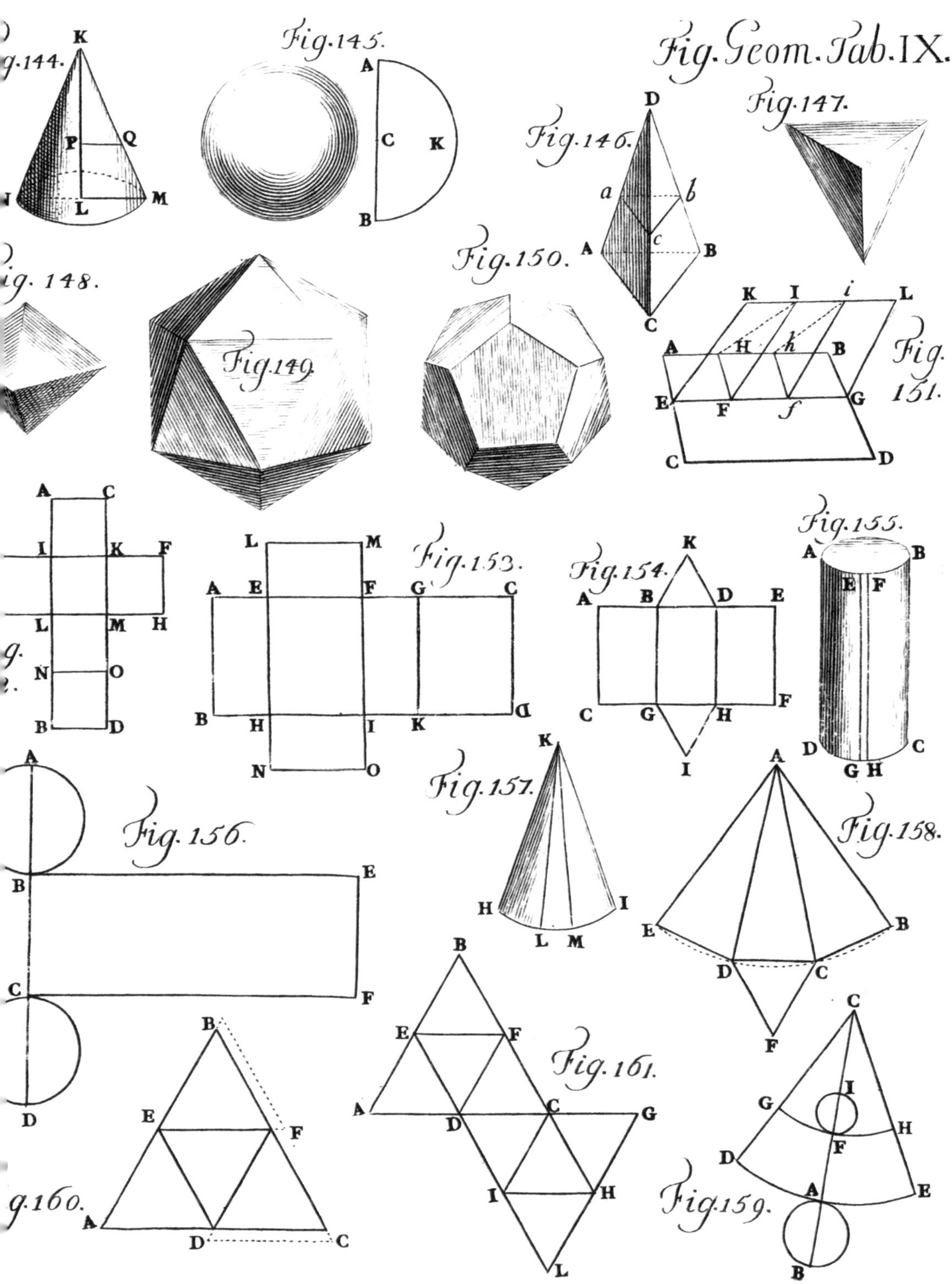

Geometric figures, from Christian Wolff, *Elementa matheseos universae*, vol. 1 (Geneva, 1743), Tab. IX. Staats- und Stadtbibliothek Augsburg (4 Math 641-1).

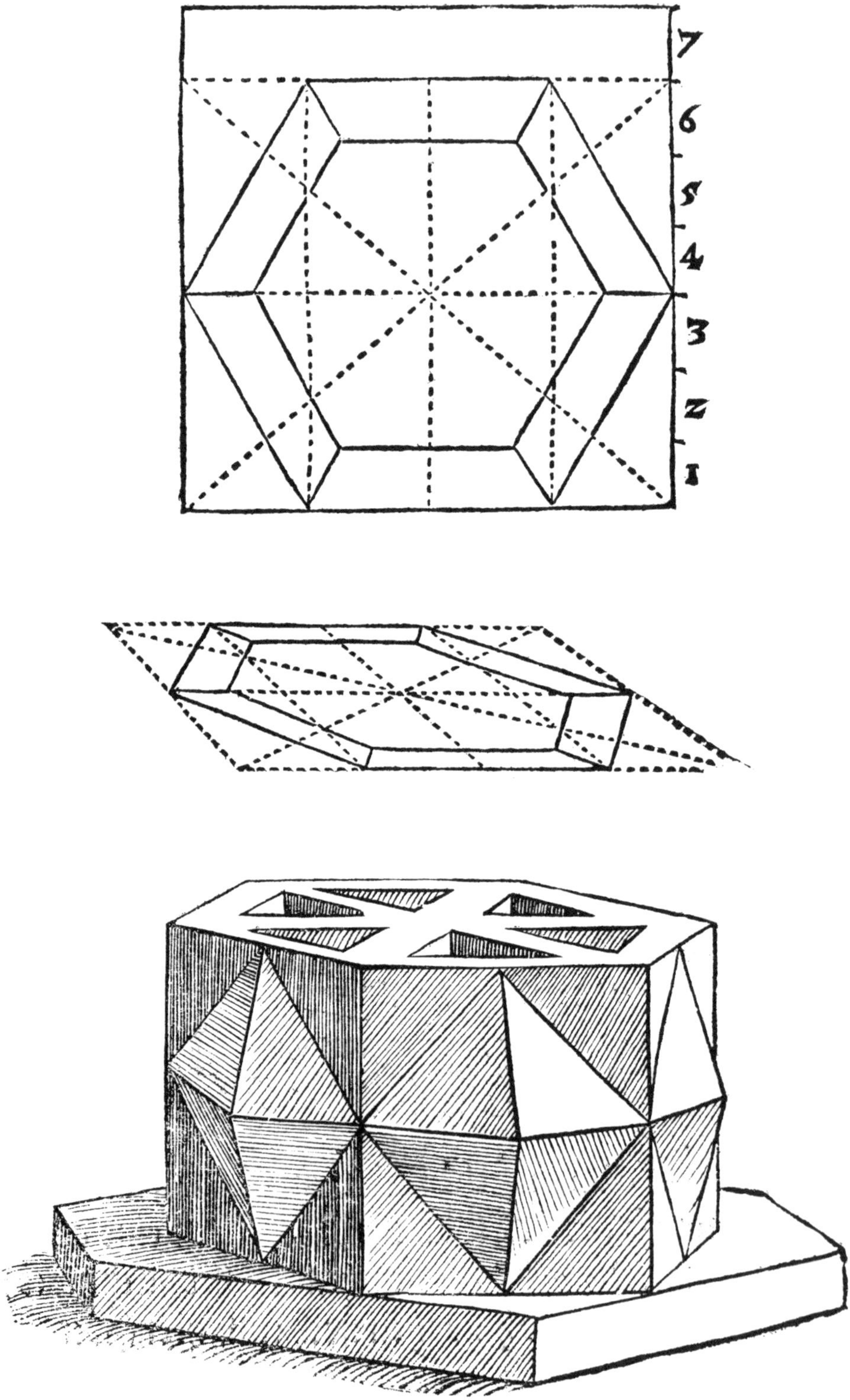
7
6
5
4
3
2
1

> Afterward, make a cross inside it, and also a straight cross. Thus the straight cross gives lengthwise two segments on both ends. If you then divide each half of the original tetragon widthwise into two parts, you would have on the top and bottom two segments on both sides, as it is drawn. 48

Unlike with Hirschvogel, Lautensack's conceptual ground is constructed perpendicular to the surface of the book—it sits within the space of the page. The sentence preceding the hexagonal figure translates as "Here I have, in a poor manner, fitted the ground/plan into perspective" (*aus dem grundt auffgezogen*). For Lautensack, form is created 49
through extrusion—the stretching out of a new three-dimensional object from a plan—that could then be selectively modified by graphic techniques, as in this instance, where the extruded hexagon has been filled in with triangular cuts excavated from its interior and diamond shapes ringing its sides. Lautensack was not concerned with the buildability of his forms as paper models, and provides no instructions about how to unfold them, assemble them from paper, or how to draw them. Instead, the hexagon is the geometrical inspiration for the rendered perspectival figure, clearly hexagonal in origin, which lies beneath it on the page.

Because for Lautensack perspectival geometry was "fitted from the ground/plane/base" (*aus dem grundt auffgezogen*)—out from the plan and into perspective—he depicts his geometrical inventions as models situated on an abstract ground and casting printed shadows onto the page. Using this graphic technique in the reverse direction in order to construct two-dimensional geometries, a pattern could theoretically be projected onto the ground plane (*in den grundt gelegt*) from any number of sectional "cuts" through a three-dimensional figure. Although these cuts would not allow one to fold a solid out of paper, the implication was that if one took enough horizontal section cuts through a solid, it would be possible to join the slices together, one on top of the other, to reconstruct the solid graphically, much as gluing together sequential slices through the trunk of a tree might allow one to reconstruct the profile of the entire trunk. Thus one solid object (perforated with triangular holes for example) could theoretically generate an infinite number of potential Swiss-cheese-like two-dimensional slices, much as one slice (or plan) could be used

Diagrams depicting the construction of an irregular body, from Heinrich Lautensack, *Des Circkels vnnd Richtscheyts* (Frankfurt am Main, 1564). Staatsbibliothek Bamberg (JH.Ma.f.7, fol. 20 recto). Photo: G. Raab.

to generate forms through extrusion, which could then be infinitely modified through cutting, filling, and appliqué.

Still the onus was on the student, or apprentice, to bridge the large gaps between theory and practice. While Lautensack taught two-dimensional construction, he did not provide enough information to draw the geometry he depicted in perspective. His images of geometrical figures stand as aspirational models of skill—products of his research into perspectival invention. Lautensack probably found it to be as important to train the eye to recognize correct (or nearly correct) perspectival constructions; to have readers be able to leap agilely between the plan and the third dimension—and thus also to develop the skill to think three-dimensionally from a two-dimensional drawing. *Des Circkels vnnd Richtscheyts* conjures a space of imaginative development between the ground plan and Lautensack's perspectival objects that one could only hope to surpass through sustained labor with and study of geometry.

Given the myriad potential ways to invent form, Lautensack's commitment to extrusion cut through the morass of creative possibility and contributed a strand of graphical experimentation, though one without the conceptual rigor of Hirschvogel or Dürer. Limitless geometrical permutations were possible from the same polygonal base. In danger of toppling over physically and intellectually, the geometrical body was no longer a stable, singular solid, nor a stable if foldable surface, but contained within itself the latent techniques of its own transformation. This was a further shift in emphasis from the strictures of Euclid to a text predicated upon developing its readers' capabilities to generate geometrical variation—variation that was palpably irregular, overtly *not* in *Elements*, nor, frankly, capable of being described in text alone. Lautensack encourages bounded geometrical play based on related difference, in which a repertoire of form is experimentally developed through two main techniques: projecting a planimetric "cut" of a polyhedron onto a "ground" and extruding a solid polyhedral object out of a plan that could then be tweaked, incised, and developed to achieve a range of related geometries. Once modeled graphically in three dimensions, "physicalized" movement animates Lautensack's geometrical constructions, as in many of the geometrical *Lehrbücher* —a twisting and stretching, stars teetering on their vertices, the properties of materials translated into a graphic world.

Perhaps it was Petrus Ramus who most passionately espoused
the benefits of the practical application of and tactile relation to geom-
etry as utilized in southern German workshops. In his *Prooemium* 50
mathematicum (1567), he defended geometry against charges of use-
lessness by emphasizing its importance in mixed mathematics and
the arts. In Ramus's view, France could above all learn from the 51
Germans about how to unify theory and practice, a goal he consid-
ered integral to mathematical pedagogy. A visit in 1568 to the work-
shops of Jamnitzer and Lencker substantiated his belief that the
combination of geometry with the artisanal activity occurring in
Nuremberg was evidence of a new method of practical reasoning that
was producing exciting and innovative results. According to Ramus's 52
contemporary biographer, Jamnitzer and Lencker, who lived near
each other on Zisselgasse, had constructed a machine of such rarity,
perchance the *Perspectiu Tisch* later depicted by Pfinzing, that the two
artists demanded money from Ramus and his companion, Frédéric
Reisner, to see it. Ramus was so intrigued that he offered in return 53
to have one of his treatises on geometry translated into German (by
Reisner) if the two Nurembergers would let him inspect the machine. 54

It is unarguable that precision measurement and design draw-
ing skills were critical for artisans, both for leading names like
Jamnitzer as well as for the many specialized craftsmen involved in
the production of mathematical instruments such as sundials, astro-
labes, angle-measuring instruments for navigation, and surveying
tools. Less clear, however, is the exact role played by these *Lehrbücher* 55
in the workshop and to what extent they actually fulfilled their
intended functions of transforming geometrical habits of mind. The
Hirschvogel and Lautensack texts stand as pedagogical touchstones
that cannot be taken at face value as evidence of their owners hav-
ing acquired the proffered skills. Unlike highly ornamented, decora-
tive prints, which were valued as base motifs to transpose onto other
media, the surviving geometrical *Lehrbücher* are almost uniformly
clear of marginalia. Is this a testament to their disposability, or
were their pages never the sites of active reworking, their diagrams
extracted for further copying and experimentation, calculations cas-
cading along the edges as can be found in many editions of *Elements*?
What then are we to make of their purported graphic innovations, in
the absence of concrete substantiation of their use?

The very existence of the *Lehrbücher* was emblematic of a shift in priorities away from the transcendent proportions of a universal order of a "pure" mathematics toward a set of "physicalized" geometrical operations, albeit visualized in two dimensions. In service of this goal, Hirschvogel and Lautensack might well have been substituted for Lencker, Rodler, or any of the other southern German artists who began publishing geometric tractates at the same time. Each one unique in the graphic accent it provided to geometric instruction, *Lehrbücher* were responses to the evolution of a market for books of geometrical demonstration, the physicalist tradition of French synthetic geometry, developed and popularized by Gaspard
56 Monge (1746–1818) and Théodore Olivier (1793–1853). Conspicuously straddling the boundaries between mathematics and mechanics, the eighteenth-century synthetic geometers used physical operations as the inspiration for equivalent graphic-mathematical operations, an approach that was picked up in the teaching of mathematics in
57 Germany during the nineteenth and early twentieth centuries.

Yet unlike these later schools of thought, which utilized the textbook as integral to courses of study the Renaissance precursors were collected at the discretion of the purchaser, seemingly without an overarching structure policing use. In the absence of such a structure, the *Lehrbücher* may have occupied, a more amorphous status. In contradistinction to the geometrical exercises and definitions of Euclid (an abstraction in which geometry was wholly divorced from real-world application), *Lehrbücher* used a theoretical geometry to teach perspectival construction (the precise plotting of imaginary geometry on paper) that explicitly made use of inventive manipulations of visual information. The assembly strategy of geometrical forms in Hirschvogel—out of paper and in and out of planimetric and perspectival views—is one example promoting physical constructability as a key property of geometry. Lautensack's *Lehrbuch* is but one other, showing how a singular base geometry could produce limitless, and irregular, related permutations.

The self-reflexiveness demonstrated by these authors, in which geometry was made the explicit subject of images, underscores the importance of training precision drawing skills, much as Dürer had hoped would in Germany. Just as geometry established the core knowledge essential to all forms of mixed mathematics, so too was

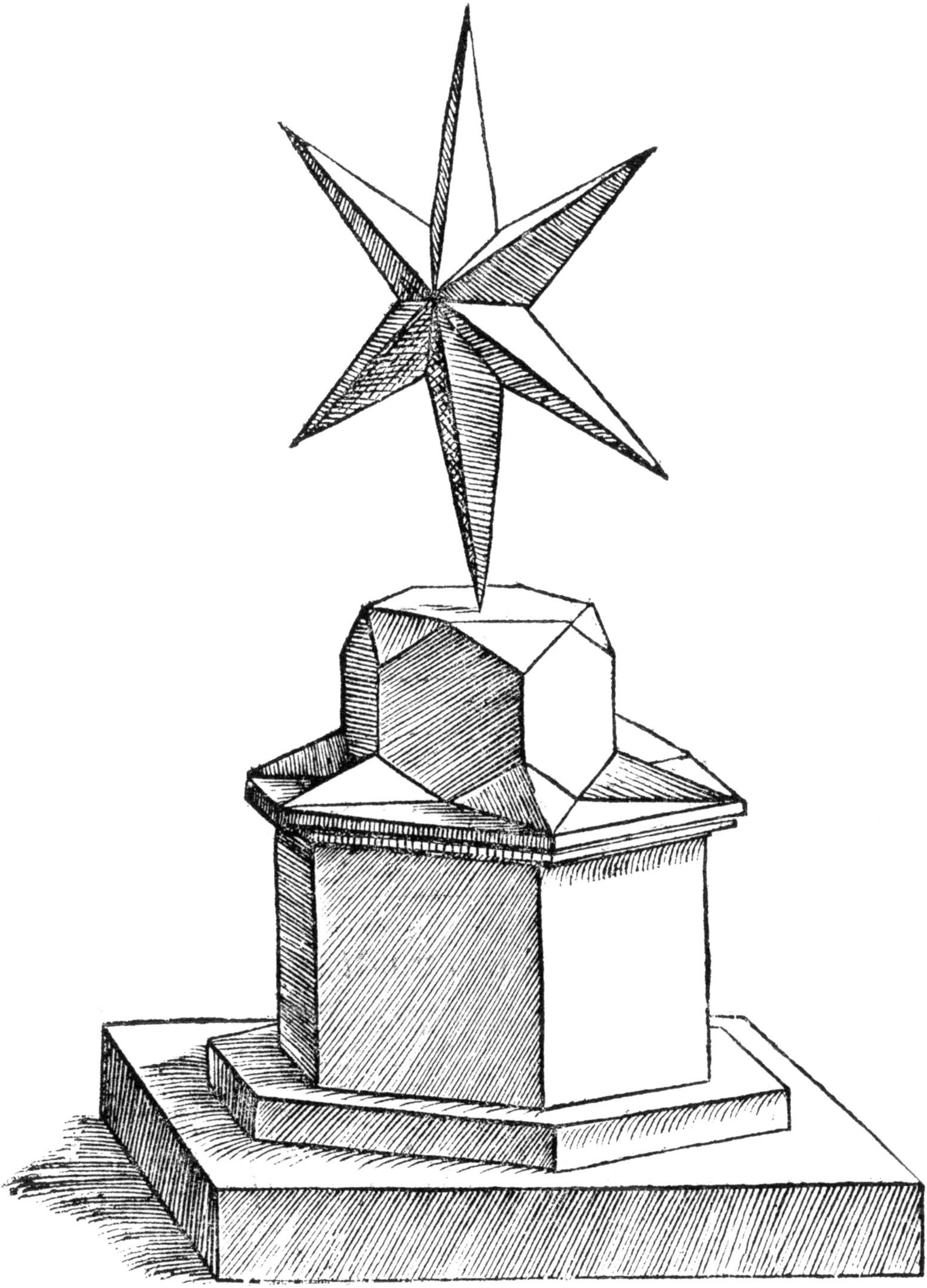

Perspectival model, from Heinrich Lautensack, *Des Circkels vnnd Richtscheyts* (Frankfurt am Main, 1564). Staatsbibliothek Bamberg (JH.Ma.f.7, fol. 20 verso). Photo: G. Raab.

it fundamental for learning how to cultivate the imagination and for the strict and self-conscious translation of design ideas from the mind to the page. What had begun as training exercises in Dürer had shifted into the realm of boundless, graphic invention, a further loosening of the already permeable perimeter between theory and practice, the teaching of knowledge and the selling of mastery.

Artisans harnessed geometry in all areas, pioneering a new range of practical techniques intended to make the world more comprehensible. And yet much remains to be discovered, hidden by the vagaries of reconstructing historical modes of thought dedicated to teaching thought to others. Were *Lehrbücher* read continuously or selectively, alone or in conjunction with other practical geometry books? Were they handled only by master artisans or by apprentices (or both)? While they were known to have been purchased by elite audiences, as testified to by the inventories of numerous ducal libraries including at Wolfenbüttel, Stuttgart, and Dresden, did they also appeal to the industrious common man or *Hausvater*, as the German publisher of numerous practical tractates, Walther Hermann Ryff (ca. 1500–1548), identified his intended audience? One is left to ponder the content of the *Lehrbuch*, eschewing the binary distinction between the textual and the visual, and the evaluation of these texts' success in espousing graphic stratagems for bridging manual and mental practices. Having sprung from the fruits of Dürer's originary meditation on geometrical materiality, these books multiplied in abundance and circulated through every major German city and princely collection. This dispersal of a uniquely artisanal engagement with geometry, imprinted upon a new genre of generously illustrated and, importantly, affordable books of instruction, coincided with artisans' deepening affinity for geometry and geometrical forms. This critical momentum would culminate in the fantastical, irregular creations of the latter half of the century, creations that could only have been conceptualized on the back of the techniques espoused by the likes of Lautensack and Hirschvogel. Begotten of and stimulating perspectival mastery, cycled into medial pathways linking practical texts to workshop practices, perspective machines to domestic spaces, and the pressures of the urban environment to the environment of the mind, *Lehrbücher* demarcated the expanding borders of an expansive and unmapped territory for the early modern mathematical imagination.

Georges Reverdy (active 1529–1557). *The Architect* (detail). The Metropolitan Museum of Art, New York, The Elisha Whittelsey Collection, The Elisha Whittelsey Fund, 1964 (64.568.10).

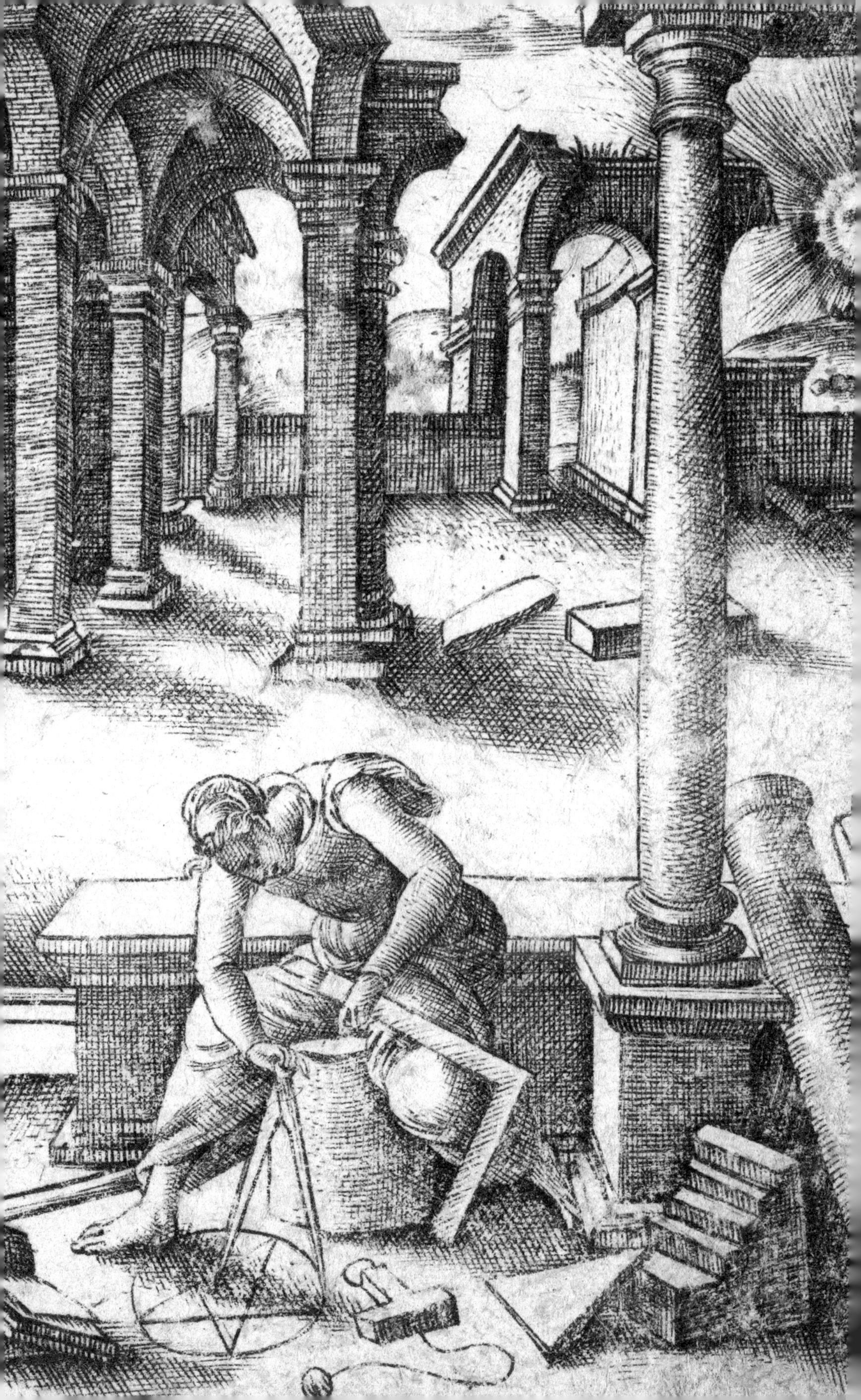

LOVE
THE A
PERSPE

R OF

RT OF

CTIVE

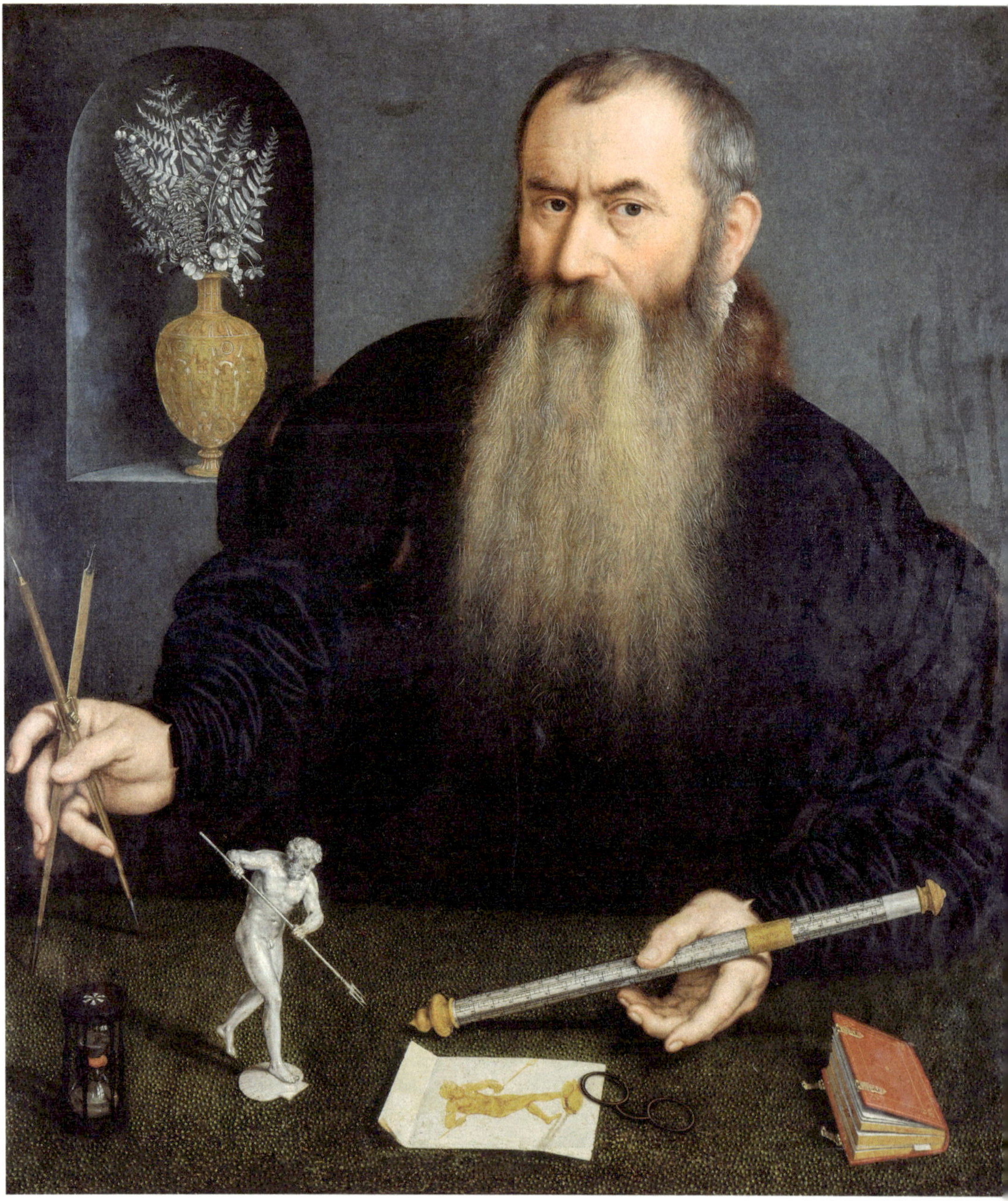

Nicolas de Neufchâtel (ca. 1524 – ca. 1567/87). *Portrait of the Goldsmith Wenzel Jamnitzer*. 1562–1563. Musée d'Art et d'Histoire, Geneva (1825–0023).

It is the beard of the Viennese-born Wenzel Jamnitzer (1508–1585), the most famous master goldsmith in Nuremberg, a city of goldsmiths, and one of the two most famous living goldsmiths at the time of its creation along with the Florentine Benvenuto Cellini (1500–1571), which first commands attention in this self-commissioned portrait. A rich ombre of silver and gold, its colors echoed by the bronze vase of metallic flowers behind him, the beard bifurcates into seven wispy tresses motioning toward each of the seven items chosen to represent the epistemic aspirations of the art and science of goldsmithing at its mid-sixteenth-century zenith. Jamnitzer grasps a silver measuring scale (*Maßstab*) in his left hand and a variable proportional compass (*Reduktionszirkel*) in his right; a prayer book (piety), spectacles (wisdom), and hourglass (mortality) flank a doubled figure of Neptune
preserved in a cyclical state of intermedial translation. Modeled as 1
a silver statuette and sketched at a reduced size for an iteration in gold, the Neptunes together cast the personage of the goldsmith in terms of his unparalleled control of the second and third dimensions. Goldsmithing, the portrait declares, enjoys all the tools necessary to lay exceptional claim to the economies of measurement and transformation, to the choreography of material, matter, and data that tie casting to drawing and back again.

Johann Neudörffer's *Nachrichten von Künstlern und Werkleuten* (Proceedings on artists and craftsmen) (Nuremberg, 1547) described Wenzel, a member of Nuremberg's goldsmith association from 1534, and his brother Albrecht as working both with silver and gold, having a great understanding of perspective and proportion, and having cut coats-of-arms and seal dies in silver, stone, and iron. By the 1730 publication of Johann Gabriel Doppelmayr's (1677–1750) *Historische Nachricht von den nürnbergischen Mathematicis und Künstlern* ... (Historical proceeding of Nuremberg mathematicians and artists), Jamnitzer's reputation as the most renowned of sixteenth-century goldsmiths had long been established, not least because of the work he had produced for four Habsburg emperors—Charles V, Ferdinand I, Maximilian II, and Rudolf II. His prodigious skill and success had also been recognized by his contemporaries in the city's influential community of goldsmiths. As an indication of his prominence, the new rules governing the submission of masterpieces in Nuremberg, issued in 1572, held that all applicants for the title

of master goldsmith were to produce masterpieces working from a
2 pattern cup by Jamnitzer as a base model.

On the basis of his graphic work with geometry, Jamnitzer
appears twice in *Historische Nachricht*, as both a goldsmith and an
accomplished mathematician. "A lover of the art of perspective," he
had represented more than one hundred and sixty bodies in *Perspectiva
corporum regularium* (1568), his luxuriously illustrated meditation on
3 the Platonic solids. The unabashed opulence of *Perspectiva* stood out
immediately from a polyhedral field increasingly crowded with com-
petitors, distinguished by its immaculate engraving, generous folio
size, and pristine impressions. Reflecting the unique prominence that
explorations of polyhedral geometry had attained in the years since
their unfolding in Dürer's *Underweysung*, *Perspectiva* covered no other
topic or geometrical application. The first five of its six sections were
devoted to the presentation of twenty-three floating geometrical vari-
ants based upon the five Platonic solids, while the last section included
twenty-three pages of freestanding, freeform geometrical models.
That *Perspectiva* was deeply important to Jamnitzer is evident, not least
of all, from the time, patience, and training it would have taken to
produce the intricate underlying drawings. Though the introduction
laments the "long and boring path" he had followed during his forty
years of practicing the "very lovely and gracious art (which is called
Optica by the learned and otherwise commonly *Perspectiva*)," his grave-
stone in Nuremberg's Johannisfriedhof bears a plaque based on the
4 book's title page. At its four corners, the female allegories Arithmetic,
Geometry, Perspective, and Architecture are rendered in relief. The
lower figure of Perspective holds up a cube while the posture of
Geometry and Arithmetic have been modified from the original print
5 so as to better present a dodecahedron and icosahedron.

The grave presents Jamnitzer as a theorist of geometry rather
than the most famous northern European goldsmith of his time,
known for the seamless integration of naturalistic foliage with
delicate life casts in crowning achievements like the Merckelsche
6 Tafelaufsatz (centerpiece) from 1549. That a goldsmith renowned for
such virtuosic mimicry of flora and fauna chose to invest the time and
energy it would have taken to produce *Perspectiva corporum regularium*,
arguably the epitome of the polyhedral genre, and, critically, that he
had the drawing skills to pull it off, testifies to the new notoriety of

geometry in Reformation-era decorative arts in Germany and the place of drawing generally within the goldsmith's workshop. To be sure, the ability to think through and represent forms on paper was critical to all manner of craftsmanship, like goldsmithing, dedicated to manipulating expensive material into more expensive objects, as it was much cheaper and more efficient to experiment on paper than with gold and silver. *Perspectiva* implicates the status of the drawing in the workshop, and suggests how the realities of the workshop—both commercial and tactile—might have informed the graphic, if not literally in form then in an approach to form making conditioned by the realities and materialities of artisanal practice.

Unlike the iconic mathematical equation, an $E = mc^2$ or, closer to our subject matter, Leonhard Euler's (1707–1783) polyhedral formula, $V - E + T = 2$, virtuosity for the Nuremberg artisans was never expressed through simplicity and the discarding of the extraneous, or indeed in the striving for beauty predicated on reducing an equality to its most irreducible core. Though a robust culture of consumption for luxury 7
goods had existed since the fifteenth century in Europe, patronage in 8
sixteenth-century Germany came to equate beauty and desirability with the image or impression of intricacy and thus positioned artisans to produce ever more detailed items. The universal had little place in the production of masterpieces, which necessarily needed to be distinguished by their *particularity*. Divorced as they were in *Perspectiva* from any second-order function they might be capable of fulfilling (in telling time, for example), polyhedra bore the weight of communicating these values on the basis of their image or form alone. Geometry as a field of enquiry, released from its Euclidean confines, had absorbed the priorities of artisanship and with it an ulterior set of culturally specific conventions for determining success. It was simply no longer sufficient to draw the regular bodies in three dimensions. This was a task for students. In order to convey mastery over geometry and perspective, the theoretical backbone of artisanship, the bodies had to evolve. Here invention reigned supreme, less was not more, and visualizability in three dimensions was key to the communication and preservation of geometrical knowledge.

Perspectiva is a direct response to Nuremberg's culture of complexity, its high standards of manufacturing—where nothing less than perfection was acceptable—the distributed networks of

assembly-line-like, highly specialized labor that could be activated en masse to produce the ornamental masterpieces for which the city was renowned, and the prominence the solids widely enjoyed as the
9 premier emblems of early modern mathematical knowledge. This was a vibrant culture of hyperactively articulated luxury objects, instruments, and surfaces, often embellished to the extreme—a predilection for intricacy that changed everyday furniture, compasses, lock mechanisms, terrestrial and celestial globes, sundials, surveying instruments, or suits of armor into opportunities for the abundant
10 ornamentation of form and function. In turn, patrons both developed a taste for and drove the market for these luxury objects, which served also as flamboyant showpieces of the skill and time required for their creation. And lead workshops formed networks of professional associations with agile groups of highly skilled makers in order to satisfy the desire for *Meisterstücke*, which often required mul-
11 tiple skill sets that could not be found under one roof. The interplay between highly precise, scientific objects which were often lavishly decorated, and decorative art objects that displayed precise geometrical and technical expertise, mutually reinforced the reliability of Nuremberg artisans to potential buyers.

Most importantly, however, *Perspectiva* was a reaction to the premium placed on the role of drawing and measurement skills. At the prosaic level, skill in the accurate and measurable representation of objects or components of objects, a skill cultivated through the use of geometrical *Lehrbücher*, was critical to the successful manufacture of the *Meisterstücke* produced in Nuremberg. Goldsmiths, intarsia specialists, and other craftsmen who dealt with ornament and ornamentation worked through drawing to train their subordinates, develop their own thoughts, coordinate design components, perfect their proposals for commissions, and present progress or presentation designs to clients. Printed ornament books were in high demand, offering up quotable repertoires of motifs in series that found their way onto all types of designed surfaces, from building facades to fabrics and precious metals—cultivating an aesthetic dedicated to communicating artistic virtuosity through intricacy and ornamentation.

The practicalities of producing secular, decorative art, in which biblical themes were often replaced by geometrical or naturalistic motifs, also led to novel typologies of drawings and of approaches

to drawing that were uniquely informed by artisanal experience. As
the spread of Reformation-appropriate subject matter contracted
away from the depiction of expressive biblical themes, the decora-
tive arts, replete with secular, geometrical motifs and unique graphic
practices, surged in popularity and investment. Drawings by gold- 12
smiths for instance, the kind of graphic practices to which *Perspectiva*
is indebted, became metamorphic and iterative, concerned with sur-
face transformations and seriality. Bringing to bear the wealth of 13
their experience as makers, goldsmiths distinguished themselves
as innovators in the graphic realm, communicating their intimate
familiarity with the performance of matter in flux and their emi-
nently pragmatic ideas about the generation of form.

It was not unusual for precious metal workers to keep sketch-
books of observations on human and animal figures alongside future
design ideas and graphic experiments with ornament and geome-
try. One of the greatest collections of early modern goldsmith draw-
ings belonged to Basilius Amerbach (1533–1591), a prominent lawyer
and art collector from Basel, Switzerland, who amassed over his life-
time a substantial compendium of drawings and objects termed the
Amerbach-Kabinett. Purchased by the city of Basel in 1661 and since 14
distributed across the city's museums, among its many treasures the
Kabinett holds items from different goldsmith workshops contem-
porary to Amerbach, including several rare molds and design draw-
ings from Wenzel Jamnitzer's workshop, which Amerbach acquired
after Jamnitzer died from the plague in 1585. The Historisches
Museum holds a number of lead patterns used by Jamnitzer that
have been matched to parts of the splendid coffer he was commis-
sioned by the city of Nuremberg to make for Maximilian II in 1570. 15
These patterns are small strips of ornament, the largest 9.4 cm by
2.5 cm, depicting *Rollwerk* (strapwork ornament), a miniature row of
columns, and interlocking ropelike motifs. Traced onto a soft wood 16
like lime or pear, the molds had first been carved by *Bildschnitzer* or
Formscheider (model makers) and then cast in wax and covered in ter-
racotta to make the final molds for the lead patterns. To build up
the visual density of Maximilian's coffer, or indeed any high degree
of ornamental metalwork, a goldsmith like Jamnitzer would have
relied upon multiple lead patterns derived directly from drawings
which could be combined together as a kit of interchangeable parts. 17

overleaf Wenzel Jamnitzer (1507/08–1585). Design options, from the "Berlin Sketchbook." 1545–1546. Bpk/Kunstbibliothek, Staatliche Museen zu Berlin (OZ 1,12 and OZ 1,21).

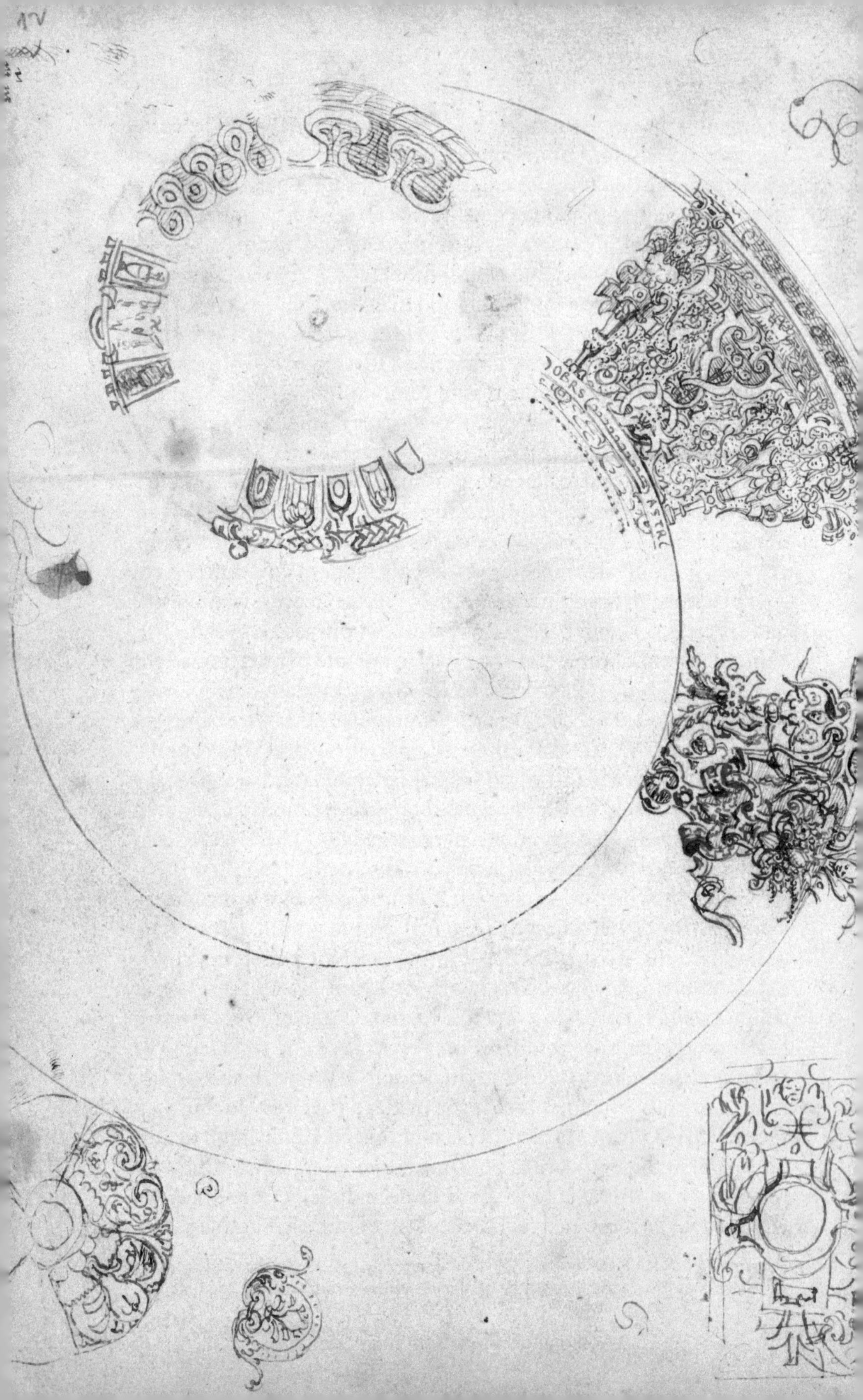

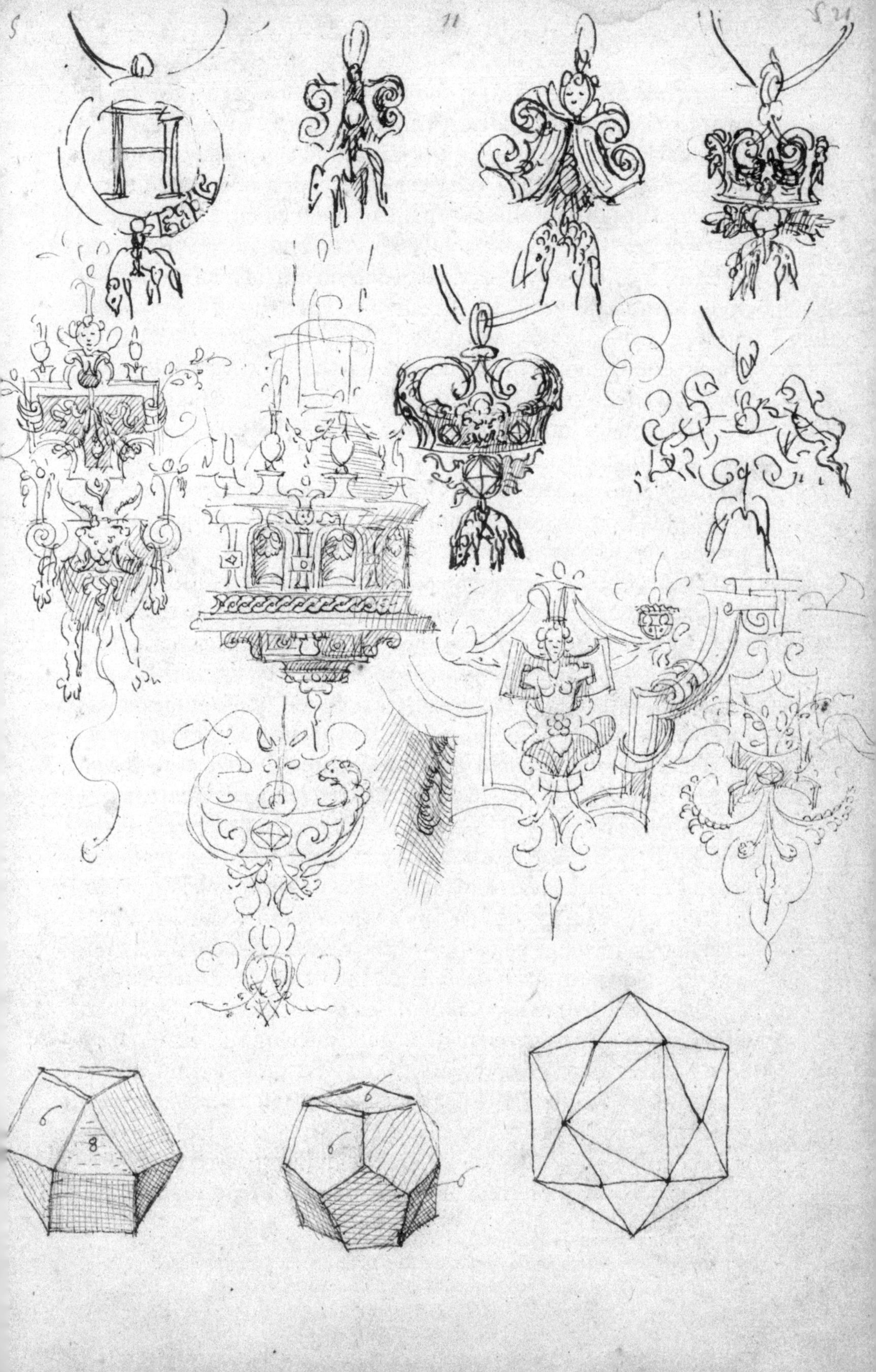

While others disguised the inevitable breaks in a composite
ornamental pattern by soldering tiny metal parts or covering the seam
in special elements, gilding, embossing, and enameling, Jamnitzer
was the first goldsmith to make use of revolving stamps in order to
18 impress strips of continuous ornament upon metal. The surface of
masterpieces like Maximilian's coffer was composed of numerous
patterns, continuously cast and based upon ornamental motifs that
would have been compiled over years of inspiration and design explo-
19 ration, or copied from printed *Musterbücher* (pattern books). Design
schemes from Jamnitzer's own notebook at the Berlin Kunstbibliothek
record his experimentation with different designs and his search for
an ornamental solution that would work given the parameters of a
commission, further confirming that he used the medium of draw-
ing in order to explore options for continuous decorative elements
to be cast in lead. For this reason, lead patterns were among the most
precious and important commodities that a goldsmith could own, as
20 they could be reused, recombined, and repurposed for future objects.
To acquire the patterns of another goldsmith meant that one could
replicate the ornament that defined a goldsmith's personal style, a
problem that the Nuremberg city council took very seriously. Council
records show that on December 19, 1549, goldsmith Peter Kuster was
threatened with jail if he did not report how he had come into pos-
session of Jamnitzer's mold (*mödel*) and patterns (*kunsteisen*), as well
21 as the names of those who had given the offending articles to him.

The great variety of uses to which drawing could be put in the
goldsmith's workshop resulted in a vast array of drawing types and
graphic epistemologies, a knowledge provided by the making, view-
ing, and use of images. As evidenced from thethree-hundred-plus
anonymous drawings belonging to the Amerbach-Kabinett, a min-
iature collection in itself designated the Basler Goldschmiedrisse
(Basel goldsmith drawings), skill in portraying diverse subject mat-
ter was an essential prerequisite to goldsmithing, and would have
been tightly bound up both with the creative process and with the
22 selection of a design to pursue in precious metal. Inserted between
basic stereometric figures, *mazzocchi*, and perspectival cubes drawn
in pen with a gray wash were high-end drawings used as "photo-
graphic" records of precious artworks, for instance to augment the

Workshop of Hans Holbein the Younger (1497–1543). Design for a lidded goblet with precious stones. Ca. 1535. Kunstmuseum Basel, Kupferstichkabinett, Depositum der Gottfried Keller-Stiftung 1924 (1924.64).

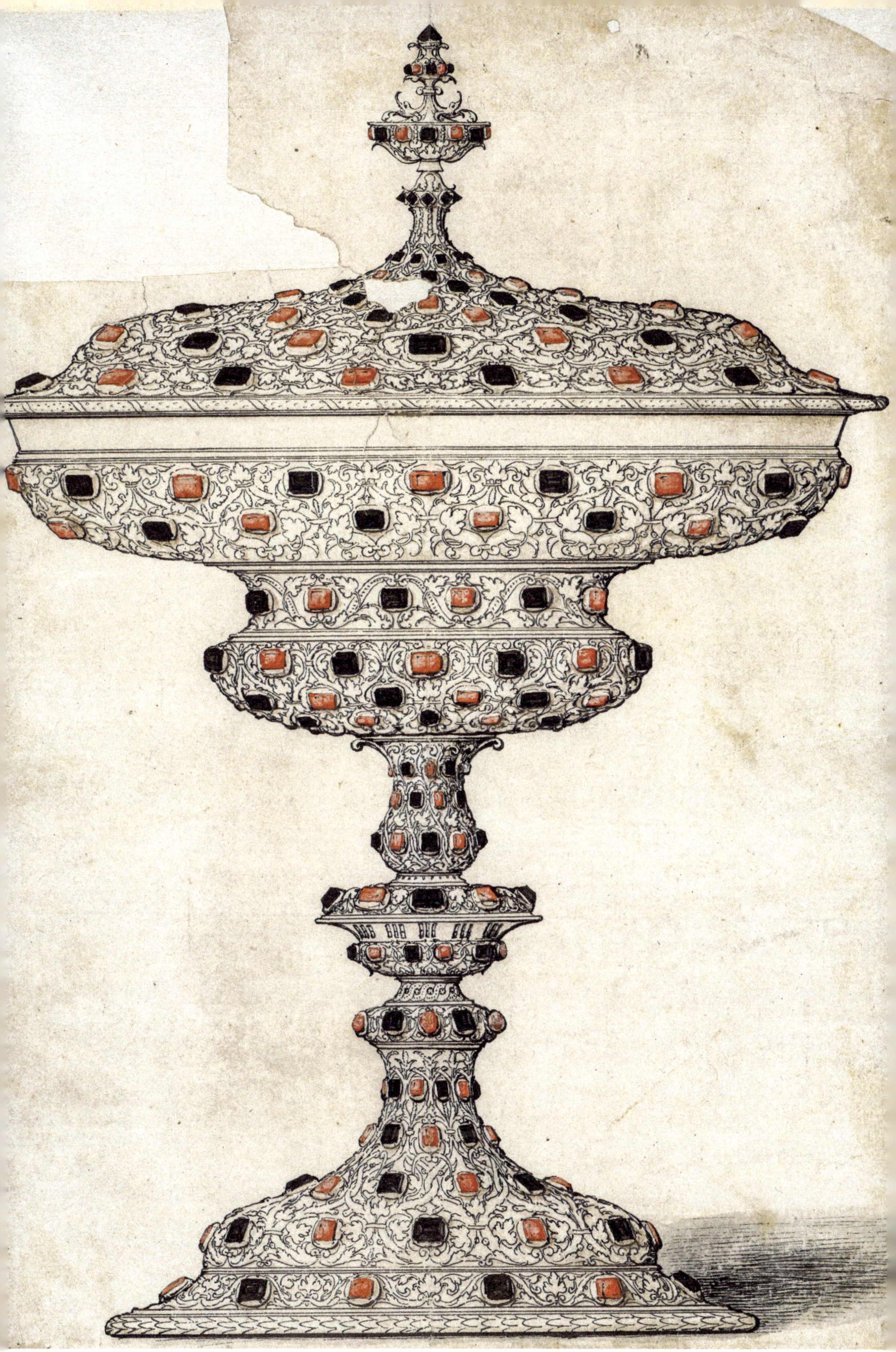

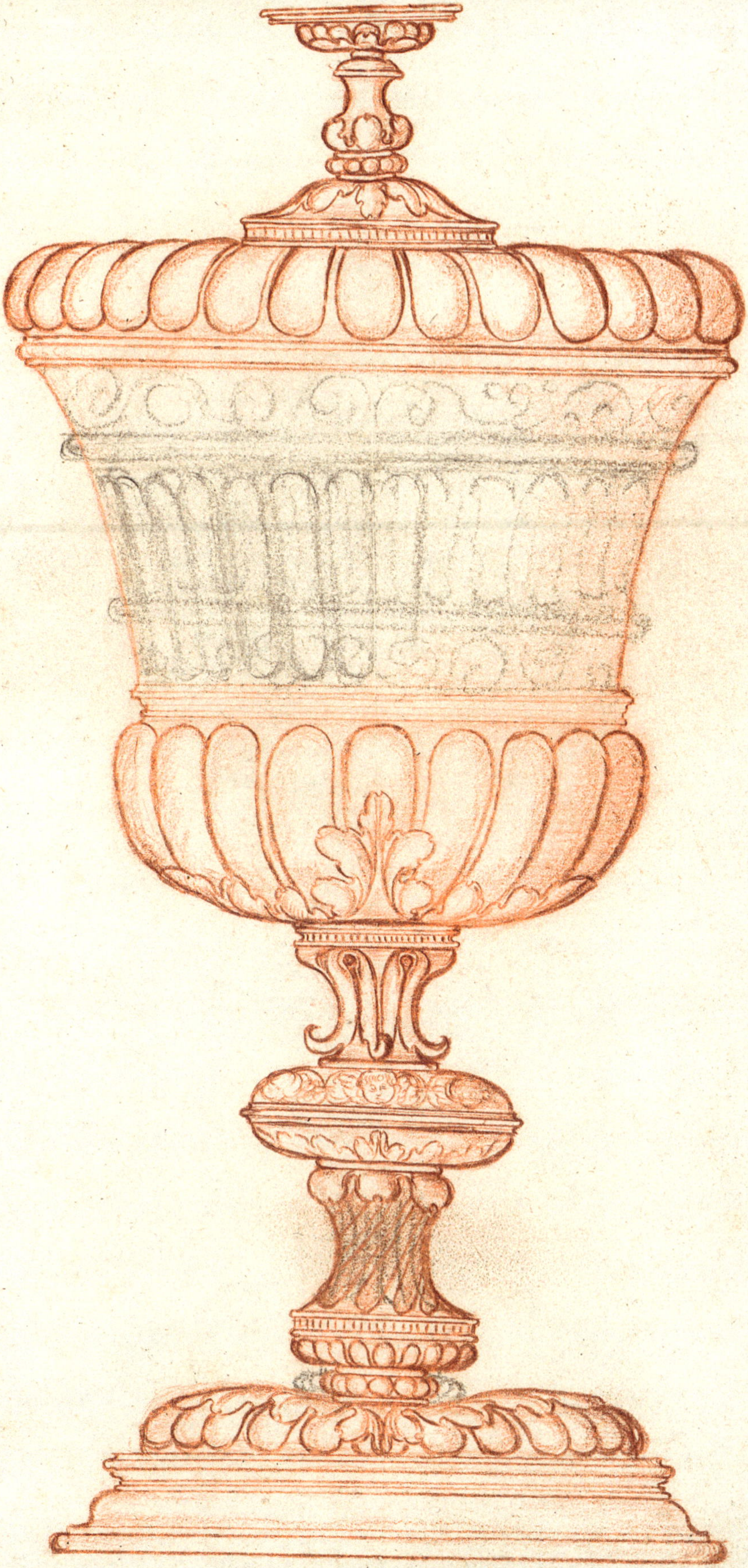

inventory of an estate. The design for a covered goblet from Hans Holbein the Younger's (1497/8–1543) workshop may have been one such example, given that the exactness of the lines and the hyperreality of the rendering far exceeded what would have been required from a working drawing. 23

The production of accurate drawings was also essential to conveying the sense of a future design, prior to receiving funding or a commitment from a potential patron. On December 22, 1556, Italian artist Jakob Strada reported to Archduke Ferdinand II on Jamnitzer's desire to produce for him a vessel modeled after the creation of Adam and Eve, and Jamnitzer's response to Ferdinand's request for a design drawing. Jamnitzer apparently replied, via Strada, that he could make the drawing as detailed as Ferdinand would want, but that he first had to know Ferdinand's opinion as to the final size of the piece. Furthermore, that it would also be difficult to understand the work solely from a drawing as the final piece would take up more space (would be at a larger scale) than the drawing, and thus that it would be prudent to make a model out of plaster, as was the common practice. 24

Drawing was the primary medium used to work out ideas, convey design intentions to clients, and train the aesthetic sensibilities of studio staff through theoretical exercises. In stark contradistinction to Holbein's, other drawings from the Basler Goldschmiedrisse are clearly amateur in quality and executed by apprentices. A goblet (*Deckelpokal*) drawn in red chalk is incomplete, revealing an uncertain pencil outline in its midsection—quite likely the result of an unfinished design idea. The artist can be seen struggling to represent the elongated, vertical ovals in perspective while the protruding candelabra-like cover of the goblet is awkwardly similar to the base, as if the artist had not been able to come up with a uniquc capstone element for the piece and had opted instead to simply scale down the goblet's base for reuse at the top. Another drawing, a clumsy sketch of a pitcher drawn with a dull pencil, is clearly the work of someone just learning how to design on paper. The lack of precision makes it difficult to imagine either drawing as having been used as a final design for an object, while the absence of ornament likely attests to the author's deliberate concentration on the pitcher's outline. The ponderous, inelegant proportions of the pitcher, with its thin spout appended to the thick body of its container, and the more proficient if still amateur

Design for a covered goblet. First half of sixteenth century. Kunstmuseum Basel, Kupferstichkabinett, Amerbach-Kabinett 1662 (U.XII.65).

goblet in chalk are examples of invisible, “internal” documents created within the workshop by its personnel: abortive design studies, not intended for public consumption, using drawing to develop and refine the design aesthetic of the workshop’s assistants.

Goldsmithing was thus intimately tied to graphic habits that facilitated the development of sophisticated surfaces to be conceptualized prior to the production of objects. Consequently, knowledge given through drawing and site-specific to the goldsmith workshop emerged to enable designs to be created and assessed, in sequence and relative to each other. At the level of the sketch and much as in Jamnitzer’s Berlin notebook, variations in ornament from the *Jeweller’s Pocket Book* (ca. 1550, south Germany), a small calfskin bound book stuffed with design ideas (including sketches based upon Holbein and Virgil Solis) and intended for ease of portability, reveal many different permutations of the same idea through shorthand drawing,
25 each iteration only very slightly modified from its predecessor. It follows that, for example, a drawing of a circular plate from the *Pocket Book* is not a passive rendering but is divided up into seven sections with two competing portions of ornament, in reference to what the completed plate would look like with either of two design options.

Other goldsmith drawings explicitly engaged with seriality *in perpetuum*, converting the transformative characteristics of ornamental metalwork into methodologies for generating form. Arrayed on pages in sequences of elevation options and geometric plans, the Basel collection houses sheets of linear design permutations for late Gothic-style baldachins and wimpergs (Gothic ornamental gables
26 with tracery), inked in hardline. The artist would have started with one scheme and then altered it in the next iteration. The design sequence displays an interest in progressive alteration, as the elements are graphically tweaked from one scheme to the next to create new options; first the outermost spires rotate inward, while the central element is extended up, next the two curvilinear spires rotate inside of the outer elements, and so on. That they were drawn in series would have permitted them to be seen relative to one another, allowing the master goldsmith to choose one particular option to develop through object making. While a client might not have been able to tell the difference between any of the baldachin design options once built—they all sport multiple spires, curved symmetrically

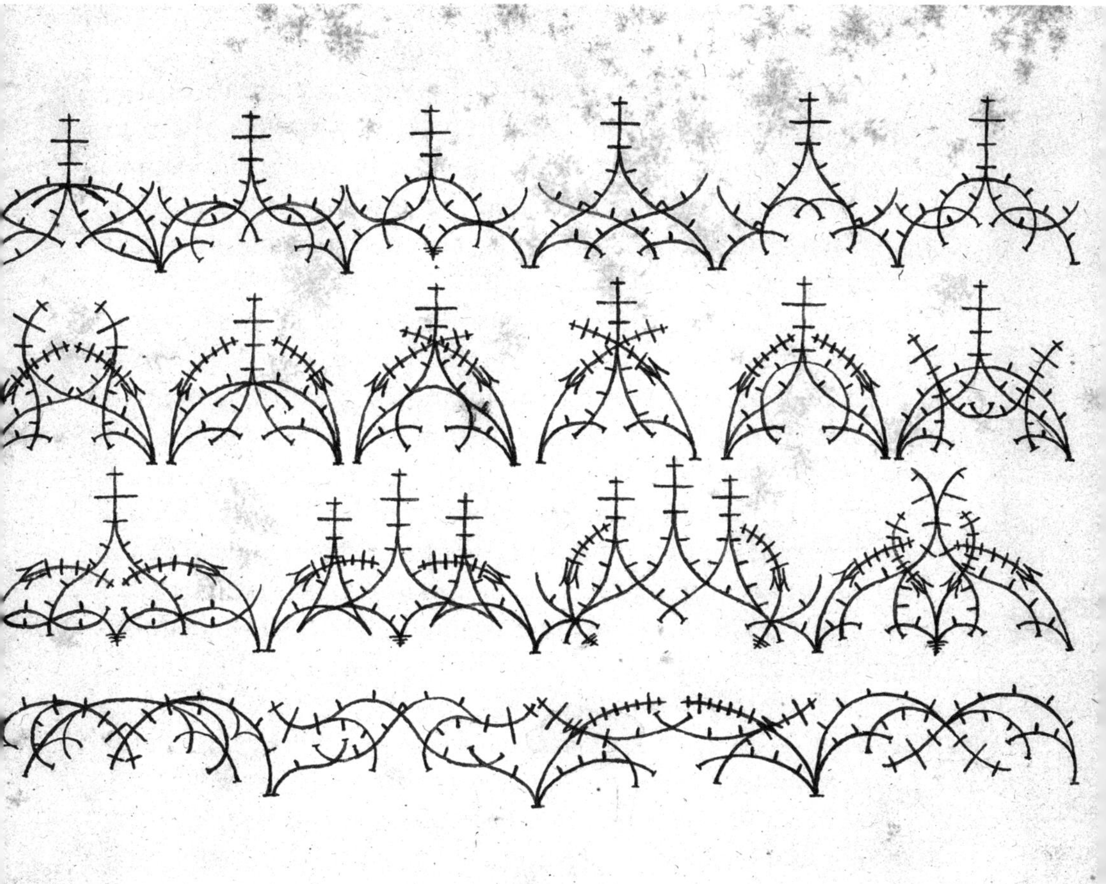

Design for baldachins and wimpergs. First half of sixteenth century. Kunstmuseum Basel, Kupferstichkabinett, Amerbach-Kabinett 1662 (U.XI.12).

with crosses—and any one of the options might well have satisfied a commission to produce a baldachin, this type of drawing would have allowed the goldsmith to decide, himself, which option he liked best. Lining the schemes up next to each other was a graphic conceit that allowed him to assess their merits and take forward the iteration that he felt best represented his design intention.

All of these designs, whether they were base motifs or fragments of ornament, finished thoughts or fleeting moments of inspiration, could then be applied to any variety of precious commissions. A plate design, for instance, required that ornament work on a flat, circular shape; a goblet, ornament that could wrap around its sides. What ornament is thus decorating is the relation between the form of the object and its surface. Form is a quality upon which character, conveyed through ornament, can be applied, engraved, or cast. In practice, it appears that this orientation toward the potentialities of ornament led to designs, and design thinking, possessed of strong metamorphic dimensions, although metamorphosis was not restricted to the figural and might well be depicted through the interplay, transformation, and packing of abstract, geometrical motifs, as in the wimperg designs in the Basler Goldschmiedrisse. Taking into account the roving movement of the eye, goldsmiths conjured up sumptuous displays of talent intended to entertain the viewer, a further iteration of the taste for the aesthetic of intricacy that swept across southern Germany, made possible by the intense, creative concentration on the surface of things.

Perspectiva inflects and is inflected by these unique drawing practices, in which the image rehearsed a geometric metamorphosis that both satisfied a prospective client's desire for design options while also manipulating the viewer into sustained contemplation of new kinds of nonfigural, visual narratives. But as much as genres of goldsmith drawings combined with decades of polyhedral expertise prepared the groundwork for Jamnitzer's magnum opus, *Perspectiva* rises above its source material in the sheer grandiosity of its pictorial ambitions. A poetic epic in artistic terms of what could be read today as filmic syntax, *Perspectiva* served as a site for Jamnitzer to explore the possibilities of print to communicate material change, reflecting his mastery of the shaping and manipulation of molten metal as much as it did his complete command of geometry and perspective.

Perspectiva was printed by Jost Amman and engraved by Hans
Sachs (one of the main characters in Richard Wagner's *Die Meistersinger
von Nürnberg*), who published his own *Ständebuch* (Book of Trades) in
Frankfurt that same year, and was acquired by the princely *Kunst-
kammern* of Prague, Dresden, and Ambras Castle in Innsbruck along-
side Jamnitzer's commissions in gold. Jamnitzer had begun work 27
on *Perspectiva* at least ten years before it was published. We know this
because in 1558 he had sent a letter to Archduke Ferdinand II (1529–
1595) in Innsbruck claiming to have "invented the laudable, useful
and ingenious art of perspective, the likes of which have never been
seen before." Dedicated to Emperor Maximilian II (1527–1576), older 28
brother of Ferdinand II, its full title translates as "a diligent expo-
sition of how the five regular solids of which Plato writes in the
Timaeus and Euclid in his *Elements* are artfully brought into perspec-
tive using a particularly new, thorough and proper method never
before employed. And appended to this a fine introduction on how
out of the same five bodies one can go on endlessly making many
other bodies of various kinds and shapes." 29

Unlike Italians like Piero della Francesca or Alberti who worked
in similar fields and typically made use of extensive textual commen-
tary, in German geometrical and perspectival treatises the relation
between text and image tended to lean heavily toward the visual,
relying on the image to speak, if not completely on its own, then pri-
marily so. Envisioning his book as a collectible, a coffee table book and
work of art in its own right rather than a working treatise, Jamnitzer
does not define any rules of geometrical transformation that might
be repeated or expanded upon by others, a strategic withholding
of the kind of information that both Barbaro in *La pratica della per-
spettiva* and the Flemish mathematician Simon Stevin (1548–1620)
in his *Problemata geometrica* (1583) would include. And although 30
Jamnitzer did state in the foreword his unrealized intention to pub-
lish a second more writerly companion treatise on geometry, the
volume's explanatory text is limited to the elaborate German/Latin
frontispieces that preface each section of Platonic solids and link the
solid to the element associated with it by Plato in the *Timaeus*. For
instance, the frontispiece for the first solid, the pyramid, reads: "A.1;
IGNIS. Das Fewer [Fire]; TETRAEDRON; Siue Pyramis trilaterata. Ein
trianglichter Kegel [A triangular cone]." The title appears suspended

in a burst of flame rising up from a roaring cauldron and is followed by a description of how to geometrically construct a pyramid, similar in tone and content to the many *Lehrbücher* to which Jamnitzer presumably had access.

> The first of the five regular bodies is a body made from four equilateral triangles, surfaces or intersections of straight lines on which it may be placed. It has six edges or straight lines, twelve flat angles and four corners to its geometry. From this triangular body or cone further originate twenty-three other bodies, made in a variety of
> 31 different ways. As is shown hereafter.

The five sections consist of twenty-four polyhedral variants apiece beginning with an initial Platonic solid, six per individual page, four pages in total. The polyhedra are represented as floating objects inside six circular vessels whose rims are connected by a continuous strip of material wrapped with ornamental frames. Each of the four pages proceeds from the simplest polyhedral variant in the top left position to the solid that Jamnitzer considered to have been most completely transformed in the lower right position. Nearly each section devotes the first page to transformations that required a graphic excavation of the solid, the second page adds a progressive chopping up of the solids' surfaces, while the last two pages explore the intersection of two or more solids and their stellation. Following the exploration of the polyhedral variants is a section consisting of forty freestanding models, two per page, in clear mastery of the genre of unbuilt polyhedral models populating the German *Lehrbücher*, and three final pages depicting tableaus of *mazzocchi*, stars, crosses, and pyramids leaning on sundry supporting structures.

Perspectiva was no common *Lehrbuch*, published to supplement Jamnitzer's income. In the degree of its graphic-philosophical aspirations, it resoundingly surpassed the records of geometrical investigations in the various *Lehrbücher* circulating around Nuremberg's workshops. Interspersed throughout the *Rollwerk* that makes up each section's frontispiece are objects and figures that testify to the associations of the Platonic solids with the virtues and instruments of measurement as well as the classic association of the Platonic solids with the elements. Jamnitzer stressed these allusions, which had been

Jost Amman (before 1539–1591), after Wenzel Jamnitzer (1507/1508–1585). Polyhedral bodies, from *Perspectiva corporum regularium* (Nuremberg, 1568). The Metropolitan Museum of Art, New York, Harris Brisbane Dick Fund, 1924 (24.45.1, fols. B III, C V, and D V).

E

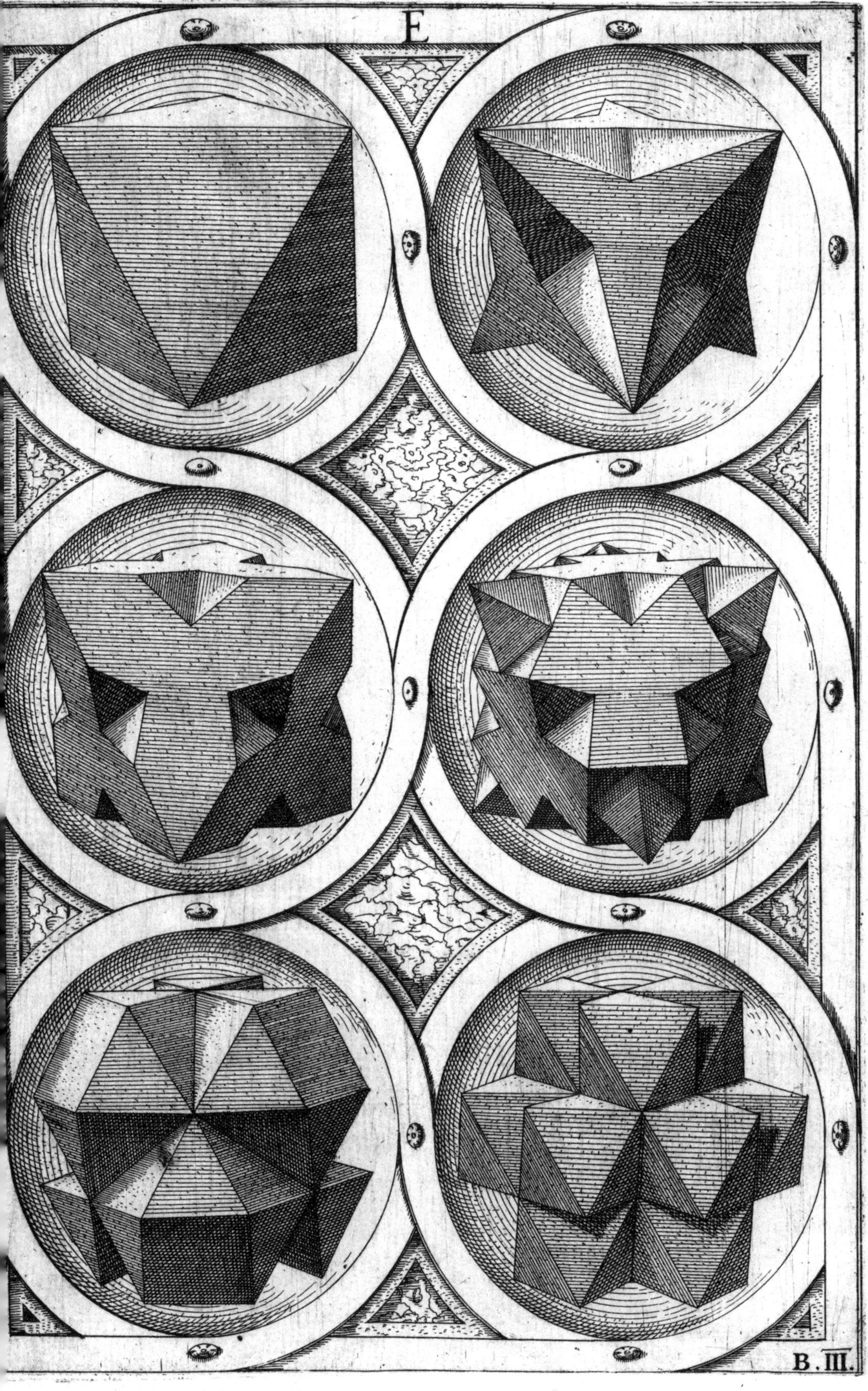

B.III.

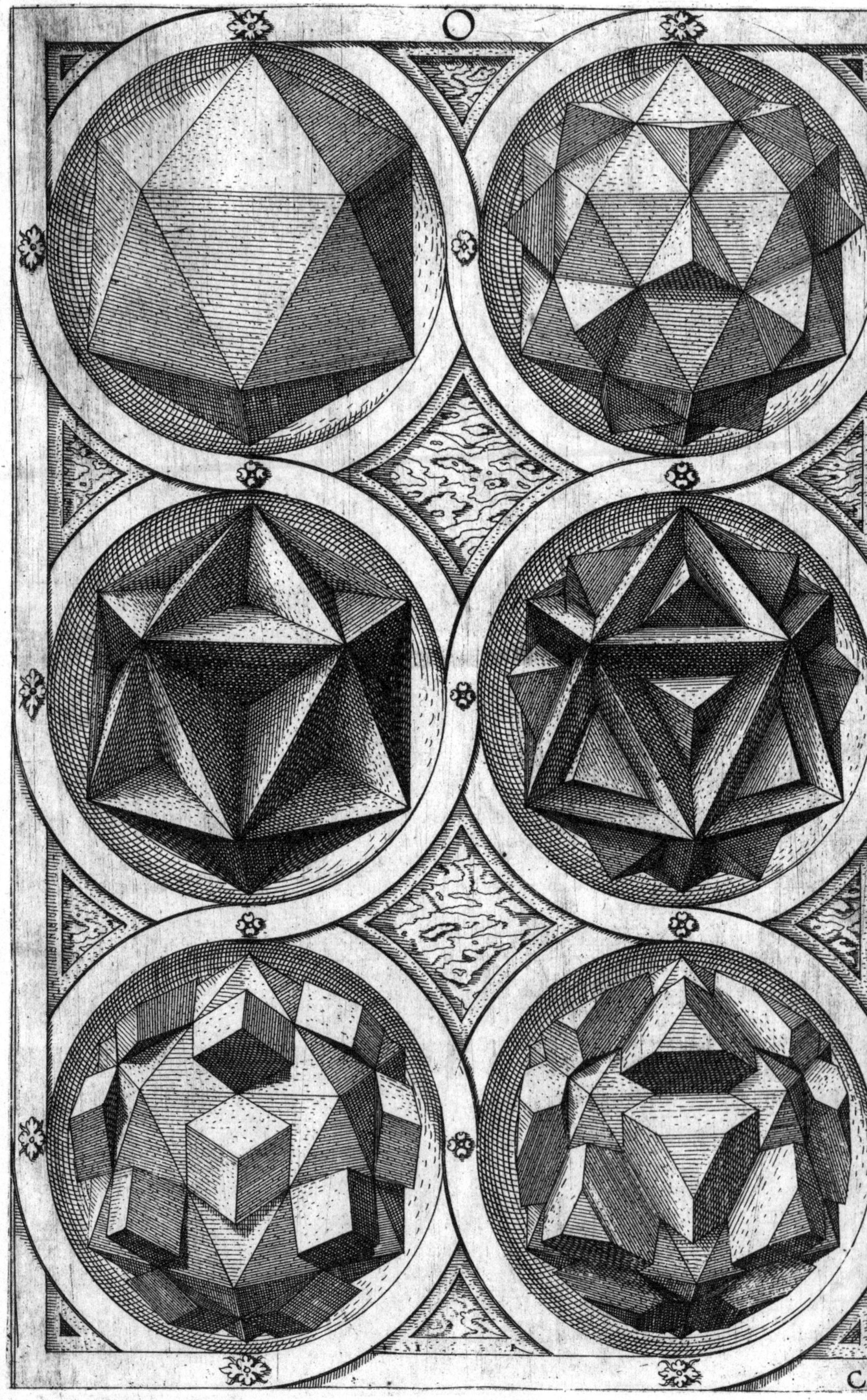
O

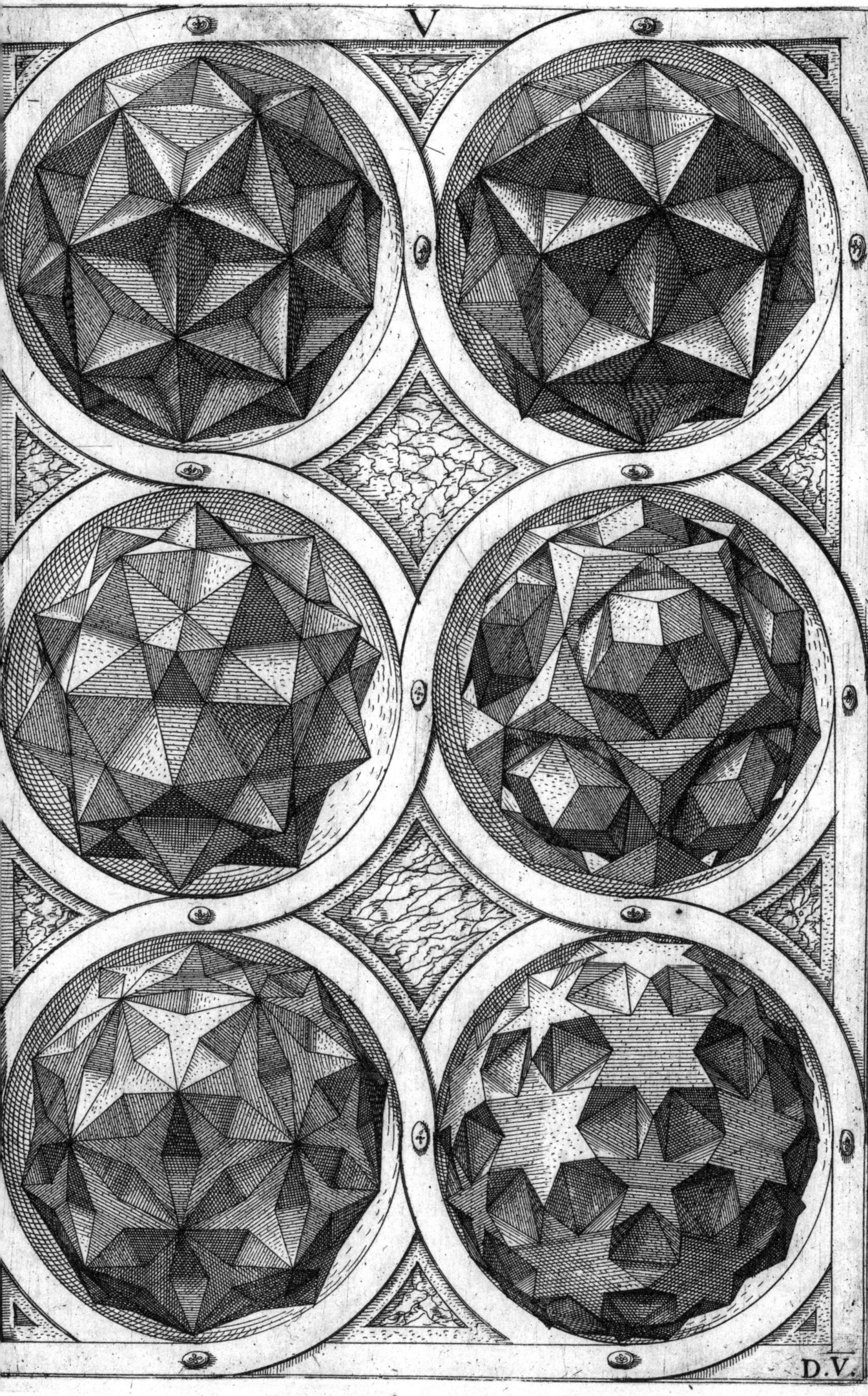

abandoned by Dürer and the *Lehrbuch* authors, in order to elevate his claims for the significance of his geometrical inventions. Hence the frontispiece for the pyramid, associated with fire in Plato, is decorated with dragons, guns, lanterns, incense, and torches, while the icosahedron, represented by water, comprises crabs, squid, tridents, shells, turtles, and sea serpents. Each subsequent body is similarly adorned with references that speak to their elemental designations. The octahedron signifies air; the hexahedron, earth; and the dodecahedron, heaven (the fifth element). The polyhedra are further linked with vowels depicted above each page of designs: "Tetraedron" with A, "Octaedron" with E, "Hexaedron" with I, "Icosaedron" with O, and "Dodecaedron" with U. The Alpha-Omega coupling (represented by Jamnitzer as the vowels A-U), the first and last letters in the classical Greek alphabet, corresponds to the first section of *Perspectiva*, signified by the simplest polyhedron, the tetrahedron, and the last
32 section, signified by the dodecahedron, the most complex.

The geometrical construction of the alphabet has its own parallel trajectory in the Renaissance. Pacioli appended letters to the appendix of *De divina proportione* and there is a section covering the drafting of the alphabet in the section preceding Albrecht Dürer's unfolding of the Platonic solids in *Underweysung*. But Jamnitzer's use of vowels as a classificatory mechanism for his polyhedral permutations does not focus upon their constructability as figures and has more in common with Lucretius's *De rerum natura* (first printed edition 1473), which addresses the recombinability of letters, the building blocks of language, as mirroring the changing states of matter.

> For the same letters signify sky, sea, earth, rivers, sun, the same too crops, trees, living creatures ...but it is by position that things sound different. So in things themselves likewise when meetings of matter, its motions, order, position, shapes are changed, things too are bound
> 33 to be changed.

Later artists, such as Peter Halt in the title page to his *Perspektivische Reiß Kunst*, would further explicate the correspondence between the solids and the vowels. In Halt, the vowels are drawn as three-dimensional objects and are balanced atop the Platonic solids, the building blocks of perspectival and geometrical knowledge. As the preface to the second, posthumous edition of *Perspectiva literaria* (1595) by

Jamnitzer's contemporary, the goldsmith Johannes Lencker, put it,
"wie man ohne Vokale nicht sprechen könne, ohne die regulären
Körper nichts in der perspektivischen Reisskunst erreiche"—as one
cannot speak without the use of vowels, so without the regular bod-
ies no knowledge of the art of perspective is attainable. 34

In the foreword to *Perspectiva*, Jamnitzer spells out his vision of
the natural order of the universe. "God … is good and true and has
everything beautifully and artfully ordered … Heaven and earth and
the magnificent lights, the sun, moon, and stars, with which he has
adorned the heavens." Within this order, the infinite variety of "earthly 35
bodies" (*irdische Coͤrper*), including mankind, possess their own indi-
vidual dispositions, distinctions that may be revealed through inter-
relationship, much as combining the primary elements like fire and
water reveals their fundamental characteristics. In Jamnitzer's view, 36
nature is defined by these degrees of variation and change, a view that
dovetailed with a distinctly Aristotelian natural philosophy.

> In addition, I want to show if at all possible that there are still more elements in nature than the four natural elements and the selfsame five universal essences. Although there can be no more than five regular bodies, there are other kinds of bodies, which are formed in the same way and can be produced from the same geometrical bases [*Boͤden*]. These may still be compared to the selfsame five elemental bodies of nature, because namely the trilateral pyramid or the cone both terminate into points as do fire and flames. … And like all the other earthly bodies, all living creatures both men and plants, namely foliage and grass, have been created from a mixture of these four elements of nature placed together, as well as have also other geometrical bodies been unendingly mixed and set together, as is the intention of my work in which 140 different bodies can be seen, both whole and skeletal. I have rather wanted to bring these [geometrical] bodies into my new perspective because they have been transformed [*gekehrt*] inward and outward and into many kinds of different corners, sides, angles, and points. 37

More than any other Renaissance artist, Jamnitzer's goldsmithing
was legendary for breaching the art-nature divide. Literally making

use of plants and animals as the base for his life-cast-covered compositions, Jamnitzer's work demanded to be seen as a direct copy from nature, in which nature imprinted its form onto material, not solely
38 a process whereby an artist had striven to replicate (or perfect) nature. But what Jamnitzer's goldsmithing could not do was dramatize the process of gaining the deductive and propositional knowledge that came from the manipulation of matter. Jamnitzer's foreword to *Perspectiva* makes evident that he intends to use the Platonic solids as emblems to the transformation and potentiality inherent in nature, and that he sees geometrical recombinability, "ohn endtlich miscirt" (unendingly mixed), as an analogous process of invention and construction. The transmutation of the Platonic solids into a hypothetically infinite stream of multifaceted geometry signified, for Jamnitzer, the natural act of creation that mirrored the construction of language and the divine order of God. By showing "wie auß denselbigen Fůnff Cŏrpern one Endt / gar viel andere Cŏrper / mancherley Art vnd gestalt / gemacht" (how out of the same five bodies one can go on endlessly making many other bodies of various kinds and shapes), Jamnitzer raised the intellectual stakes of perspectival work with geometry. Perspective was not just a teachable art for representing objects, and geometrical instruction not just a stepping-stone on the way to more advanced forms of applied geometrical knowledge. Accurately and measurably working with perspectival geometry was akin to bringing new bodies into existence, to forming sounds into words, and to creating life on paper where before there had been none, even if Jamnitzer, bucking the semantic field in the treatise's title, deemed his many inventions "regular" on account of their symmetry.

By viewing his work as performing the principles of natural variation, Jamnitzer was implicitly laying claim to knowledge of natural philosophy supported by years of practice and observation. But the graphic liquidity of *Perspectiva* could only have been conceptualized by someone versed in the behavior of molten materials. The precedents for these graphic transformations lie not only in goldsmith workshop drawings but also in the relationship of goldsmithing to the manipulation of form and the "keen pleasure aroused by strange
39 transmutations in the late Renaissance." Beginning from the Platonic solids as base patterns, *Perspectiva* performs the state changes of metal

through a series of highly articulated variants. This is an alchemical transformation of the graphic emblems of perspectival geometry, one that showed that the nature of the solids as conceived by Plato and Euclid was transitory and part of a liquid continuum of movement.

The use of the word "alchemical" is more than just descriptive,
as it has been established that knowledge of the formation and mani-
pulation of metals affected the working processes of both goldsmiths
and alchemists alike. The collection of theories compiled in the
German minister Johannes Mathesius's (1504–1565) "Third Sermon.
On the Origin, Growing and Reduction of Metals and Minerals and
Ores" glosses not only the overlap of alchemical and metalworking
knowledge but also "how strongly early modern natural philosophy
and metallurgic research are influenced by the Christian episteme." 40
Base metals, like lead or zinc, were treated by alchemists in the manner
that the sick were treated by physicians, by purging them of impuri-
ties in the heat of the forge. "For just as gold is tested in fire a seventh
time, the physician must be proven by fire a seventh time and more." 41
Moreover, the purification of metal in the forge was an established
biblical allusion to the quality of God's work. In his final sermon, 42
"Death's Duel," the cleric John Donne (1572–1631) uses the language
of alchemy, and goldsmithing, to describe spiritual rebirth—from
death to resurrection, and man's relationship to the Christ sacrifice. 43
"The melting process (passing from a solid state to a fluid one) is a
metaphor for the mystical opening of the soul." It is into the space 44
created by the intersection of metallurgy, alchemical transformation,
and Christian theology that we must insert Jamnitzer the goldsmith
and author of *Perspectiva*, who, like God in Mathesius's *Sarepta*, "hat
mancherly schmeltzwerck inn seinem laboratorio" (had all kinds of
melted products in his workshop). 45

Jamnitzer illustrates "a sixteenth century swept up in change
and fascinated by genesis and metamorphosis," albeit in geometrical
fashion. *Perspectiva* is, ultimately, the graphic product of an artistic 46
culture seeped in the pragmatics of making decorated objects and elevated to the level of a printed masterpiece; born of the same creative reality as the many drawings that would have been floating around Jamnitzer's workshop and informed by workshop practices and the innate properties of the material handled there. His patrons were meant to consume the images he etched onto the surface of his

masterworks over long periods of time, as conditioned by the shape of the object. Here, geometry could be just as engrossing as figures, particularly if viewers understood geometry in terms of its transformation from one thing to another. Jamnitzer understood too that the shape of the underlying object would define the types of narrative he could portray and whether it had a start or an end or ran on a perpetual loop. And that enigmatic, visual texture would cause his viewers to slow down and allow themselves to become absorbed in the weight of the detail, growing joyful then fearful as their gaze darted across the different scenes etched in gold, losing themselves in a rapture of wonderment.

The flow of form and bodies in ornament held within itself a series of psychological or perceptual effects communicating to the early modern viewer a sequence of meanings in the time designed for their reception by its creator. The goldsmith workshop was not limited to the production of objects; it also produced narratives wrapped around objects, or, in the case of *Perspectiva*, endless transformation reproduced as wondrous filmstrips bound together. Echoing Hephaestus's forging of the shield of Achilles in the *Iliad*, the *locus classicus* for the narrative skill of the craftsman, these *Meisterstücke* decelerated seeing for connoisseurs of the decorative arts in an ekphrasis charged and recharged through the consumption of layers
47 and lines of surface ornament.

The time and labor embedded in the production of *Perspectiva*'s pages—Jamnitzer's certainly, but also the printer Jost Amman's contributions as well as those of a host of invisible workshop technicians and makers of paper or wood polyhedra—were intended to provoke admiration and wonder in *Perspectiva* as a portable, and purchasable, mathematical *Wunderkammer* that exhibited the exoticism for which *Wunderkammern* were known. To create "difficult" or impossible objects that graphically performed the discipline and labor required to produce them was to justify to buyers the cultural worth and asking price. If "philosophers traditionally measured nature's skill by
48 the elegant economy with which she had fitted form to function," *Perspectiva* was supernatural, captivating spectators with the extent of its engrossing precision while capitalizing on the taste for the complexification of surfaces, so prized by patrons, that had enabled Nuremberg's unrivaled precision manufacturing industry to thrive.

The unique drawing culture that supported the production of decorative art objects in crafts workshops engendered a level of design literacy of which *Perspectiva* is exemplary, directly conditioning Jamnitzer's experience with, and predilection for, material transformation. It remains an epitome of artisanal contributions to geometry, intimately tied to pervasive cultures and practices of drawing, to an evolving aesthetic that valued abstraction and complexity, and to the influence of ornament making as an epistemic practice; a new, artisanal investment in the theoretical, inflected by the materialities of practice and the cinematics of viewing ornament. Jamnitzer was understandably proud of his accomplishments in visualizing and inventing geometry, perhaps believing that his work could rival the humanist contributions to geometrical knowledge. *Perspectiva* thematizes the entire design process, as one polyhedral shape morphs into the next from page to page. Graphic form is unstable, elastic, prone to change and metamorphosis, not unlike the precious metals that Jamnitzer spent his days reworking. For gold, in whatever form or shape it arrived at the workshop, would necessarily be pounded into gold leaf, subjected to the heat of the forge and melted down to be made workable and to remove impurities. A lifetime of melting down the work of other goldsmiths and transforming the precious, raw material into new, exquisite forms must have impressed in Jamnitzer a bittersweet appreciation for the fragility and temporality of objects. Hence, a palpable violence quivers at the edges of even the most fragile of his physical creations. Only by daring to achieve prodigious heights of complexity might a goldsmith prevent, as much as possible, the eventual melting down and reuse of his work. To epitomize, aesthetically, a moment in the collective taste, to be representative of all that a particular style had to offer, was as much as he could do to guarantee the preservation of his labor.

overleaf Jost Amman (before 1539–1591), after Wenzel Jamnitzer (1507/08–1585). Spherical models, from *Perspectiva corporum regularium* (Nuremberg, 1568). The Metropolitan Museum of Art, New York, Harris Brisbane Dick Fund, 1924 (24.45.1, fol. G1).

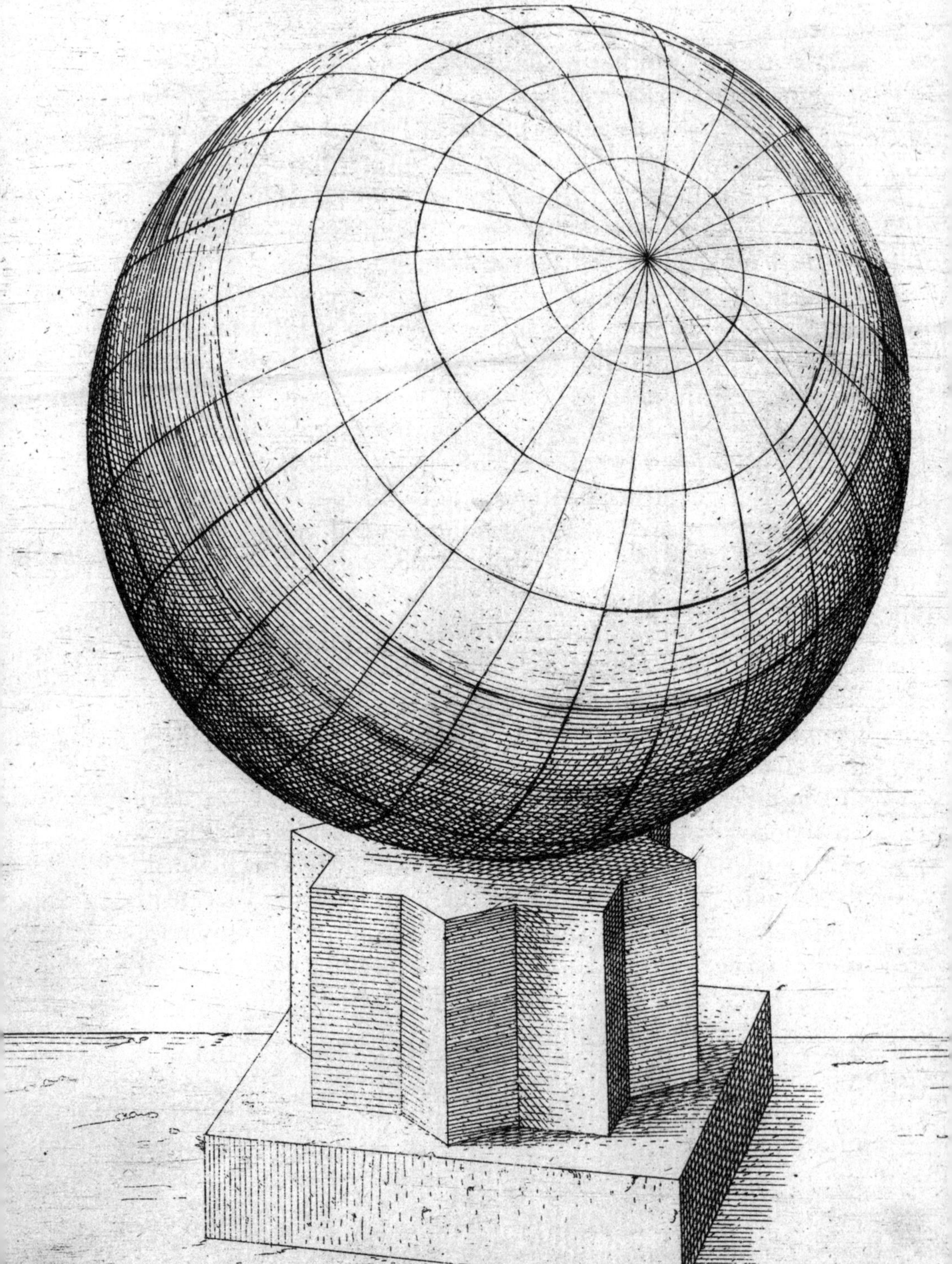

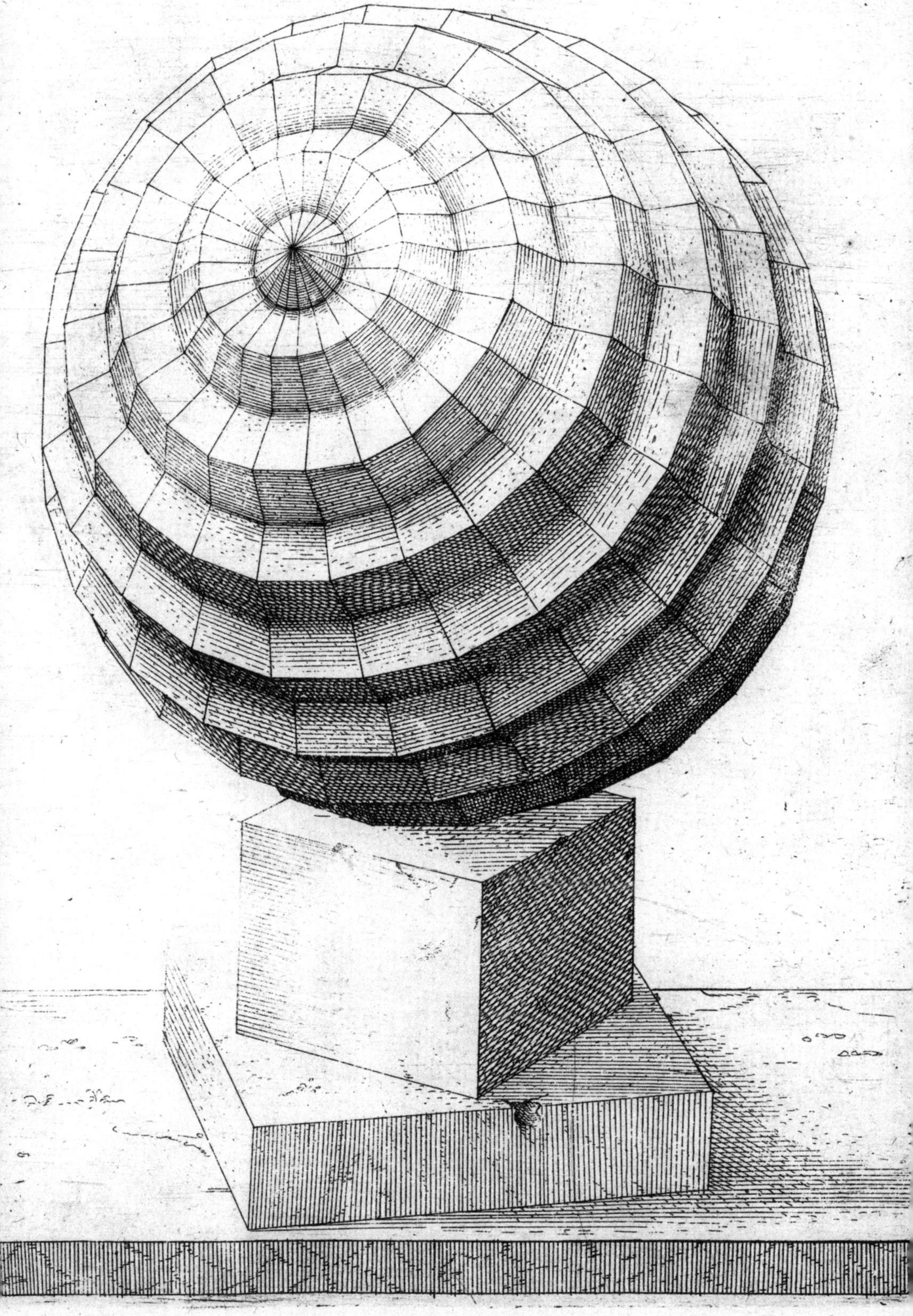

CVTA

IN MAN

PART

Y WAYS

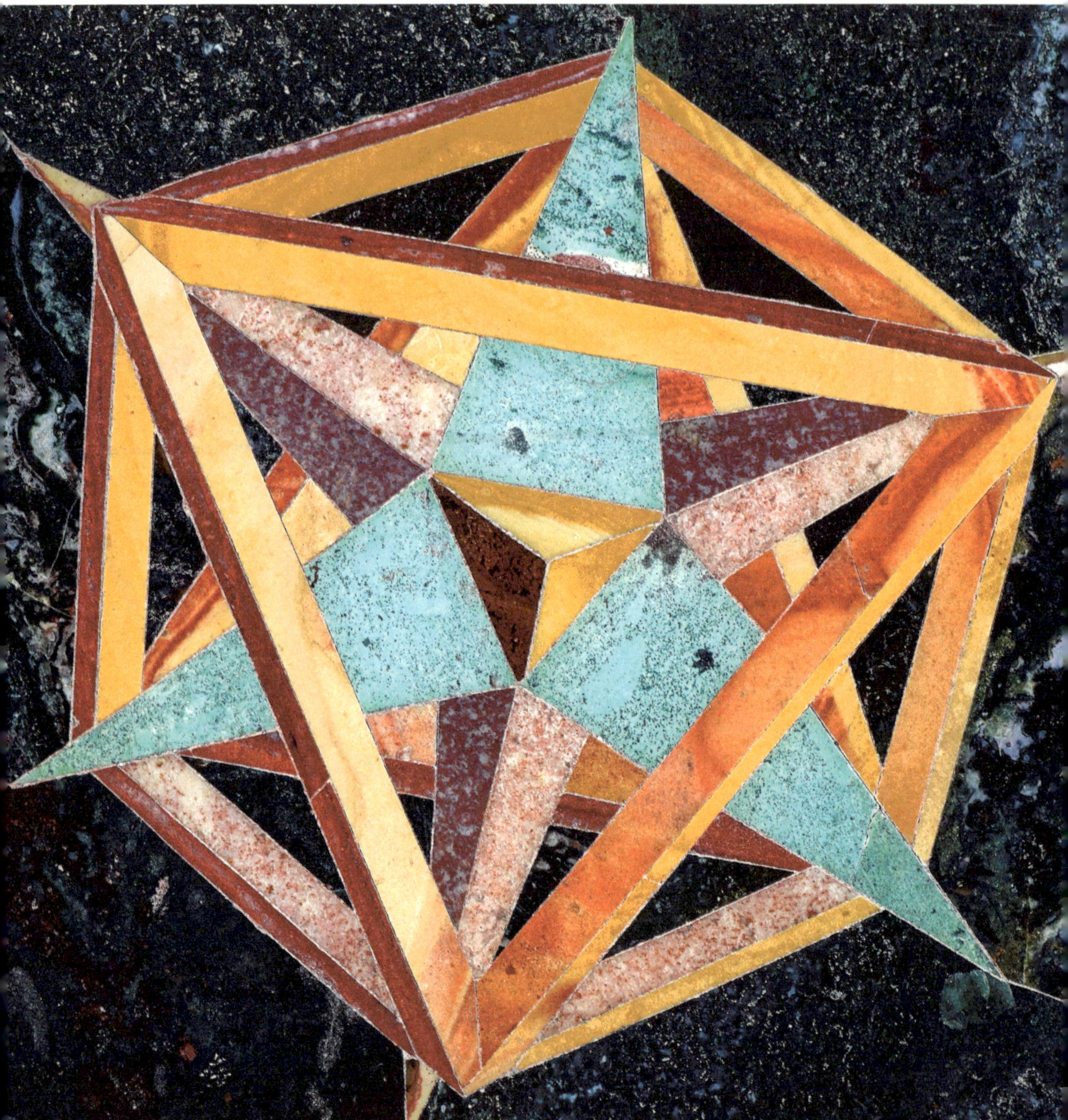

Cosimo di Giovanni Castrucci (1590–1619); Dionisio Miseroni (1607–1661); Workshop of Giuliano di Piero Pandolfini (1615–1637). Detail of nested polyhedra from tabletop belonging to Prince Karl I of Liechtenstein. Ca. 1620–1623. Liechtenstein, The Princely Collections, Vaduz–Vienna (SK 1401). Photo: ©2020 Liechtenstein, The Princely Collections, Vaduz–Vienna/SCALA, Florence.

In the upper foyer of the Hôtel de Ville de Mulhouse, a distinctively pink building adorned with tromp-l'oeil and an expressive branching staircase, rests a mammoth wooden cabinet from the southern German city of Augsburg, its many marquetry panels inlaid with a dazzlingly modern, urban vision. A labyrinth of terraced planes hover above each other, seemingly without support and connected only by small bridges and floating stairways. The scene runs continuously behind the miniature wooden columns that mark the borders between the cabinet's compartments, giving the impression that the space extends beyond the cabinet's surface. The urban scene is eerily devoid of life but multiperspectival, as if a composite image stitched together from the eyes of a crowd. Tracing the contours of this city, under porticos, up the stairs and in and out of the jet-black space beyond, gives the impression that the city is as tenuous as it is ever-expanding, floating in nothingness, a utopian purveyor of information—both a place devoid of imperfections (an *eutopos*) illustrating perspectival principles and an abstract site (an *outopos*) existing nowhere.

Resuscitated by Florentine artists in the fourteenth century, by the mid-sixteenth century the art of intarsia, later developing into marquetry, had evolved into a natural outlet for the kind of geometrical creativity increasingly associated with German craftsmen. 1
Requiring the ability to work between media, dissecting and recomposing drawn or printed figures in small slabs (3–10 millimeters thick) of veneer wood elements or tesserae, intarsia retained an intimate connection to the circulation of combinable ornament prints produced by and for workshops. And if there was one place that specialized in the art of luxury cabinetmaking it was Augsburg, a bustling commercial hub some 150 kilometers south of Nuremberg and home to the Fuggers, a mercantile family conglomerate with transcontinental interests spanning banking, mining, and the spice trade. With connections at courts, banks, and ports across Europe, it was the Fuggers that loaned money to Magellan to finance his global circumnavigation. But this was only one of their many overseas ventures. The Portuguese trade ship *Bom Jesus*, which sank off the coast of southern Namibia in 1533 en route to western India from Lisbon, was laden with copper ingots from Central Europe imprinted with the trademark of the Fugger company. Indeed Jakob Fugger (1459–1525) bears 2

the distinction of having been the world's richest-ever man, with a
3 reported net worth equivalent to $400 billion in today's currency.

The Augsburg *Schreiner* (cabinetmakers) were known for a local specialty of opulent *Schreibtische* (cabinets that folded down into writing desks) and *Kunstkabinette* resplendent with marquetry (*Geschrottenwerk* as per Johann Neudörffer or *legni tinti* as per Vasari), much like the exquisite example in Mulhouse. Cabinet surfaces, following what evolved into a characteristic Augsburg or southern German style, were opportunities for visual density, whether floral, mythological, architectural, geometric, or some combination thereof, and were prized across the courts of Europe. King Charles I of Spain, better known as Emperor Charles V, commissioned the Augsburger Lorenz Strohmeir in 1554; his fellow citizen Bartholmä Weishaupt made an ornate cabinet for King Philip II that was delivered to Spain; and in 1600 Hieronymus Fleischer worked for the
4 Archduchess Maria of Austria and her daughter the Queen of Spain. A petition to the Augsburg city council in 1568 enunciated the success that the city had begun to enjoy and the extent of its self-identification with the art of "cabinet making, which earns praise abroad and
5 in this city for how all sorts of things can be so skillfully inlaid, which no painter can match in the colors."

The Augsburg council had reason to defend cabinetmaking and the craft of marquetry against the incursions of an Italianate art history that privileged the fine over the decorative arts. While discussing the work of Italian artists, Vasari in *On Painting* had fired a broadside against intarsia, which according to him was "less than painting" and of "short duration because of worms and fire," "a line of work consist[ing] only in the choice of designs that may be adapted to it." For Vasari, the motifs that distinguished the Augsburg cabinets—the abstracted, alien urbanity in a perpetual cycle of decay and renewal, the collage of multiperspectival and foreshortened viewpoints, of architecture, geometry, and overgrown ornament collapsing into and out of each other—did not amount to any grand urban or philosophical vision. Rather, "buildings and objects with rectangular outlines" were just more easily adaptable to the innate methods of intarsia production, namely, the technical skills required to create "Compositions that are made in Tinted Woods, fitted together after the manner of
6 a picture." Indeed, in a withering critique of Paolo Uccello's oeuvre,

Master with House Monogram. "Wrangelschrank" (detail). Augsburg, 1566. LWL-Museum für Kunst und Kultur, Westfälisches Landesmuseum, Münster (K-605 LM). Photo: H. Neander.

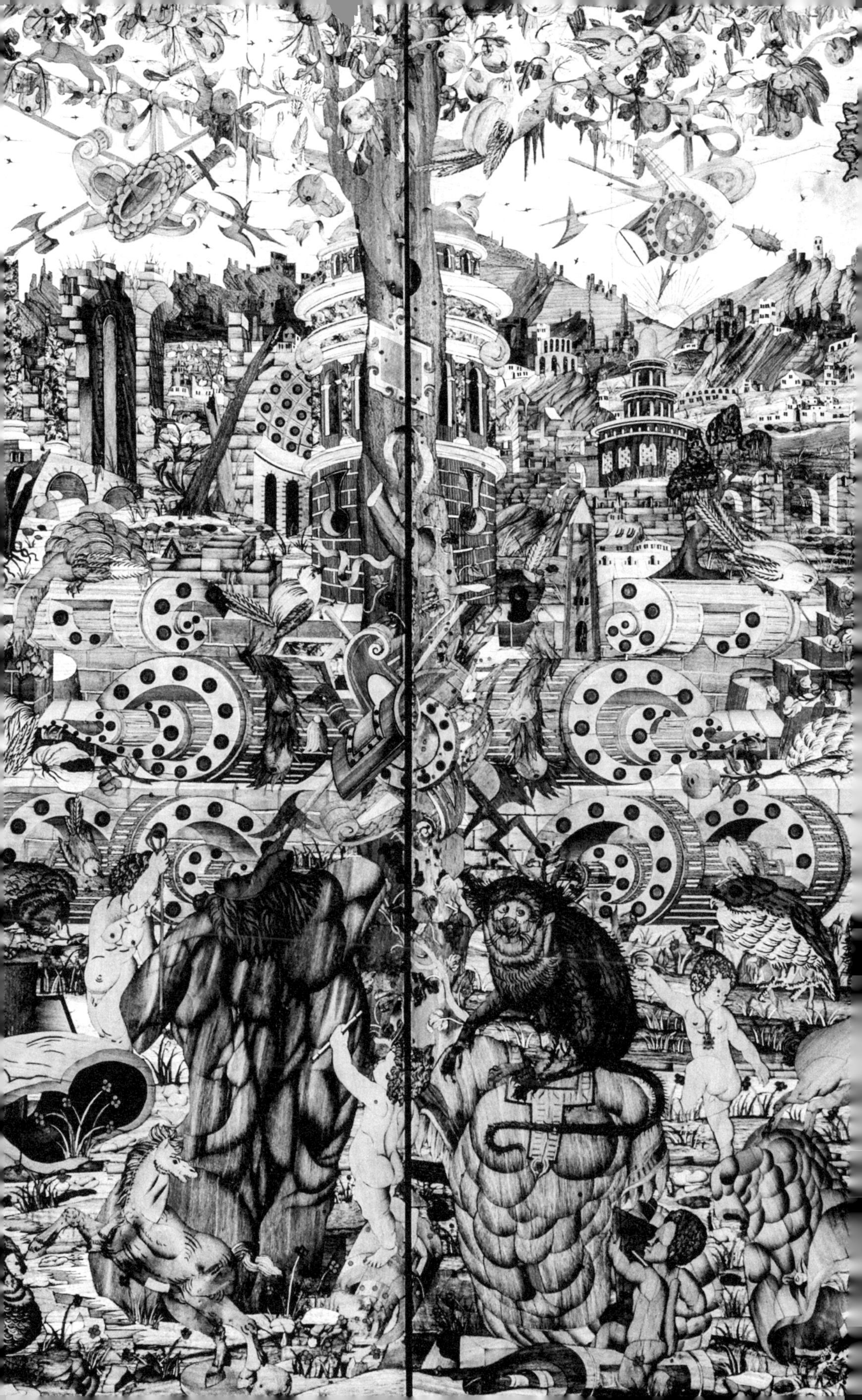

a critique he claimed was seconded by none other than Donatello, Vasari deemed intarsia to be the only appropriate medium for the imaging of geometrical solids such as "mazzocchi with pointed ornaments, and squares drawn in perspective from diverse aspects; spheres with seventy-two diamond-shaped facets, with wood shavings wound round sticks on each facet." According to Vasari, Donatello had apparently conveyed the message personally to Uccello that "these are things that are only useful to men who work at the inlaying of wood, seeing that they fill their borders with chips and shavings,
7 with spirals both round and square, and with other similar things."

And yet Vasari's dismissal of these representational spaces—and by extension the network of prints and drawings connected to their conception and fabrication—as too reductive and too defined by material constraint to evidence artistic intention risks overlooking an exciting chapter in the history of three-dimensional thinking. In the Mulhouse cabinet or the so-called *Wrangelschrank* (Augsburg, 1566) now in Münster, as with many of the cabinets originating in Augsburg, intarsiated surfaces depicting postapocalyptic cityscapes, embellished with strange creatures and monstrous polyhedra, staged a new kind of spatial complexity beyond the search for any kind of order, decorum, divine proportion, or stable classical referent long abandoned. Bearing little relation to the ideals of *The Ideal City* and expressly designed to toy with space, exploit scalar ambiguity, and elude the primacy of the human gaze, these artifacts capitalized on the tastes of a nascent 3D culture in northern Europe that favored visual complexity over restraint and states of transformation over the staid perspectival objects of the *studiolo*. Standing at the vibrant intersection of the decorative arts, geometrical perspective, and print culture, the Augsburg cabinet did not stage subtle tweakings of the Renaissance picture plane to create more realistic settings, but masterful geometrical assemblages demonstrating the artist's command of the third dimension, collected together for permanent display and tethered to the network of paper ephemera that sustained them.

In one respect, however, Vasari may have been correct. Outside of the graphic images of the German polyhedrists and their southern neighbors like Daniele Barbaro and Lorenzo Sirigatti, *corpora* appear most frequently as elements in intarsia, marquetry, and *pietre dure* (the companion art of cutting and arranging polished stone). Polyhedra

appear on a grand coin chest for Archduke Ferdinand II by the court
carpenter Conrad Gottlieb (1513–1620) for Ambras Castle in Innsbruck. 8
A sumptuous tabletop in *pietre dure* made by the Castrucci workshop
in Prague for Prince Karl von Liechtenstein (1569–1627) depicts a tab-
leau of compartmentalized images of pastoral scenes, musical instru-
ments, weaponry, and Prince Karl's monogram in the center. Its
surface is inlaid with four identical geometrical assemblages consist-
ing of a star octahedron nested inside a skeletal octahedron each of
which points in the direction of the centrally located royal crest, while
the outer border is ringed by alternating motifs of polyhedra (two
interpenetrating tetrahedra), salamanders, snails, and frogs. A pair 9
of miniature cabinets in finely gilded leather cases (1560–1570) by the
Master of Perspective, one in the Metropolitan Museum of Art and
the other currently in a private collection, each fold open to reveal sev-
eral polyhedra on their inside covers and inlaid ivory scenes referring
to the liberal arts on their drawers. The cabinet in the Metropolitan 10
Museum of Art was originally a gift from the Nuremberg town coun-
cil to an important visitor, possibly a Hapsburg, given the presence
of the Hapsburg eagle insignia, and likely Albrecht VII, Archduke
of Austria (1559–1621), who counted the cabinet among his posses-
sions. The Museum Angewandte Kunst in Frankfurt has in its pos-
session a small lectern covered in twenty different polyhedra, while a
chessboard in the *Kunstkammer* of the Kunsthistorisches Museum in
Vienna, originally mentioned as well in the 1596 inventory of Ambras
Castle in Innsbruck and made from various woods, ebony, ivory, and
mother-of-pearl, depicts all manner of polyhedra. From the former 11
Medici collections, a pair of early seventeenth-century bedpost fini-
als in the shape of stellated polyhedra, mounted in gilded brass and
composed of triangular pieces of lapus lazuli and jasper, are tantaliz-
ing reminders of the grand ducal workshops for *pietre dure* in Florence
and the royal market for polyhedra in the domestic setting. 12

The most extravagant showpiece of polyhedral geometry in
existence is surely a seventeenth-century southern German cabinet
in the Museum für Angewandte Kunst in Cologne. When closed, the 13
images adorning its upper compartment depict a woman, possibly
an allegory of nature, seated in a fertile landscape, feeding a dog with
her right hand and a putto with her left. Surrounding the cabinet's

overleaf Cabinet with inlaid polyhedra. Nuremberg (?), first quarter of sixteenth century. Museum für Angewandte Kunst, Cologne (A1451). Photo: ©Rheinisches Bildarchiv, rba_c002484.

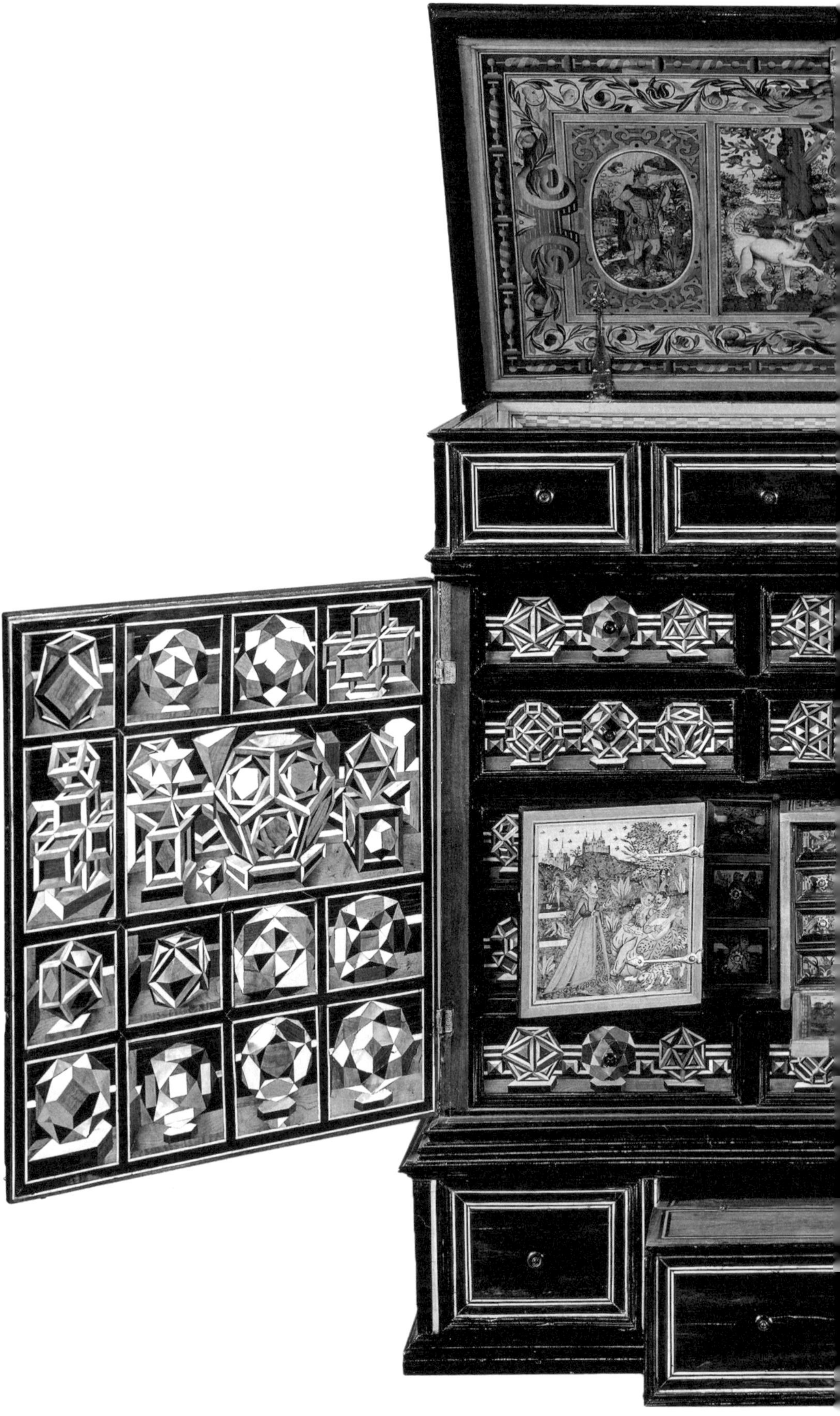

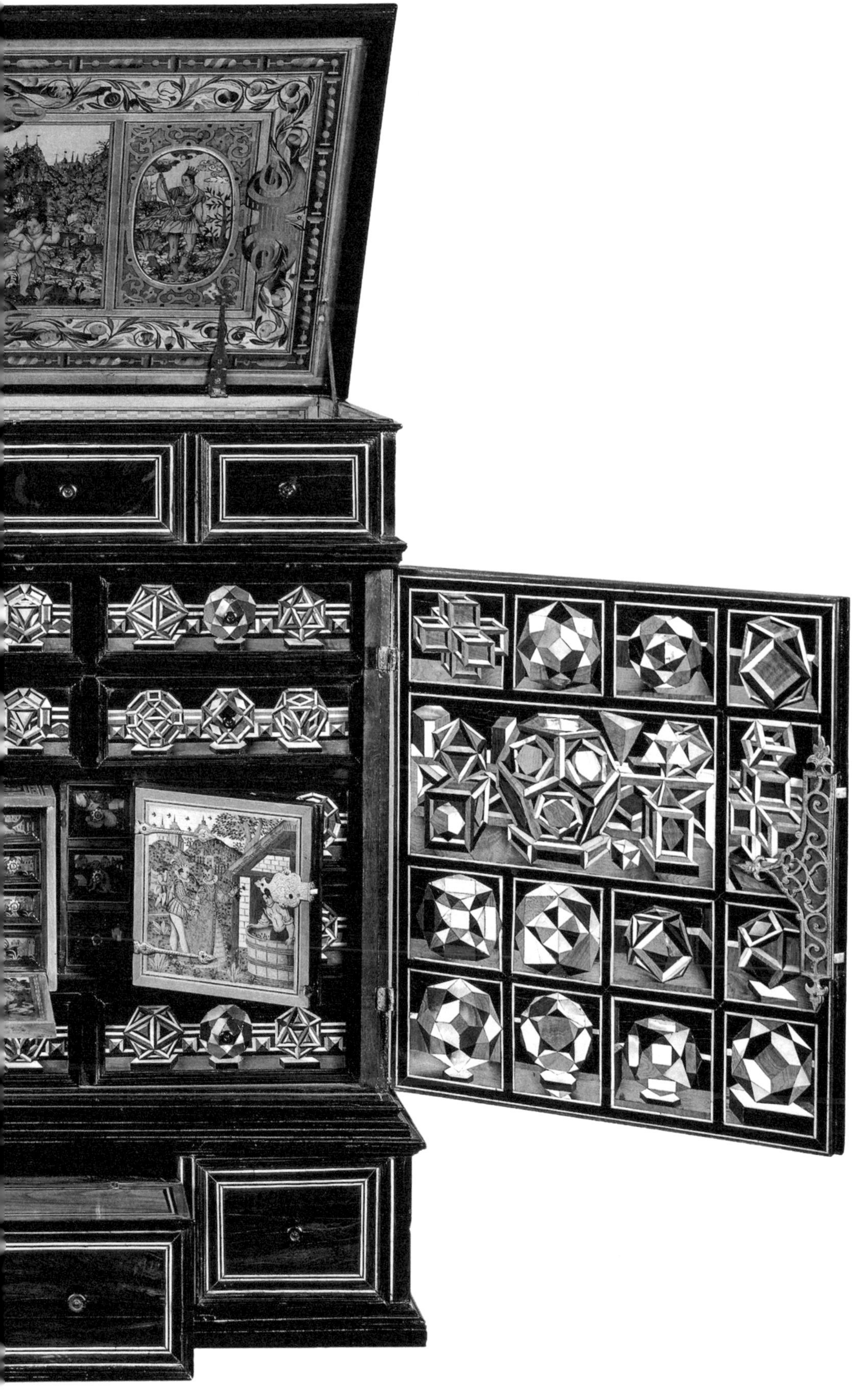

exterior are inlaid scenes of courtly life in contemporary or lightly exotic dress, played out against pastoral settings. Yet once thrown open, the cabinet's doors reveal a geometrical *vanitas*, a sensory overload of polyhedra. Fifteen drawers with three polyhedra each are flanked by two central doors, themselves covered in a grid of polyhedra. A further set of hidden compartments conceals more polyhedra, themselves framed by two doors showing noblemen wandering through a pleasure garden. The whole poses a disjunction between the philosophical passivity of contemplating existence and the hive of fermenting shapes and irregular forms bubbling within the mind of the cabinet's creator and beneath the surface of things.

Supported through precise techniques of making and drawing polyhedra on paper, by the mid-sixteenth century the Platonic solids had become firmly established as the premier arena for generating new geometrical forms by mixed-mathematicians and artists, much of this due to the high print quality and success enjoyed by Jamnitzer's *Perspectiva corporum regularium* (1568). It is evident that decorative artists were quick to take notice of the burgeoning taste for, and marketability of, perspectival *corpora*; in their hands, the solids leapt off the page into more precious materials, coming to adorn domestic furniture and select interiors of some of Europe's finest residences and collections. In this way, *corpora*, multiplying and metamorphosing, were deployed within new visual contexts that balanced their status as emblems of knowledge and precision with a mounting sense of their irrationality and monstrosity. The rudimentary diagrams in Euclid had faded into distant memory, and the elemental iconology Jamnitzer had exploited to great effect in *Perspectiva* loosened its allegorical hold.

In their stead, artists collaged together references from books of perspective, architecture, and geometry to create disorienting landscapes in which polyhedra played prominent roles as emissaries of strangeness and disorder. Instructive is one of two alternative cover pages to *Geometria et perspectiva* (1567, republished 1617), a striking collection of eleven prints by the little-known artist Lorentz Stöer (active 1557–1620/21) and one of the most popular graphic resources
14 for the Augsburg cabinets. The cover depicts an archetypal northern European city in which the buildings have been conspicuously stripped of ornament; several citizens gaze fearfully at a floating

skeletal dodecahedron that hovers above the city like a UFO. On an overgrown outcropping among the ruins rests another skeletal form enclosing a solid, presumably having just landed. The three men gaze at it in wonderment, so much so that one of them holds back his two companions lest they be overwhelmed entirely.

Geometry, as portrayed on the cover of *Geometria et perspectiva*, is a specialized body of knowledge, alien to the common man but capable of structuring a relation between the ancient and the modern city. This is a conceit borne out and developed on most if not all of the plates in the series. Plate 1, for instance, depicts a motif of modern ornamental scrollwork popular in the late sixteenth century on the left-hand side of the page. The scrollwork has been scaled up to the size of a wall, making it function as an elaborate screen veiling the view of a city in the distance. In the foreground, a diamond balances on point atop a cube resting on a rectangular base, while the edge of a ruined complex of classical architecture looms in the middle distance. The solids in plate 2 similarly separate two ornamental structures—a life-size scrollwork installation and the decrepit remains of a grand ancient building whose curvilinear contusions echo one another.

The juxtaposition of modern ornament with historic, classical architecture which has fallen into disrepair implies at once a mutual affinity, and that Stöer's ornament has taken inspiration and even emergence from urban ruination. Its function abandoned, its color faded, its form degraded, the ruin that has become ornamental to the modern city is finally made to translate into ornament itself. The mirroring of modern ornament and the ornamental ruins of antiquity hinge on Stöer's own mastery of geometry which bridges the visual and conceptual gap between the old and the new, calling attention to the utility of geometry for the analysis of ancient architecture and its transformation back into ornament. Geometry, Stöer implies, enables one to operate from a position of authority and knowledge upon the bland and generic urban surfaces of the northern European city on his cover page. Architecture, on the brink of death and in a perpetual state of collapse, gives birth to a new geometry arising from the ashes of a broken world.

Though the geometrical accent of Stöer's oeuvre was eccentric, it belonged squarely to a field of ornamental prints dedicated to the depiction of architectural ruin and often used as the basis for wood

15 inlay in high-profile commissions. Supplying design vocabularies that could be used to adorn objects in any way that the purchaser of the print desired, prints could be cut and pasted upon the surface of furniture as cheap decorative options or used in combination by *Schreiner* to construct the urban scenography that distinguished the faces of the Augsburg cabinet. In addition to Stöer, oft-copied sixteenth-century visual references were *Büchlein von den alten Gebewen* (ca. 1555), a set of prints after Jacques Androuet du Cerceau's engravings from 1550 by Léonard Thiry, the work of Virgil Solis (1514–1562), Hieronymus Cock's *Praecipua aliquot romanae antiquitatis ruinarum monumenta* (1551), Walter Rivius's *Vitruvius Teutsch* (Nuremberg, 1548), Wendel Dietterlin's *Architectvra: Von Außtheilung, Symmetria vnd Proportion der Fünff Seulen* (Nuremberg, 1598), and Sebastiano Serlio's *Tutte l'opere d'architettura, et prospetiva* (volumes published from 1537 onward). The Dutch artist Hans Vredeman de Vries published two books on the ornamental orders in 1565 in which he urged readers to sample the classical forms from his books and "with what one finds
16 here one may do what one sees fit."

On the basis of the title page of *Geometria et perspectiva*, in which Stöer claims that the images of his "zerbrochne Gebew" (ruined buildings) would be useful to *Schreiner*, it appears that Stöer similarly and successfully courted the copying of his work by other craftsmen who scoured the contemporary art scene for patterns to use in the marquetry that graced the most prized southern German furniture. "Here in Geometria et Perspectiva are a number of ruined buildings, especially and pleasingly arranged and designed by Lorentz Stöer, painter and citizen of Augsburg, useful for craftsmen working in
17 intarsia and many other enthusiasts."

Though the base prints have long since been discarded, the marquetry doors for the apartment of Philip II at the Escorial near Madrid preserve motifs found in the printed oeuvre of Vredeman de Vries, Théodore de Bry, and Nicolaes de Bruyn, and feature strap-
18 work cartouches and polyhedral geometry inspired by Stöer. A contemporary intarsiated Augsburg cabinet sporting a sphere with a diamond poking through its surface holes matches a similar geometry in reverse from Stöer's *Geometria et perspectiva*, likely indicat-
19 ing direct copying. Several Stöer prints sold into private collections were the referent for intarsia panels on the lower doors of a cabinet

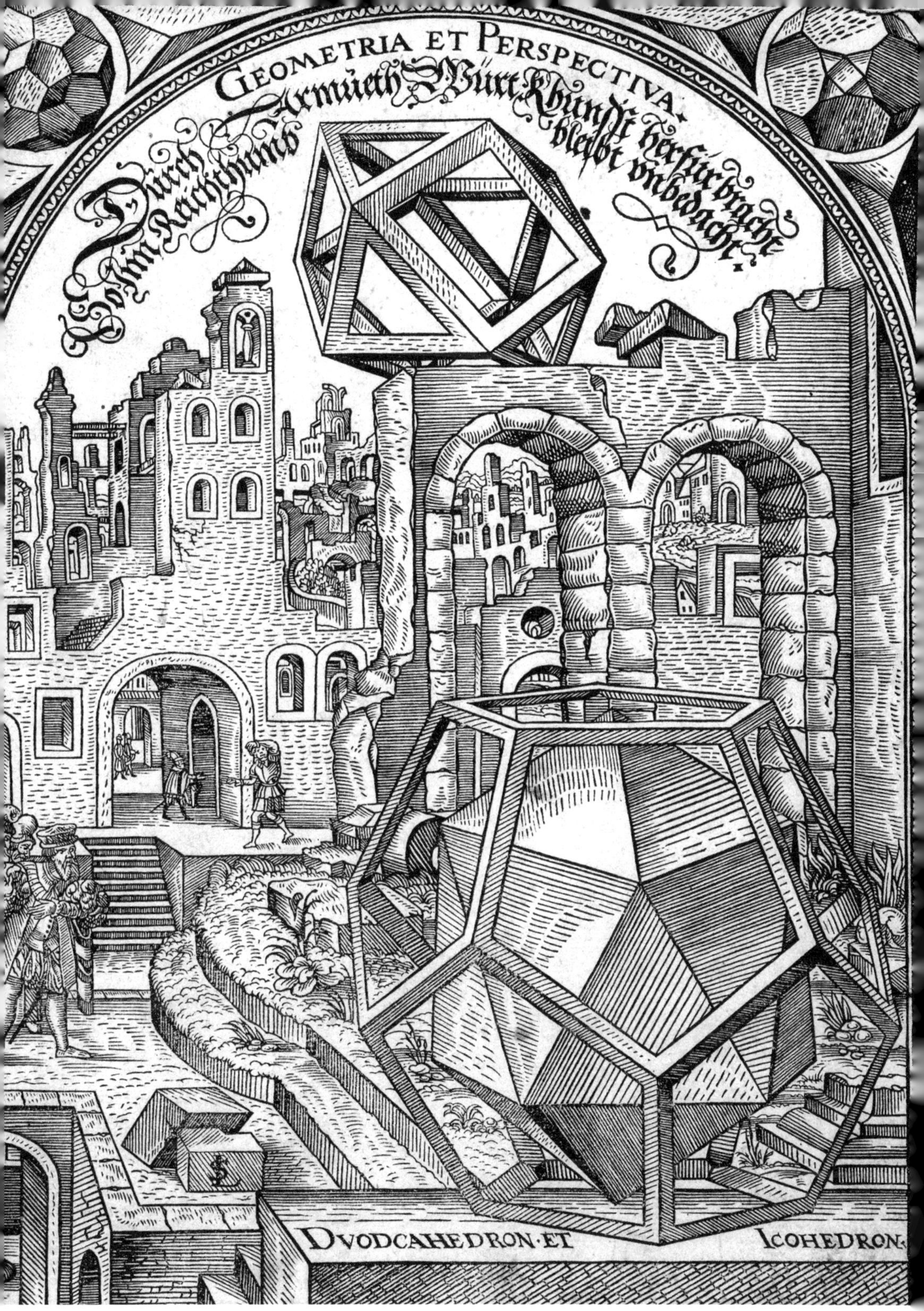

Cover page, from Lorentz Stöer, *Geometria et perspectiva* (Augsburg: Hans Rogel, 1567). The Metropolitan Museum of Art, New York, The Elisha Whittelsey Collection, The Elisha Whittelsey Fund, 1984 (1984.1085.1).

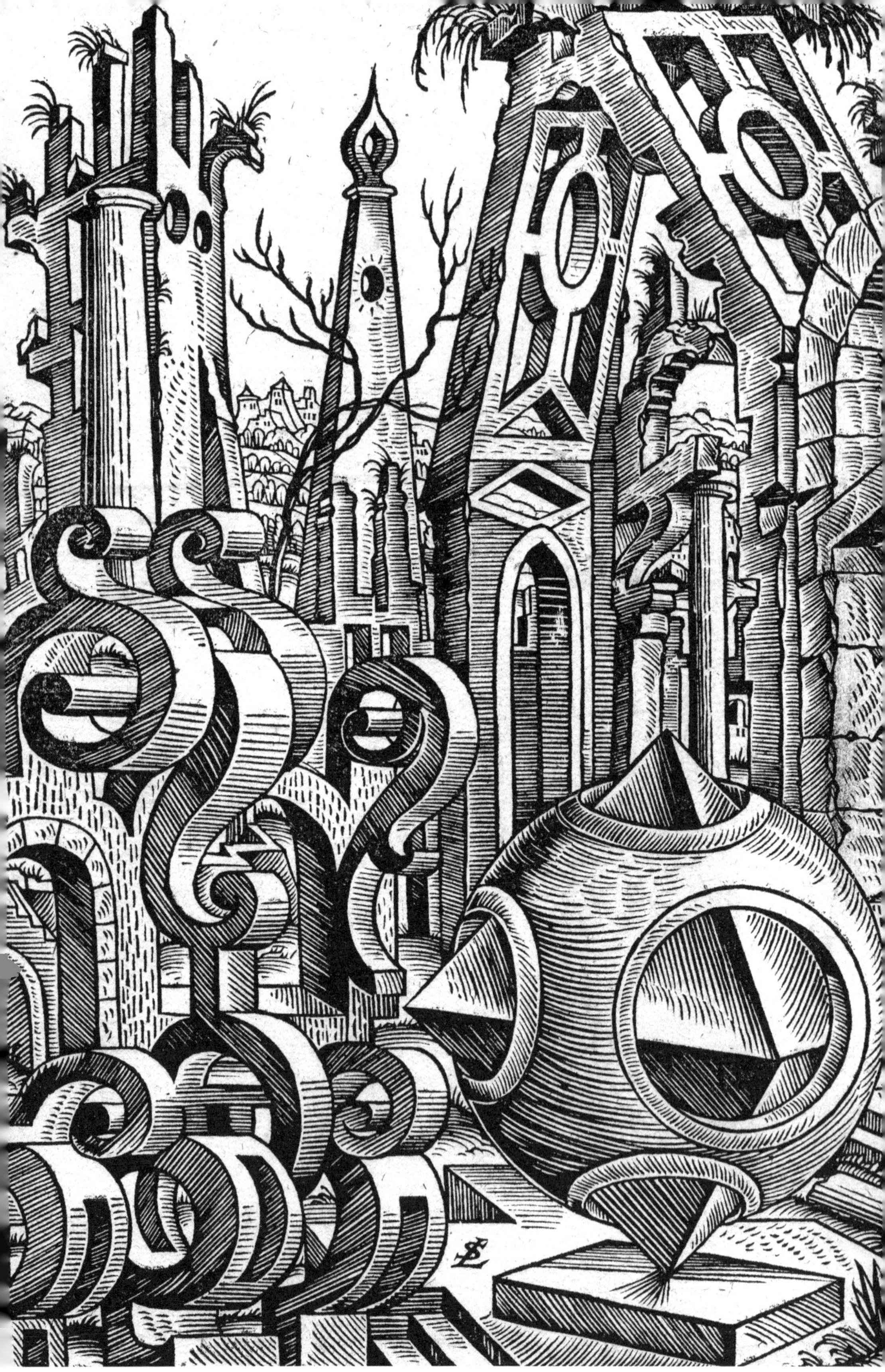

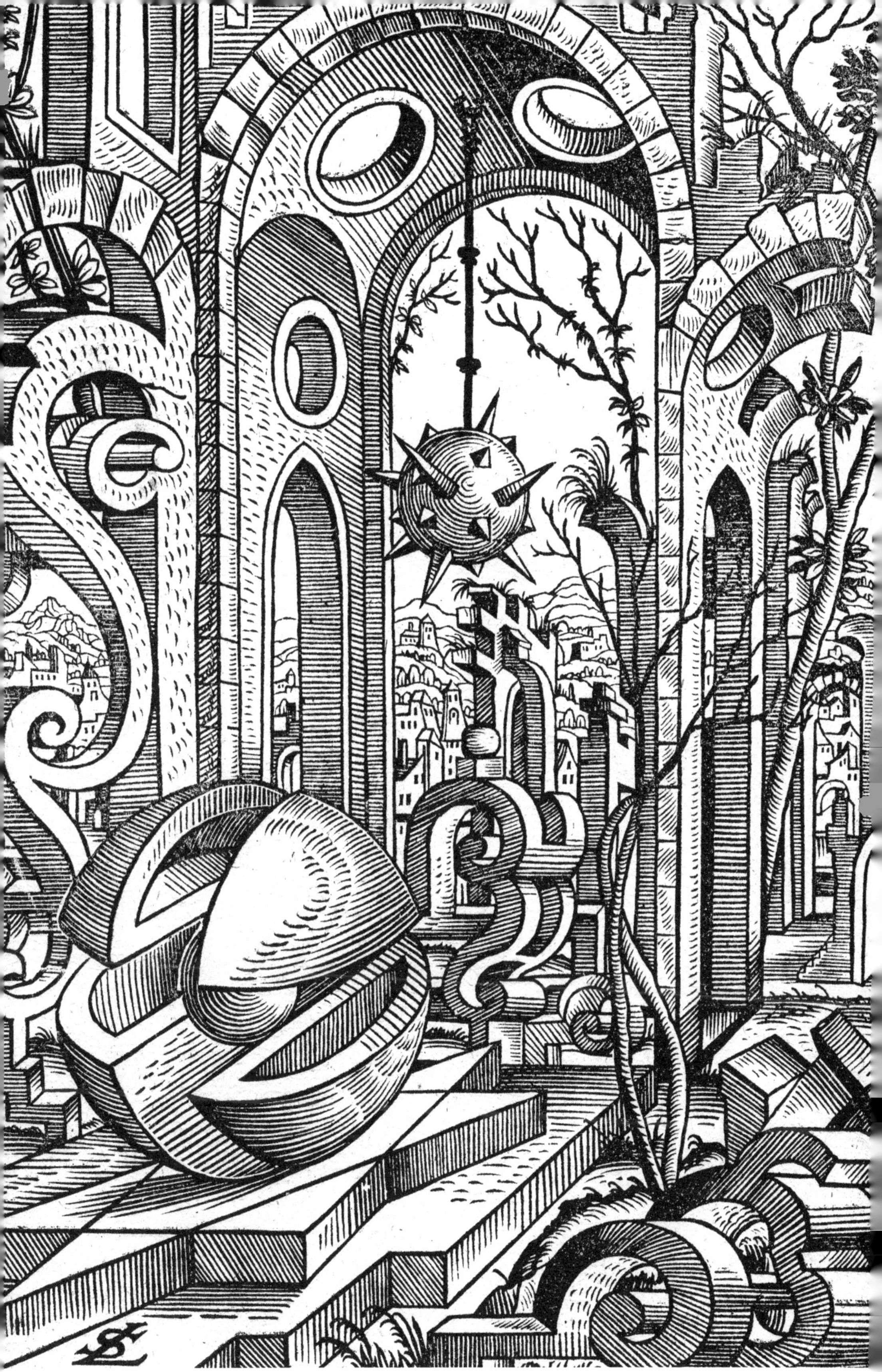

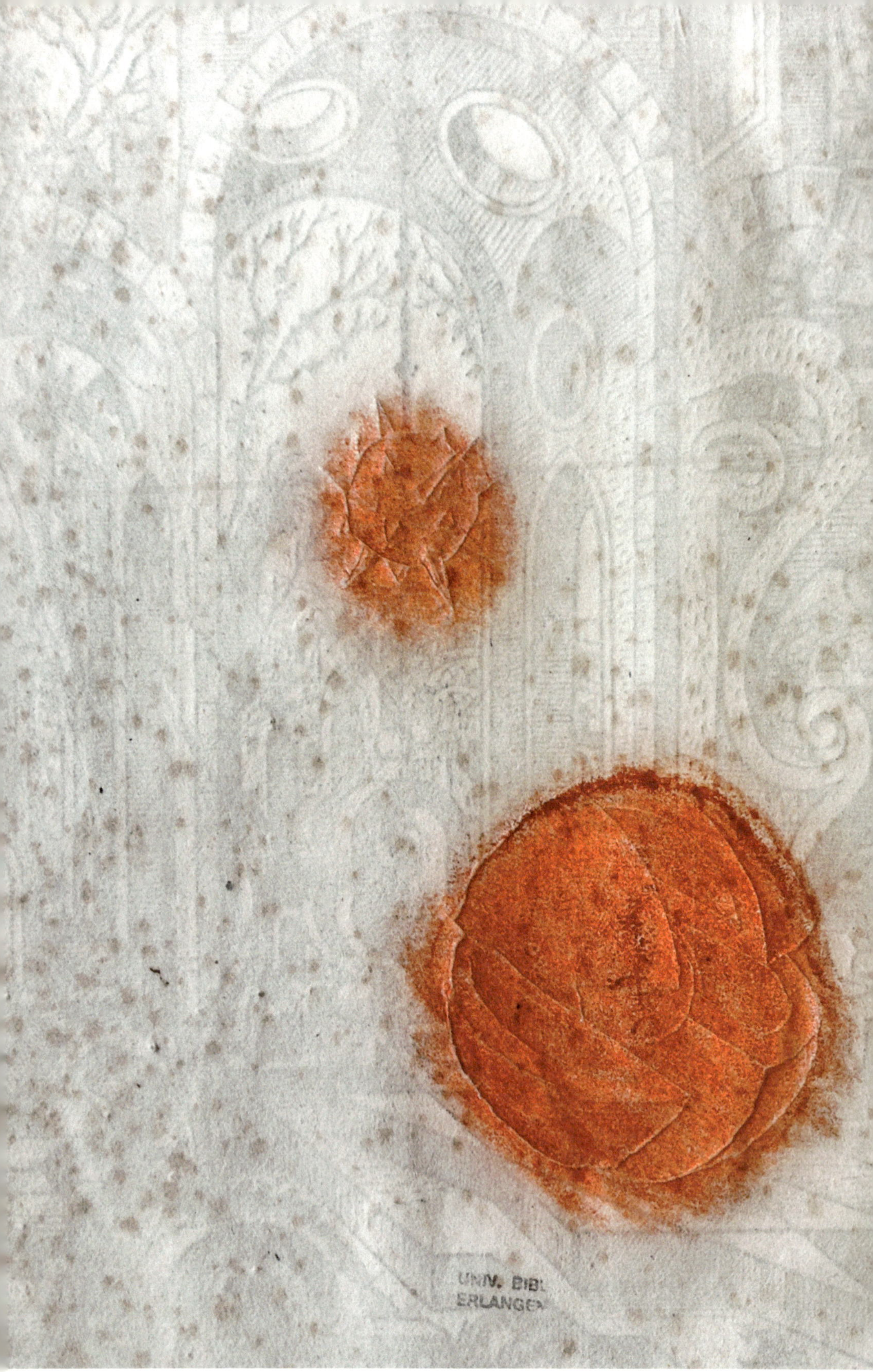
UNIV. BIBL
ERLANGE

in Ulm, located at the Ulmer Museum. And distinct similarities have been noted between Stöer's prints and the aforementioned cabinet in Cologne as well as cabinets in London and Bilbao. 20

Everyone has a Stöer, a shadowy figure few have heard of, trotted out to make broad claims that cannot otherwise be substantiated. But occasionally even a Stöer can be useful. Most of the Augsburg *Schreibtische* remain anonymous, the signatures of their makers, if they exist at all, hidden within elaborate internal joinery and requiring the conservation nightmare of a complete dismantling of the cabinet to uncover. At another level, the disappearance of Stöer from art historical genealogies, alongside his complete omission from the history of geometry, confirms a doubled schism between the histories of print culture and histories of the decorative arts, as well as the interconnected histories of the decorative arts and mathematics.

These anecdotal examples are indicative of a broader sphere of influence in which Stöer's *Geometria et perspectiva* was appreciated, not just as a printed book of fantastical images but also as a unique source of geometrical source material that could be combined onto the decorative surfaces for which Augsburg was famous. Unlike his contemporaries and apart from Jamnitzer's *Perspectiva*, which was too precious a book to serve as a direct source for marquetry, Stöer remains the artist most consistently working on the solids and their perpetual transformation. One must look to Stöer and his obsessive pursuit of the near infinite variability of irregular bodies to understand the potential of perspectival geometry taken almost to the point of madness, where any claim to rationality has evaporated in a stream of mutant hybrids and monstrous geometrical fantasies. Encompassing lessons honed from a life's work on the possibilities of geometrical form, Stöer's voluminous graphic oeuvre consists of hundreds of wildly colored drawings of regular and irregular polyhedra and testifies to a marvelous master project on the geometrical imagination arguably unrivaled until today.

“Lorentz Stöer is less well known than he ought to be.” This state- 21
ment, from Paul von Stetten's expansive Enlightenment-era history

preceding spread Geometric forms among ruins, from Lorentz Stöer, *Geometria et perspectiva* (Augsburg, 1567). Universitätsbibliothek Salzburg (G 363 II, 9 and 10).

facing Verso of pl. 10, with red chalk and light scoring, from Lorentz Stöer, *Geometria et perspectiva* (Augsburg, 1567). Bound in *Optica*. Universitätsbibliothek Erlangen-Nürnberg (MS 2774). Photograph by author.

of the applied arts in Augsburg, attests to the absence of any major,
realized commissions that could have substantiated Stöer's avowed
profession as a "Maller Burger Inn Augspurg" (painter and citizen
in Augsburg), as he described himself in the title page of *Geometria
et perspectiva*. Only fragments are known about a life that for von
Stetten seemed to have left little more than scattered collections of
22 disparate "design drafts [*Entwürfe*] betraying a bold and lively spirit."
Stöer gave up his citizenship in Nuremberg to move to Augsburg in
1557, and had already by April 8, 1555 received a royal privilege from
Ferdinand I to print a book with the title *Perspectiva a Laurentio Stöero
23 in lucem prodita*, of which no copies have survived. What is known,
however, is that his print work quickly attained a degree of renown
and that reproducing his designs without his permission was
punished by the confiscation of the reprint and a fine of ten gold
24 marks. The earliest known copy of *Geometria et perspectiva*, possibly
the *Perspectiva a Laurentio Stöero* referred to in the royal privilege,
appeared in 1567. But here too there is room for conjecture. One ver-
sion was issued by Hans Rogel (1532–1592) and another by Michael
Manger (active 1570–1603), competing publishers active in Augsburg
in this period, where Stöer himself appears in the tax records from
25 1562 until 1597. There are only two mentions of Stöer in the council
records of Nuremberg. The first, dated May 24, 1557, details the filing
of his taxes and specifies that "Lorennz Stöer, painter, recited his cit-
izen rights, declared his customary prescription [fiscal receipts] and
is known for it in the tax office." The second, dated August 13, 1595,
26 lists a dispute over payment.

Stöer's hand drawings are spread between Erlangen, Augsburg, and most substantially in Munich, with a heretofore unknown manuscript in Houghton Library at Harvard University. The several dates marked in the Munich and Harvard volumes (1562, 1564, and 1584 for Munich and 1585, 1598 for Harvard) reveal a vast intellectual and artistic endeavor that spanned at least three decades of sustained drawing, as well as an intimate knowledge and appreciation of Jamnitzer. Information on the volumes is scarce. The earliest known owner of the Munich manuscript after Stöer was Johann Franz Eckher von Kapfing (1649–1727), Fürstbischof of Freising from 1695 until his death. The manuscript is mentioned in the *ex libris* of Bishop Eckher in 1696 and was kept in the Dombibliothek in Freising until the mid-eighteenth

Lorentz Stöer (active 1557–1620/21). Irregular body, from *Geometria et perspectiva corpora regulata et irregulata*. 1584. Universitätsbibliothek der LMU München (HS Cim 103, p. 250).

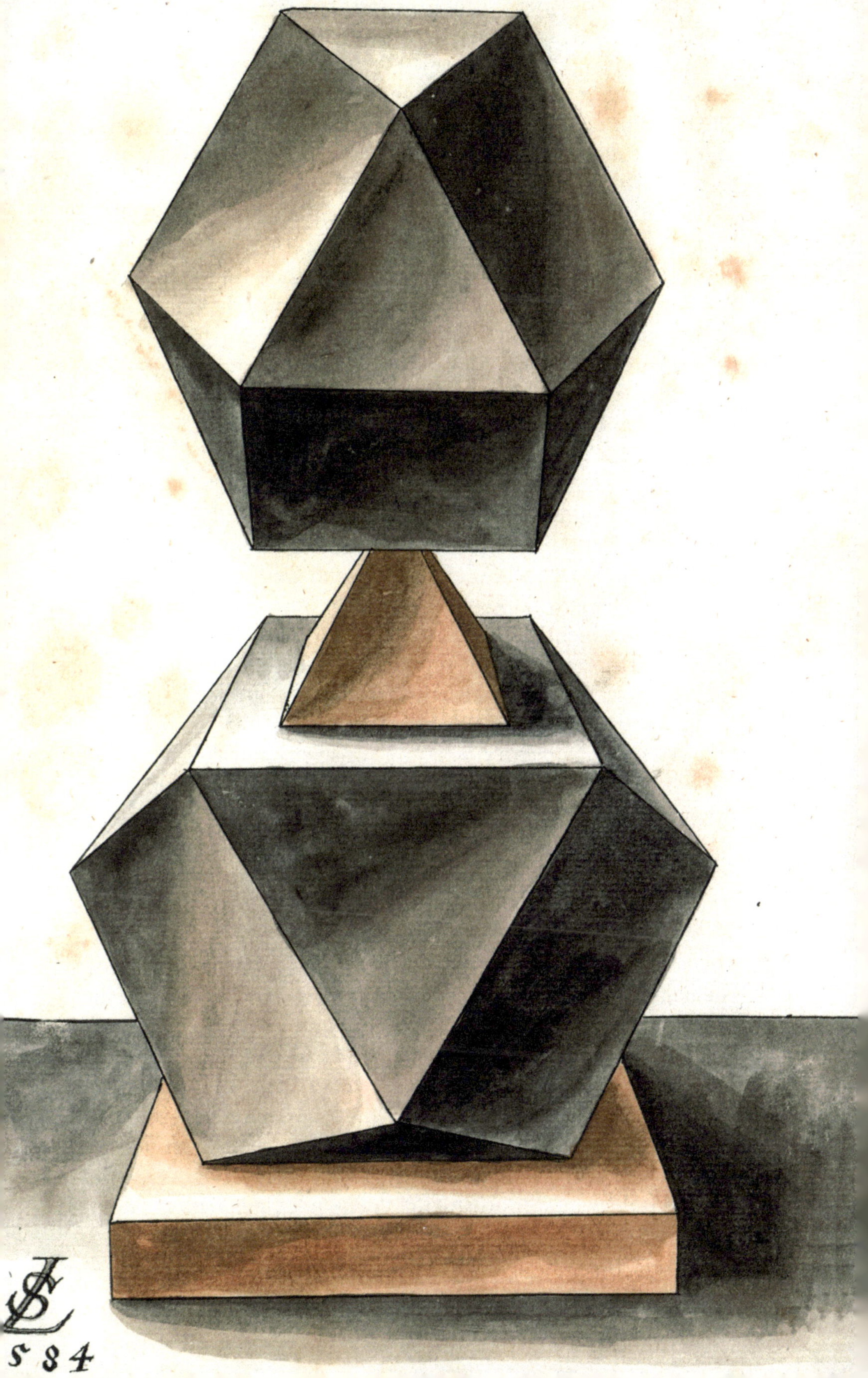
584

century when it was transferred to its present location in Munich.
The Harvard manuscript was purchased in 1950 by Philip Hofer, an
avid book collector and former curator of printing and graphic arts at
Harvard, from H. P. Kraus, a dealer in rare books who in 1948 had him-
self purchased the volume, along with twenty thousand others, from
27 Prince Franz Josef II of Lichtenstein. The earliest known provenance
of the Harvard manuscript is in the collection of the Austrian general
Frank Ritter von Hauslab (1798–1883), who donated the volume to the
28 Lichtenstein Museum in the nineteenth century.

There is enough overlap in format and presentation to suggest
that the work within the Munich and Harvard volumes would have
been created concurrently. Consisting of watercolor drawings out-
lined in pen or pencil, their colors fantastical arrays of blues, greens,
29 magentas, and oranges, substantially devoid of material or weight,
both of the volumes begin with title pages ringed by geometrical
models and repeat the phrase "Geometria et Perspectiva Corpora
Regulata et Irregulata." The cube (earth) and icosahedron (water), the
heavier elements as per Aristotle, flank the bottom of the page, with
the lighter elements represented by the tetrahedron (fire) and octa-
30 hedron (air) appearing at the top corners. While the Harvard volume
is unfinished, missing a central text, the Munich volume has been
completed and sports the title phrase "Die funff Corpora Regularia,
auff Viel und Mancherley Arth und Weis zerschnitten"—"The five
regular bodies, cut apart in many ways"—referring to the multitu-
dinous transformations undergone by the Platonic solids under
Stöer's hand. Neither preparatory models for works to be rendered
later in a more precious material nor intermediary steps integrated
into the production of more accurate drawings, many of the pages
stage two models together in perpetual dialogue with each other and
are mounted horizontally with the models' bases facing the spine of
31 the book. Nearly all of the models are situated on a long shelf that
stretches throughout the entirety of the volumes, giving the impres-
sion that the pages could be removed to create one long, continuous
presentation of the fruits of Stöer's imagination.

For Stöer, as for other northern European artists, geometers, and mixed-mathematicians, the Platonic solids were not unimpeachable Euclidean proofs to be followed, but rather served as the conceptual framework for organizing open-ended practices of graphic

experimentation, hence the epithet to Stöer's Munich drawings. The
corpora regulata had simply become too regular to resonate as suffi-
ciently exciting emblems of geometry, much as they had done previ-
ously for Pacioli, Dürer, and Caraglio since having become a prevalent
subject of artistic investment nearly one hundred years earlier. The
category of *corpora irregulata*, however, served just such a purpose,
introducing a new measure of artistic freedom from the hegemony
of "those called regular bodies by Euclid." While Winckelmann 32
famously proclaimed that there was "but one way for the moderns
to become great ... by imitating the ancients," the ancients had left
behind little if any graphic traces of their capacity to invent and visu-
alize irregular bodies. In order for Stöer to make a living and distin- 33
guish himself, geometrical arenas had to be explored that had not
been cordoned off either by ancient expertise or by the virtuosity of
previous generations of moderns.

Although the definition of irregularity was necessarily never
fixed, *corpora irregulata* maintained their relation to the Platonic sol-
ids, though the nature and strength of the relationship was deter-
mined by the individual artist. The compendium of geometrical
models on ornamental bases in Halt's *Perspektivische Reiß Kunst* are
also grouped together into sets of variations based on the Platonic
solids. More markedly, the introductions to each of the five sections 34
in Jamnitzer's *Perspectiva corporum regularium* stress states of perpetual
transformation originating with the five solids. As with the tetrahe-
dron, Jamnitzer describes the subsequent variations on the octahe-
dron as "forming 23 other different bodies," while the dodecahedron 35
"produced twenty-three different bodies of various kinds and shapes
[or models/figures/patterns] which have been brought into perspec-
tive." The implication of the generative graphic strategies employed 36
by Stöer, Halt, Jamnitzer, and others was that the same polyhedron
could yield an infinite number of irregular variations, united only by
their relationship to the originating solid.

It is apparent that the limits of geometrical morphology—what counted as the same or same enough—were able to accommodate a wide degree of visual variation. Accordingly, Stöer's notebook at the Staats- und Stadtbibliothek in Augsburg contains six drawings, all of which were classified by Stöer as six-sided polyhedra and are superficially so different that without his explicit designation it

overleaf Lorentz Stöer (active 1557–1620/21). Two "hexahedra." Ca. 1600. Staats- und Stadtbibliothek Augsburg (4 Cod. Aug. 247, fols. 14 recto and 10 recto).

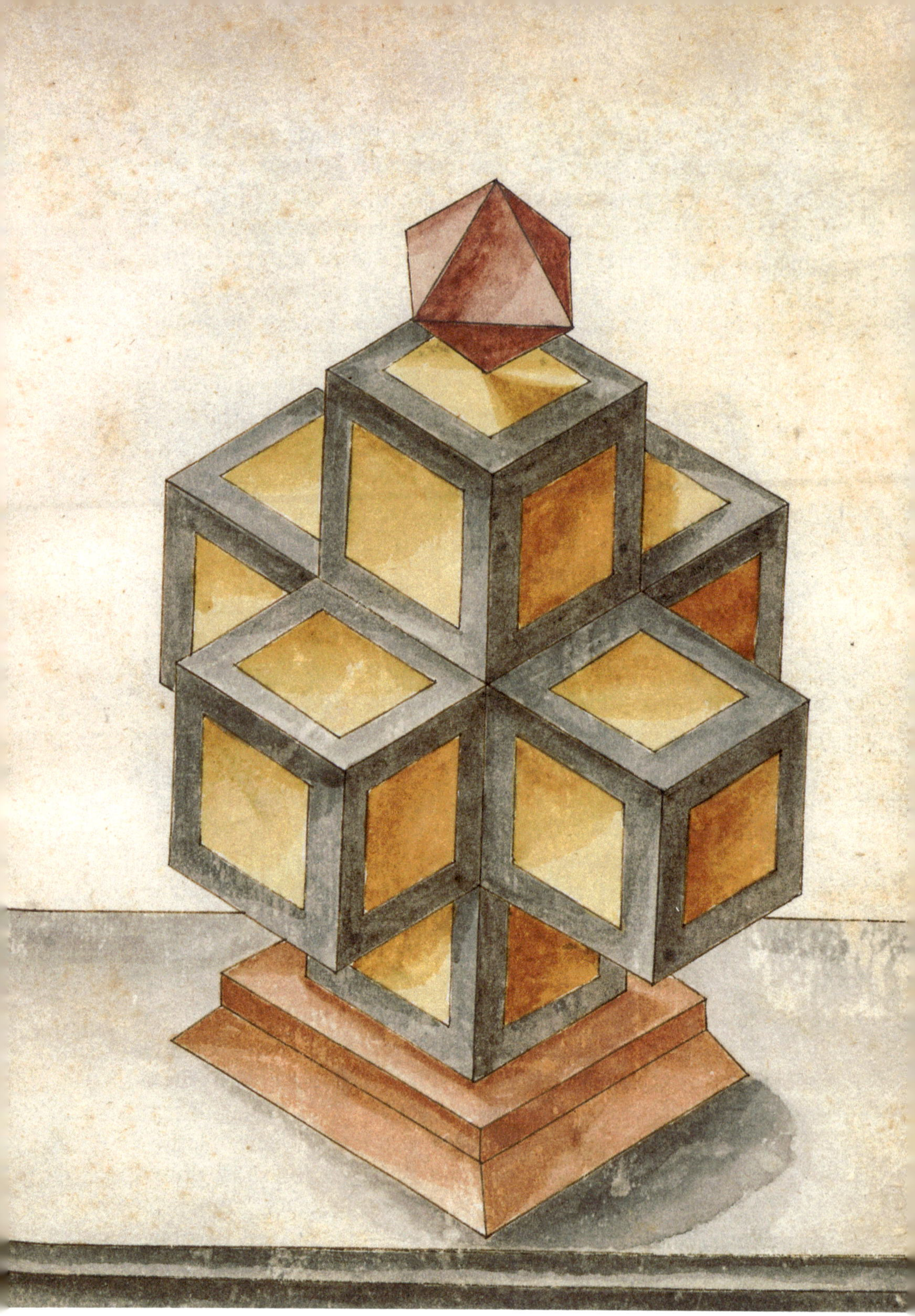

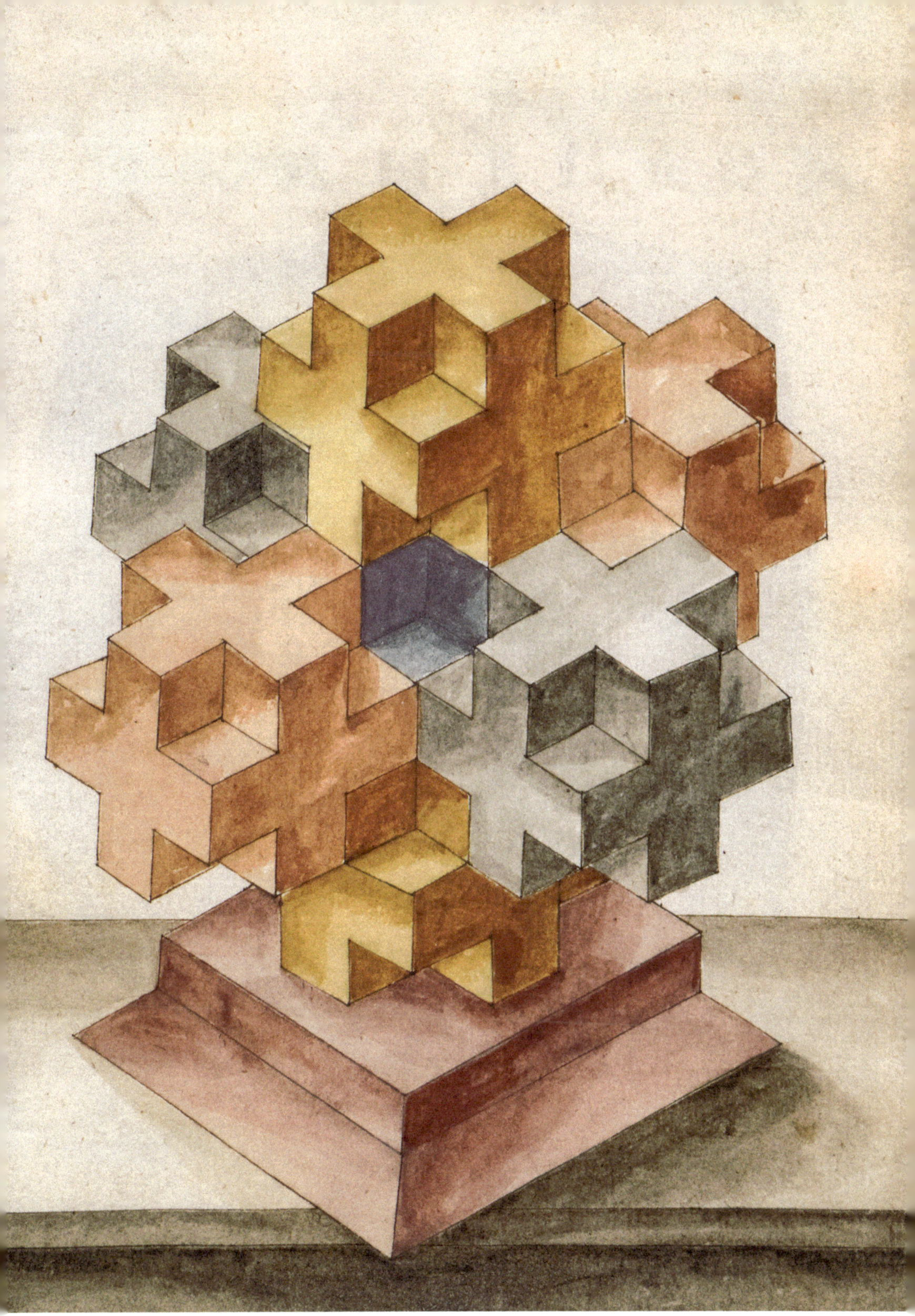

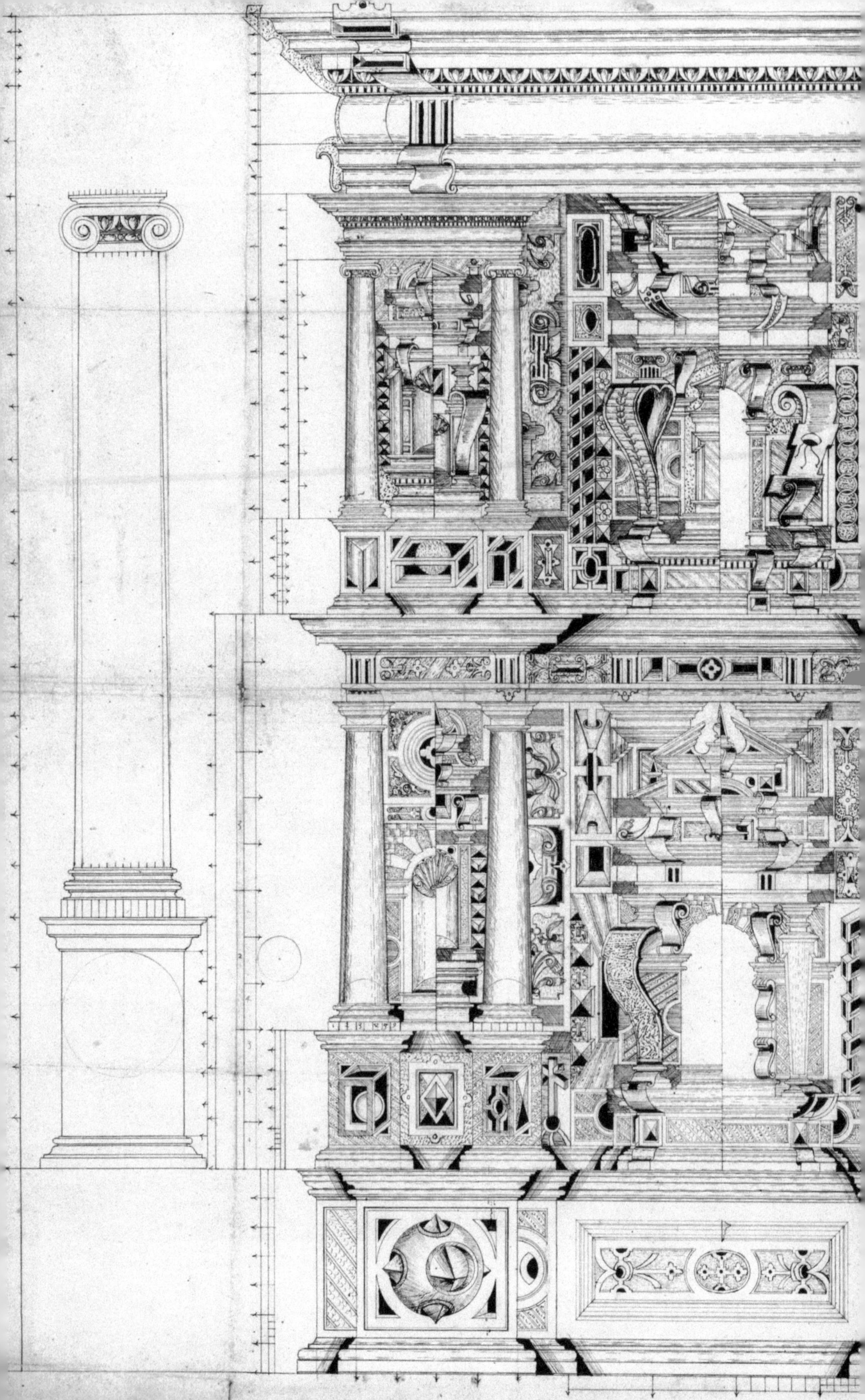

would be challenging to find any direct relationship between them,
let alone to locate them as belonging to the same geometrical category. 37
Rather, the solids served as flexible geometrical typologies helping
to organize Stöer's creative process, enabling a method to locate the
drawings within groups, and providing a conceptual framework to
organize further graphic experiments.

Although there is often little text to be read, collections of poly-
hedra were often intended to be viewed in a particular order. While 38
Stöer's volumes appear at first simply to collate hundreds of randomly
generated drawings, subsequently paginated to give them a veneer
of consistency, a closer examination suggests that the drawings were
compiled by someone with an awareness of the underlying structure
ordering Stöer's geometrical investigations, but without the patience
to make sure that the bound volume consistently reflected it. The
result gives a false impression of manic creativity, blurring together
the iterative explorations into a compounded "irregularity." Still,
an overarching progression from simpler to more complex forms,
often beginning with one of the Platonic solids as a starting point,
appears to emerge, however tentatively, from the mass of informa-
tion. For Stöer grouped his graphic experiments by Platonic solid,
much as Jamnitzer did in *Perspectiva*, and intended them to be appre-
hended in conceptual relation to the solid which inspired their gen-
eration. While the initial pages of each section are taken up by the
five Platonic solids, each drawn in six differently rotated views, there
are pairings, triplets, sequences of incrementally modified models all
tantalizingly related to each other yet spread countless pages apart
or between the two volumes in Harvard and Munich.

When Stöer, trained to work at the scale of the object, engaged architectural design explicitly, the image of architecture absorbed a secondary set of priorities or characteristics that originated in the production of ornament. His exquisite pen drawing in the Germanisches Nationalmuseum demonstrates just how architecturally innovative decorative artists could be when given the opportunity, and how a wall could serve as the ultimate "object" to adorn with hypnotic ornament. Stöer depicts a facade, likely a wall painting, bursting with embellishment and ornamentation, including one of his signature geometries at the base. He does not choose to render one proposed facade, but, much as in Jamnitzer's sketchbook or the

Lorentz Stöer (active 1557–1620/21). Drawing of facade options for a wall painting. Late sixteenth century. Germanisches Nationalmuseum, Nuremberg (Hz 6979). Photo: M. Runge.

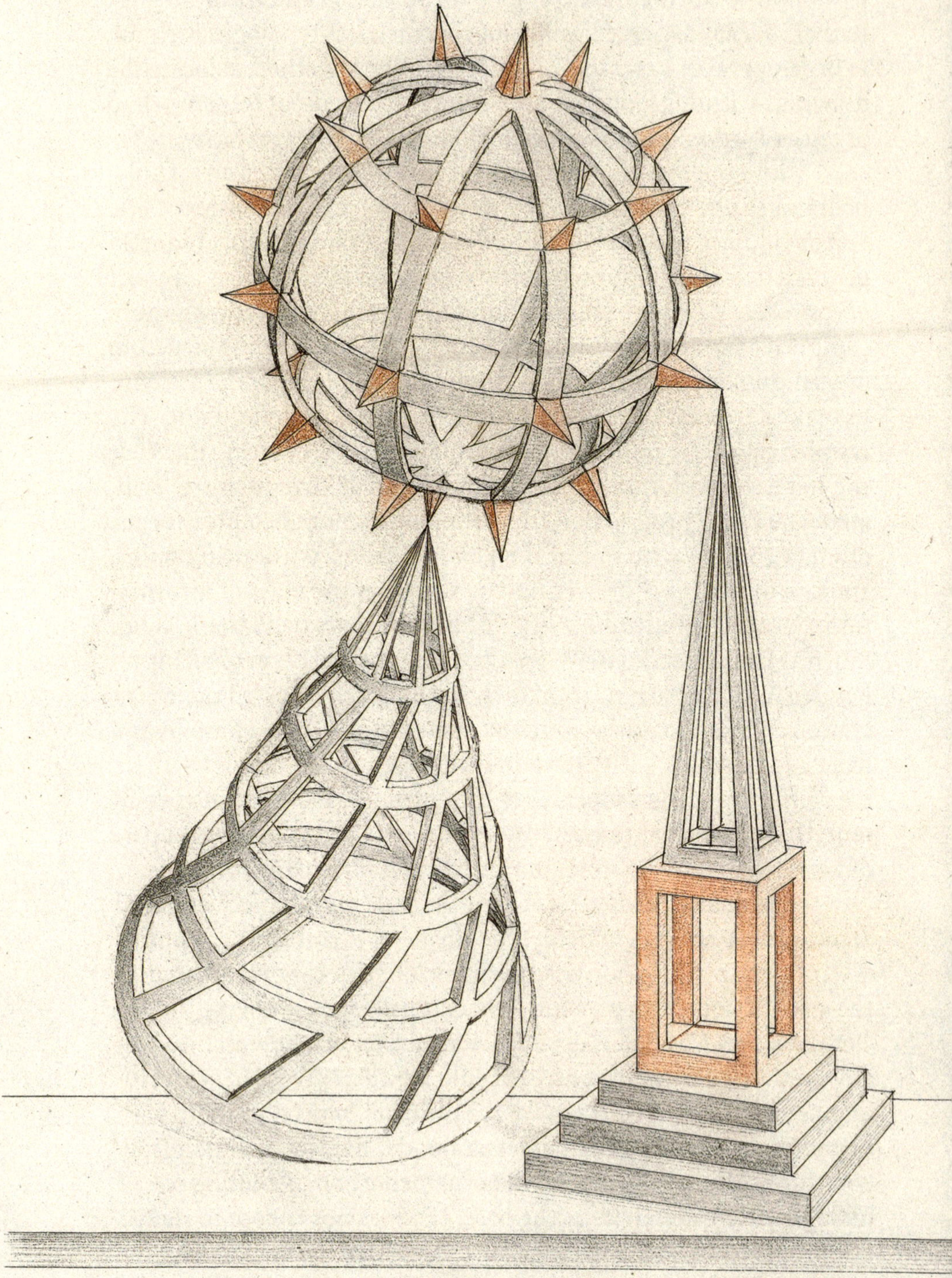

goldsmith drawings more generally, incorporates multiple possible permutations, legible by the sharp disconnects between segments of the drawing. A client would have had the option to choose elements from each of the segments to come up with a design to fit their tastes, a feat of mental acrobatics requiring the client to imaginatively connect disparate pieces of the drawing and complete the entire image in their head. The same drawing might have been used by multiple clients, given the many design options embedded in it. The result is a hypercondensed vision of architectural possibility that borrows more from the graphic conventions of the ornament print and Stöer's lifelong practice of geometrical experimentation than it does from the architectural elevation.

Beneath the bright colors of Stöer's polyhedra lurks a didactic exploration of the limits of an irregularity that could never have been achieved without the advancements in geometrical visualization and invention pioneered by the century's earlier *Lehrbücher*. Not only did these *Lehrbücher* popularize the *idea* of designing geometry, they also served as precedent and inspiration for the notion that artists could make unique and important contributions to geometrical knowledge. Perhaps this is why irregularity was never conceived in a vacuum, but always kept in bounded relation, however tenuous, to its closest Platonic solid. Exploring the constraints of these canonical forms and their combinations as opportunities for design externalized the creative potential of polyhedral, or "polyhedralish," geometry. To view one of these mathematical wonders, organized as if in a virtual *Wunderkammer*, was to be dazzled by its perspectival proficiency and also to recognize the deformations and augmentations that had altered it from the antique incarnation familiar to any student of Euclid. To transform and to invent was to be current; to build upon the old and yet to create an aesthetic of contemporary relevance was surely at the heart of Stöer's geometrical project.

The discovery of new geometrical forms and their graphic invention were folded into the same epistemological structure as learning to understand the world in geometrical-perspectival terms, accounting for the blossoming of geometrical innovation in sixteenth-century decorative arts. Copying was integral to the circulation of knowledge within and between workshops, and the transference, or implication of transference, from ornament print to wood or building surface

Christian I, Elector of Saxony (1560–1591). Geometrical models, from *Perspectief Buch*. 1576. Kupferstich-Kabinett, Staatliche Kunstsammlungen Dresden (Ca 61/30). Photo: H. Boswank.

yielded a productive frisson of geometricality thematized over and over again as a motif of metamorphosis. Representative of the power of transference between and within media to generate a limitless experimentation with geometry, the nature of artisanal innovation in geometry was the incremental newness of exploring *corpora* through serial and progressive transformation.

Ample evidence suggests that Stöer, like Dürer before him, was an active participant in a learned community of artisans who were all following each other's experiments with geometry, owned each other's books, and executed copies of geometries they found interesting or visually appealing. Take the artist Hans Jakob Ebelmann (ca. 1570–ca. 1625), who incorporated two distorted polyhedra from *Perspectiva corporum regularium* into his book of ceiling designs, *Gebäude*
39 *Bekrönungen Decken* (1609). Or an anonymous collection of drawn geometrical models at the Herzog August Bibliothek in Wolfenbüttel encompassing copies from Johannes Lencker's *Perspectiva literaria* (1567)—a collection of twelve perspectival representations of alphabetical letters and nine evocative mathematical models—one of
40 which was copied from Jamnitzer. Both the Wolfenbüttel manuscript and Stöer's Augsburg notebook indicate intimate knowledge
41 of *Perspectiva,* alongside many other examples of mutual exchange.

Whereas working from a folded-up paper or wooden polyhedral model was helpful for drawing new solids in perspective that took their basic shape from the Platonic solids, *irregulata* branched into territories with little or no reference to Plato's iconic forms. As such, they were subject to a range of reproductive strategies—inscribing, pricking, tracing from under and over—indicative of their desirability and perceived difficulty. Christian I (1560–1591), Elector of Saxony from 1586 until his death at age thirty, was only sixteen when he finished working "with his own hand" on his *Perspectief Buch* (1576), a collection of hand drawings of miniature architectures and geometrical objects preserved in the Kupferstichkabinett of
42 the Grünes Gewölbe in Dresden, undertaken under the tutelage of Lencker, who earned a salary of 117 *gulden* and 9 *kreutzern* in 1572 for giving lessons to Christian for "kunststücke aus der perspectiua," or
43 feats of perspective. Inscribed in hardline with ink, and then carefully shaded with pencil and chalk, the drawings are original compositions, and with two telling exceptions there is little evidence of

direct reference to other books or images of geometrical solids. Only
by examining the raised texture on the verso of two spherical *irregu-
lata* is it possible to ascertain that Christian had created them by trac-
ing over a lost reference image with a sharp instrument or sharpened
pencil, transferring the shape from below onto the page above.

The copy of *Geometria et perspectiva* bound into the collection of
Stöer's drawings and prints in the Universitätsbibliothek Erlangen,
entitled *Optica* and attributed to "Laurenzo Stöer," bears marked proof
of an analogous attempt to spirit away two more irregular solids from
Stöer's page. In the ninth print of the series, the outlines of the first 44
sphere, a sea-urchin-like concoction on a star-shaped base consisting
of four spherical segments propped open to reveal an internal sphere,
as well as the second, a spiky ball hanging from a rope tied to the top of
a truncated arcade ruin, have been scored with a utensil sharp enough
to create a network of raised ridges on the verso. These ridges, only vis-
ible on the verso side of the print, preserve their coating of red chalk,
a contemporary technique for producing low-tech and potentially
repeatable negative impressions. The technique is remarkably subtle.
Only by running one's fingers along the front surface of the image is it
possible to sense the slight indentation, indicating the prior presence
of a sharp instrument used to create the reverse impression.

The Harvard manuscript of Stöer's drawings is also replete
with evidence of copying, although only in the latter section which
is filled with less proficient geometrical models. Although the vol-
ume's two sections are continuous with each other, the line quality
of the second section is substantially less precise and the watermarks
of the two sections differ. The catalogue notes point out that while 45
several of the pages are signed sporadically with Stöer's monogram,
the latter drawings are peppered infrequently with the monogram
IAR—possibly the work of Hans Rogel, also known as Johann, the
publisher of *Geometria et perspectiva*. Tiny pinpricks carpet the back of
each of these latter drawings, with marks clustered around the edge
points of the figures, while almost none of the images in the first
section of the book have pinpricks. It appears likely that the copier
placed the pages over an already drawn graphic model by Stöer and
used a pin to transfer its outline points onto a new page. The fact that
this technique is not used in the beginning of the volume leads one
to believe that the artist in the latter part was unable to conceptualize

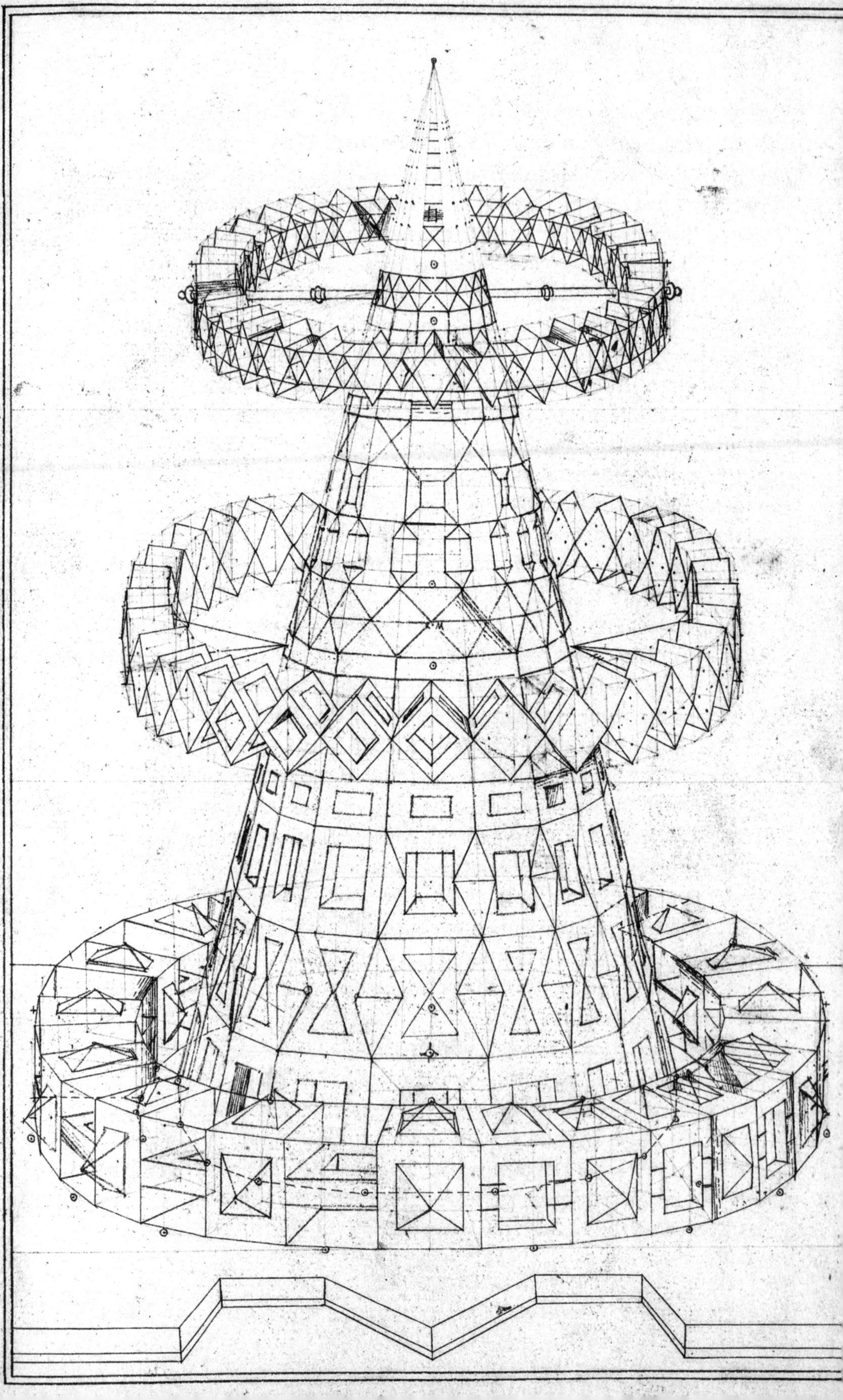

the necessary geometry on his own and that these drawings were likely produced in imitation or by a student.

Stöer's geometries were designed as three-dimensional, virtual objects to be colored in later, complete with dashed construction lines enabling a fully "transparent" view that preserve a variety of partial copying techniques and ridges on the back of drawings, as well as construction lines, basic attempts at shading in pen, and testing with less proficient lines. As a representative example, one half of a spherical irregular solid from Erlangen has been traced through—the minimal amount of inscribing necessary to copy the entire solid, given its symmetry. The intensity of practical use suggests that these pages were working drawings from Stöer's workshop or a set of models intended to be copied by Stöer or his apprentices and fleshed out in color on other sheets of paper or in other media.

It would appear that the scars of reproduction borne by his hand drawings point to numerous attempts to translate his wide stock of singular mathematical models into other media, either by apprentices or by Stöer himself. These transmissions (*Übertragungen*) preserve a certain selectivity of gaze, a roving set of geometrical predilections that fixated on unique objects while ignoring others entirely. While we don't know exactly why only parts of *Geometria et perspectiva* were copied, nor whether the copies were destined for intarsia or for other media forms, the fact that only the irregular solids were transferred likely speaks to perceptions of their specialness, whatever the copier's criteria for specialness. For dilettantes like Elector Christian, still attempting to learn the perspectival ropes, copying may have signified a response to the challenges of representing *irregulata*, and hence a demonstration of his reliance upon an original drawing or print, possibly by his prominent teacher, as a base reference.

These symphonies of pinpricks are indicative of a demand for reproductive certainty, for an exact process for translating geometrical forms as the artists had first drawn them and knowing that the copied form would be exactly the same as the original. In this sense, copying was as close as another apprentice could come to a classically mathematical proof—determined not by Euclidean logic but by the concentration and choreography of hand and eye. Artisans here did not construct geometrical figures from first principles, namely through calculation and measurement, but through the

Lorentz Stöer (active 1557–1620/21). Construction drawing for a geometrical model, from *Optica*. Universitätsbibliothek Erlangen-Nürnberg (MS 2774, fols. 60 recto and 67 recto).

same methods of graphic appropriation that animated the relations between print culture and the decorative arts. Exactness, irrevocable certainty, seriousness of purpose, a likely crop of diligent apprentices—copying here was actively pursued as a rigorous epistemological practice of "getting to know" geometry through imitation, though future iterations would be relegated to repeating the placement of the solid's vertices and edges as they had initially been conceived. The traced-over and reconstituted view of an irregular solid would thus be forever frozen in space, at once unable to be graphically rebuilt from another angle yet guaranteed to have a life beyond the confines of its original page by a culture that had grown to prize novel geometrical forms much as any other desirable subject.

Spurred by the increased possibilities born of technical specialization and prone to celebrating itself through geometrical-scientific motifs, the sixteenth century welcomed the craft of marquetry as a new genre for artisanal experimentation in which the migration of geometry into and out of architecture and ornament served as an exemplary theme colored by an unsettling moral charge. But at a certain indefinable point, geometry had slipped into disorder or confusion at the very moment its imaging had come to reflect the almost surplus craft attention for which these works were prized.

This was by no means the only instance of the Renaissance negotiating disorder. Material surviving from antiquity also had a notably disordering effect on contemporary culture, which found itself faced with the task of absorbing or discarding representations
46 on antique coins, plaquettes, and reliefs. But when the architecture of an uncannily modernist cabinet like the one in Mulhouse, or more commonly the complex of imaginary ruins emerging from du Cerceau, Stöer, and Solis, slid into *horror vacui*, it was not due to any foreign body that needed to be incorporated or expelled and had disturbed the natural order of things. Rather, anonymous *Schreinern* actively pursued the attention of the viewer, drawing him into imaginary realms made all the more engrossing by canny graphics that tipped the image from the contemplative, sublime landscape of the ruin into an optical maze, where the cities of the present are the ruins of the future and we, as viewers, are not sure whether we are looking at the cabinet from the outside or are already inside the cabinet

looking out. This was accomplished by an intensity of work that called out to be appreciated for its intricacy, just as with turned ivory, carved coconut goblets, embellished scientific instruments, or any of the other prodigious expressions of southern German craftsmanship circulating concurrently. Inasmuch as the cabinet or *Schriebtisch* was a tangible commodity that might be purchased and displayed in the home, the intangible commodity presented on its front or on its many drawers or internal facades was a collection of engrossing experiences of spatial and scalar parallax, meant to compete with the delights offered by any comparably sized cabinet of curiosities.

Geometrical models found their way from drawing and print into intarsia because they were sought after, as much for their novelty as for their austere sensuality. In this genre of media destined for transfer, polyhedra were used as representative emblems of geometry and geometricalness. Exemplified by Stöer's little-known oeuvre—itself a transpositional conduit of geometry into other media—irregular bodies proliferated in the relation space of print culture and highly technical craftsmanship. The utilitarian opulence of presenting a phalanx of polyhedra on the intarsiated surface of a cabinet in one's living room, as would have been the case with the Cologne cabinet, was intended to signal appreciation and ownership of wondrous graphic artifacts brought back from voyages into the unknown lands of perspective. The aspiration for designers and patrons must have been to present the state of modern mixed-mathematical knowledge, for which southern Germany had become a wellspring of talent, on its most precious surfaces and through its equally unrivaled abilities in carpentry. A new irregular body thus possessed a measure of significance, both because it constituted an incremental expansion of the domain of perspectival geometry, one solid at a time, and because displaying and owning a new geometrical image signified the emergence of an original "body" that had not previously existed. By redrawing the boundaries of what it meant for geometry to be "Platonic," Stöer had conjured up a promiscuous aesthetic of post-Platonic geometricality.

The uncanny pleasure of the Augsburg cabinets and all the graphic material that supported their realization is the pleasure not of order but of its absence, of allowing the possibility of being lost and of finding oneself again, reassured that the real is after all real,

even as the cabinet alludes, and remains a testament to, an ambiguous and nameless peril. At war with its surfaces, the object (or image) conjures a morality tale of the vanity of good intentions, an urban *memento mori* mediated in the space between the architecture of the cabinet and the architecture it images. The palpable unease conjured by these disjunctions, in which the ceaseless creativity of the artist is modeled, splits the city (to quote Benjamin on the *flâneur*) "into its dialectical poles. It opens up to him as a landscape, even as it closes
47 around him as a room."

Stöer's work in particular conveys a savvy understanding of its own dizzying seductiveness. The power of his drawings lies in the exquisite absence of any corresponding models, a representation that purports to depict the presence of a real object where one never existed, where there was only ever a graphic precursor, copied and reinvented through the ages. Twelve colored ink drawings of floating geometrical figures in Stöer's Augsburg notebook speak to the prevailing impulse to surrender to the iterative methods and visual qualities of mechanical production, such as the exquisite exemplars of Jamnitzer's etchings. Copied, if scaled up, from *Perspectiva*, they manage the impossible, namely to outshine the printed originals. The line weight and penmanship show incredible command, while the crosshatching strives to replicate the etching from *Perspectiva* down to the individual line. Having adopted the abstracted shading characteristics of etching, in which shadow is shown by hatched line and light is indicated by the absence of line, Stöer emulates Jamnitzer through a technological mimicry that returns the human
48 hand to the fantasy of its own erasure.

Given the endlessness of Stöer's experiments, the infrequent placement of his personal monogram raises the question of when exactly it was that incremental invention concretized into a graphic form or moment which required an explicit statement of authorship. Occasionally, we find a simple watercolor drawing, such as one depicting stacked geometrical solids in the Universitätsbibliothek München, marked with Stöer's distinctive insignia "S" intertwined around an "L," and the inscription "Lorentz Stöer Mahler [painter] in Augsburg 1567," while all of the prints in *Geometria et perspectiva* pre-
49 sent the monogram set into the plinths at the solids' bases. But more often, Stöer's monogram (frontispiece to this volume) was reserved

for what was considered to be his finest geometrical work, very selec-
tively populating his drawings with his personal attribution. Rarer
still are the few instances when his experimentation appeared to yield
a graphic result so exquisite that it required imperial privilege to
protect the skill of its invention as well as to substantiate that some-
thing truly special had been created. One such model was a dazzling 50
polyhedron located toward the end of the Munich volume (and repro-
duced as the frontispeice to this book). Set atop an *art nouveau*-like
base, two irregular and skeletal solids nested one inside the other
issue an explicit challenge to the models from *De divina proportione*, as
if proclaiming the ascension of a new master of geometry. At its base
is written the phrase "Cvm gratia et privilegio Cæs Maiest"—"by the
grace and privilege of his imperial majesty." It had taken the force of
an imperial privilege to fix Stöer's body in place, beyond the endless
transformability that otherwise threatened to subsume it beneath a
sea of variation. An endpoint, one of few, had been reached.

overleaf Lorentz Stöer (active 1557–1620/21), after Wenzel Jamnitzer (1507/08–1585). Tetrahedron and icosahedron variants. Ca. 1600. Staats- und Stadtbibliothek Augsburg (4 Cod. Aug. 247, fols. 23 recto and 32 recto).

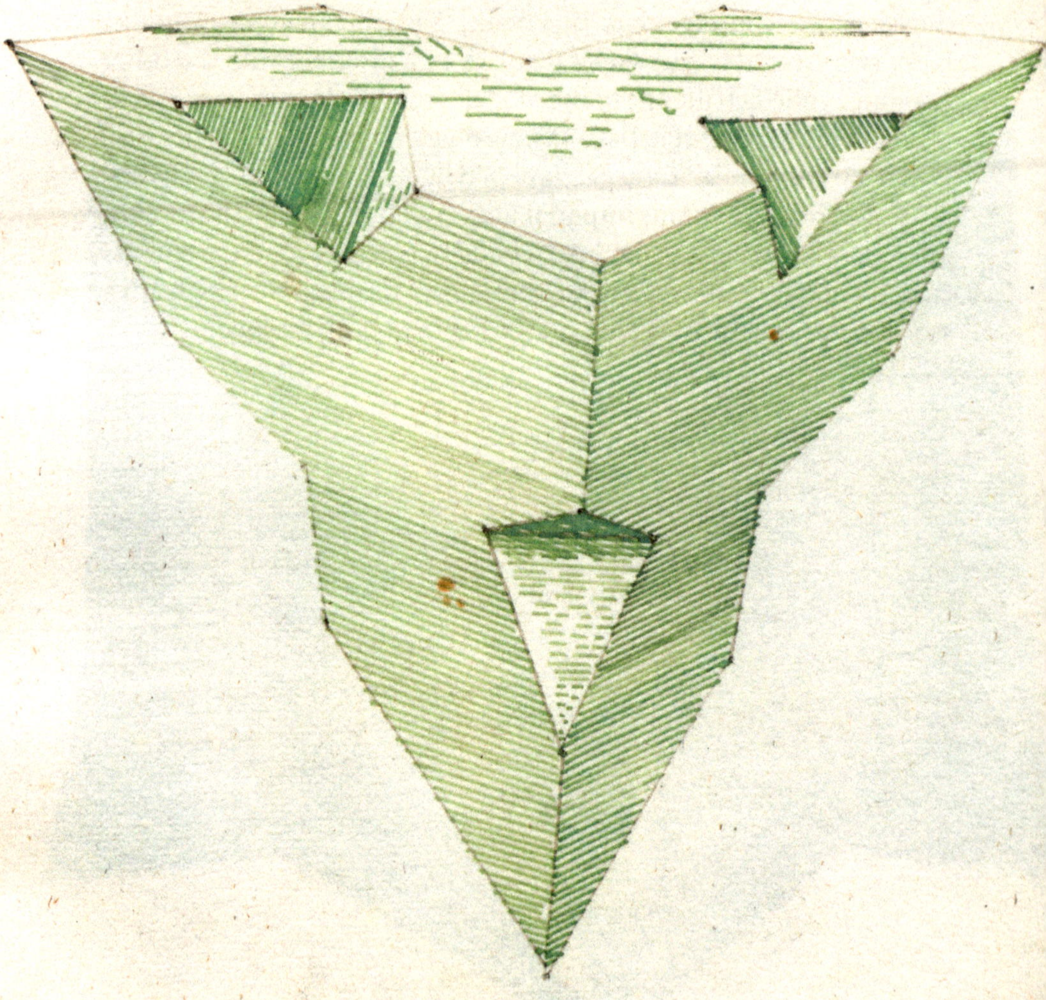

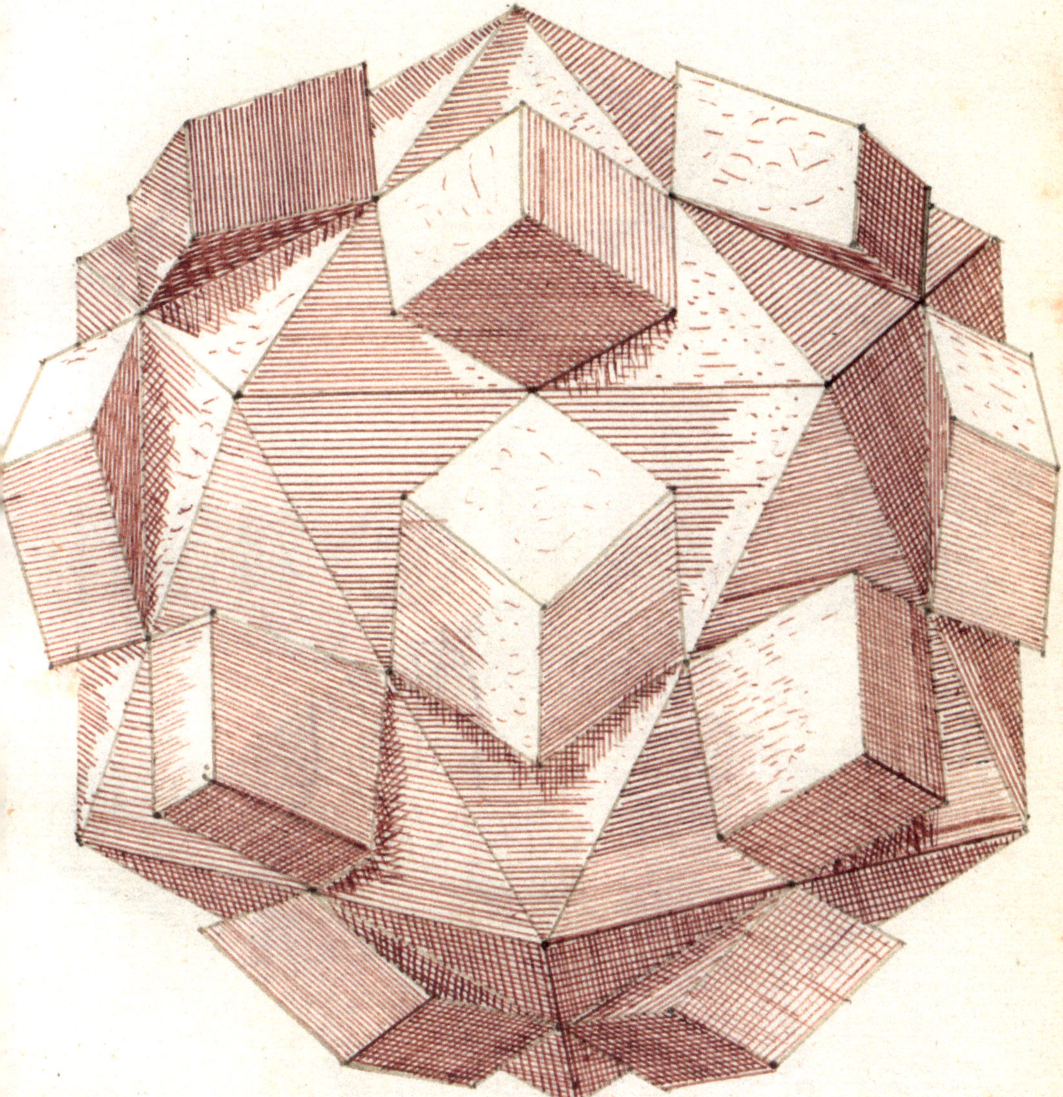

T

VIOLE

WH

E

NCE OF

MSY

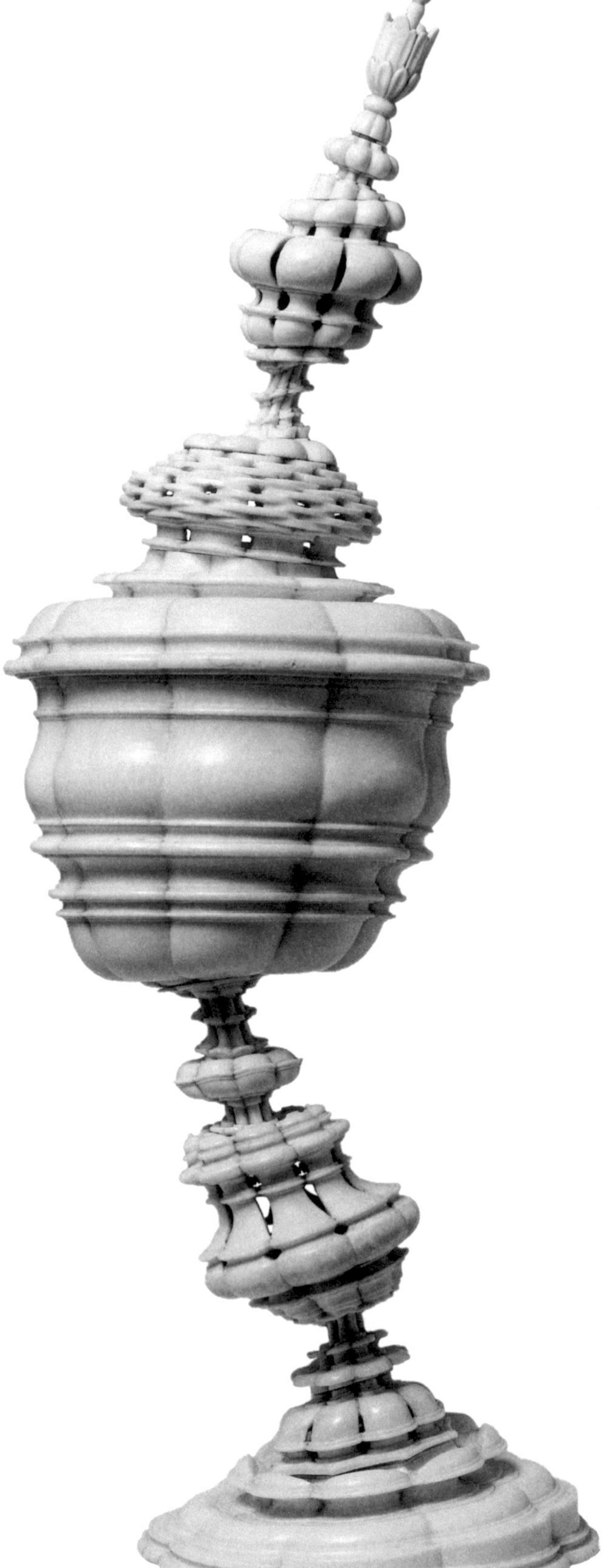

By the late sixteenth century, practical geometry in all its many forms, including prints, drawings, and books on polyhedral geometry and its perspectival construction, had come to be appreciated both as collectible artifacts and as a central component of princely learning. While *Lehrbücher* were not integrated into the curriculum of the late sixteenth-century university, the court was genuinely invested in appropriating geometrical knowledge and expertise. The library of Duke Friedrich I of Württemberg (1557–1608), for instance, contained Vredeman de Vries's *Opera mathematica* (1617), multiple editions of Euclid's *Elements, Perspectiva* (1571) by Hans Lencker, Jamnitzer's *Perspectiva*, and the first German edition of Jacques Besson's disputation on machine design, *Theatrvm instrvmentorvm et machinarum* (1578) (*Theatrvm oder Schawbuch allerley / Werckzeug und Rüstungen* [Mümbelgart, 1595]), whose translation and reissue Duke Friedrich had financed directly. August the Younger, Duke 1
of Brunswick-Lüneburg, collected nearly one thousand volumes of "Geometrie." Even the famed University of Tübingen, alma mater of the astronomer Johannes Kepler, acquired Hirschvogel's *Geometria* (1543), a hand-colored edition of Stöer's *Geometria et perspectiva* (1567), and Heinrich Lautensack's *Des Circkels vnnd Rechtseyts ...* (1564).

But in an era in which sovereigns were increasingly consumed with the authority of geometrical measurement and consequently the tools and texts used for measuring as an ever more essential component of courtly learning and the projection of power, there was no site that more wholeheartedly embraced a geometrical ethos than the Dresden court of August, Elector of Saxony (1526–1586). The 1587 inventory of the *kurfürstlich-sächsische Kunstkammer*, compiled about eighteen months after August's death, revealed that under his rule the Saxon court had accumulated one of the most singularly geometrically focused of contemporary collections, with an extensive assembly of polyhedral artifacts and literature. Recalling his 1629 visit to the *Kunstkammer*, Philipp Hainhofer (1578–1647), the famed diplomat and art advisor to Duke Philipp II of Pomerania, detailed a cornucopia of "astronomical, geometrical, geographical, scenographic, or perspectival and typographic [instruments], notably beautiful armillary spheres, celestial & terrestrial globes of wood, metal, and other material, astrolabes, quadrants, quadrata, declinatoria, nocturnalia,

Workshop of Lorenz Zick (1594–1666) or Stephen Zick (d. 1715). Standing cup. Seventeenth century. The Metropolitan Museum of Art, New York, Gift of Robert Gordon, 1910 (10.212.2a, b).

several solar clocks, cylinders, truncated [objects], various regular
2 & irregular bodies." The posthumous inventory substantiates his
description and further names many other "working" *corpora*, for
example Platonic solids by Abraham Riese, son of the mathemati-
cian Adam Riese (1492–1559), under the subheading "a regular body,
made from wood and glued [stuck together] paper," and intricate
geometrical models in wood, such as a "four-sided cube with another
3 eight-sided cube inside of it," used as prototypes for models in ivory.
Underlining the close relationship between modeling and drawing,
the subheading immediately following the collection of polyhedral
models in the inventory refers to "Ahn perspectivischen instrument
und derselben zugehörunge," suggesting a category for perspectival
4 instruments and all manner of drawing apparatuses.

Though there is no mention of how the models might have
been displayed, polyhedral models are referred to for the first time
in a later inventory from 1640 as "mathematische und stereome-
trische corpora regularia und irregularia hin undt wieder in diesem
zimmer zufinden [mathematical and stereometric regular and irreg-
5 ular bodies, which can be found here and there in the room]." It is
not inconceivable that they would have been hung on the wall, as in
the Neufchâtel painting or the Faulhaber print, or suspended with
string from the *Kunstkammer*'s ceiling as in Pacioli's *De divina propor-
tione* or Carpaccio's *studiolo*.

The inventory also recorded the accessioning of six nested *cor-
pora*, the first three of which were notable for having been produced
by women: "1 octahedron, in which a hexahedron and a tetrahedron
have been enclosed" as well as two models of hexahedra, the first
6 enclosing an octahedron and the other a tetrahedron. The last three
models were designated as "1 small cube," "1 skeletal tetrahedron
in which [there is] an octahedron," and "1 prism, which one can dis-
7 mantle into 5 pieces, made by Dr. Christof Pincker from Leipzig."
Not only had new polyhedral models been added to the collection,
but the descriptions of the older models, some of which were more
than sixty years old and originally belonged to August, had since
developed new labels that reflected their incremental articulation
and the evolution of seventeenth-century mathematics. Hence a
"Conus orthogonius" (orthogonal cube) as recorded in 1587 had
by 1640 become a "Conus orthogonius, darein ist sectio parabole

geschniedten" (an orthogonal cube, in which a parabolic section
has been cut), and Abraham Riese's earlier gift was now deemed to
be "useful for the demonstration of quadratic and cubic radiuses." 8

Inclusive of "a beautiful mathematical library for arithmetic,
geometry, architecture, perspective, and other arts," filled also, like
its neighboring spaces, with "cones and pyramids [*helffenbain gedrehte
coni vnd pyramides*], beauties of turned ivory, upon which can be seen
several regular and irregular bodies [*corpora regularia & irregularia*],"
the Dresden *Kunstkammer* consisted of seven main rooms, including
an entrance vestibule connected by stairs to the courtyard (*Großer
Schlosshof*) below and a lathe turning workshop (*Drechselkammer*)
above. Kept separate from the rest of August's library collection, one 9
room was reserved specifically for his workshop tools and research
literature, which numbered two hundred and eighty-eight "astro-
nomical, astrological, geometrical, perspectival, arithmetical, and
other art books." 10

A greatest hits of perspectival studies, the library featured
many images of geometry and geometrical models, which would
have been used as inspiration for ivory turning. While he could not 11
hope to rival the ivory stylings of his master court turners and did
not possess Christian's drawing skills, August still actively partici-
pated in turning ivory. The books shared space with "artful turned
ivory work ... which the well-regarded Elector August turned him-
self" alongside a select number of *Meisterstücke*, such as a "peculiarly 12
turned goblet, [topped by] an open-work sphere, in which are located
another 24 spheres, all of which are moveable," and a "dense sphere 13
... in which one discovers through a hole, another beautiful star and
small open-work spirals. Across it, similarly, is a box with two lids, in
which are two *contrefait* carved by Elector Christian II." 14

Though Samuel Quiccheberg, the author of one of the first treatises on museology, did not directly reference the Dresden court, he was an art advisor to Albrecht V, Duke of Bavaria (1528–1579), a near contemporary of August. His descriptions in *Inscriptiones* (1565) of a consortium of "Museums, Workshops, and Storerooms, Such as Are Meant for Furnishing Wisdom and Pleasing Arts, Which Are Sometimes Constructed Separately in Palaces and Sometimes Joined Together" anticipates Elector August's state-of-the-art setup in Dresden, which connected mathematical instruments and *corpora*, a

research library, a laboratory or working space for turning ivory, and
a space for the display of masterpieces in order to formulate the study
and practice of applied geometry produced by courtly artisans and
15 sovereigns themselves.

The impetus for the upper classes to work on mathematics
through manual procedures and to collaborate with craftsmen in
the production of objects or instruments was not uncommon during
this period, as the joins that held down the middle sciences progres-
16 sively gave way. Machines were valorized and technology gained a
prestige not heretofore enjoyed in princely circles, as did practical
geometry which was deemed to augment values important to state-
17 craft. Captured in the allegory *Labor*, from *The Four Ages of Man* (1591)
by Raphael Sadeler I (1560–1628), Minerva, goddess of crafts, com-
merce, and the arts, appears above a young man interrupted in the
act of measurement, a compass in his left hand, a large architectural
treatise in his right. In the distance, industrious men and women
engage in harvesting and weaving. On the brink of romanticizing
the virtues of physical activity and the diligence required to produce
works of art, mathematics, and science, labor had come to figure as
a princely value, whether through ivory turning or other dilettante
pastimes such as wire-drawing, minting metals, and cabinetmak-
18 ing. In a synthesis of *praxis* and *techne*, concentrating on the precise
manipulation and measurement of matter to create hybrid works
of art and mathematics directly served the development of political
19 and military faculties.

Though the lathe would undergo many modifications and iterations, its appearance in princely collections was substantially due to the technological innovations that had brought it out of the craft workshop and into the *Kunstkammern*, and later the drawings rooms, of the upper classes. Antique and premodern lathes had required turners or their assistants to manually rotate the turning material with one or both hands while the turner used tools or bows to cut into and shape the material. Though the process could result in spectacular examples of craftsmanship, the dual imperatives to rotate the material as fast as possible while intricately manipulating implements made turning a taxing physical practice. The image of lathe turning in *Theatrvm instrvmentorvm et machinarum* reveals as much. Looming over the turner, a sinuous wooden form is revealed in partial

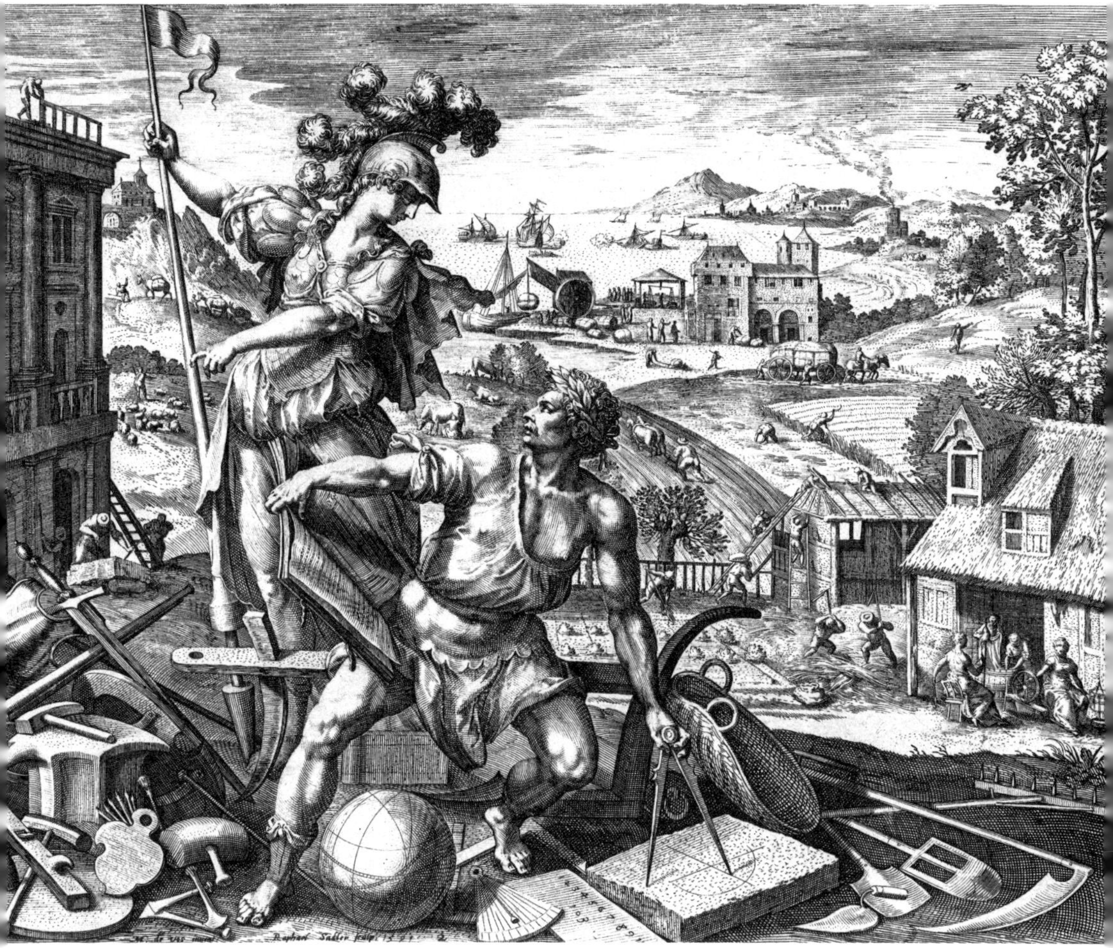

Raphael Sadeler I (1560–1628), after Maerten de Vos (1532–1603). *Labor*, from *The Four Ages of Man*. 1591. The Metropolitan Museum of Art, New York, Harris Brisbane Dick Fund, 1944 (44.62.4).

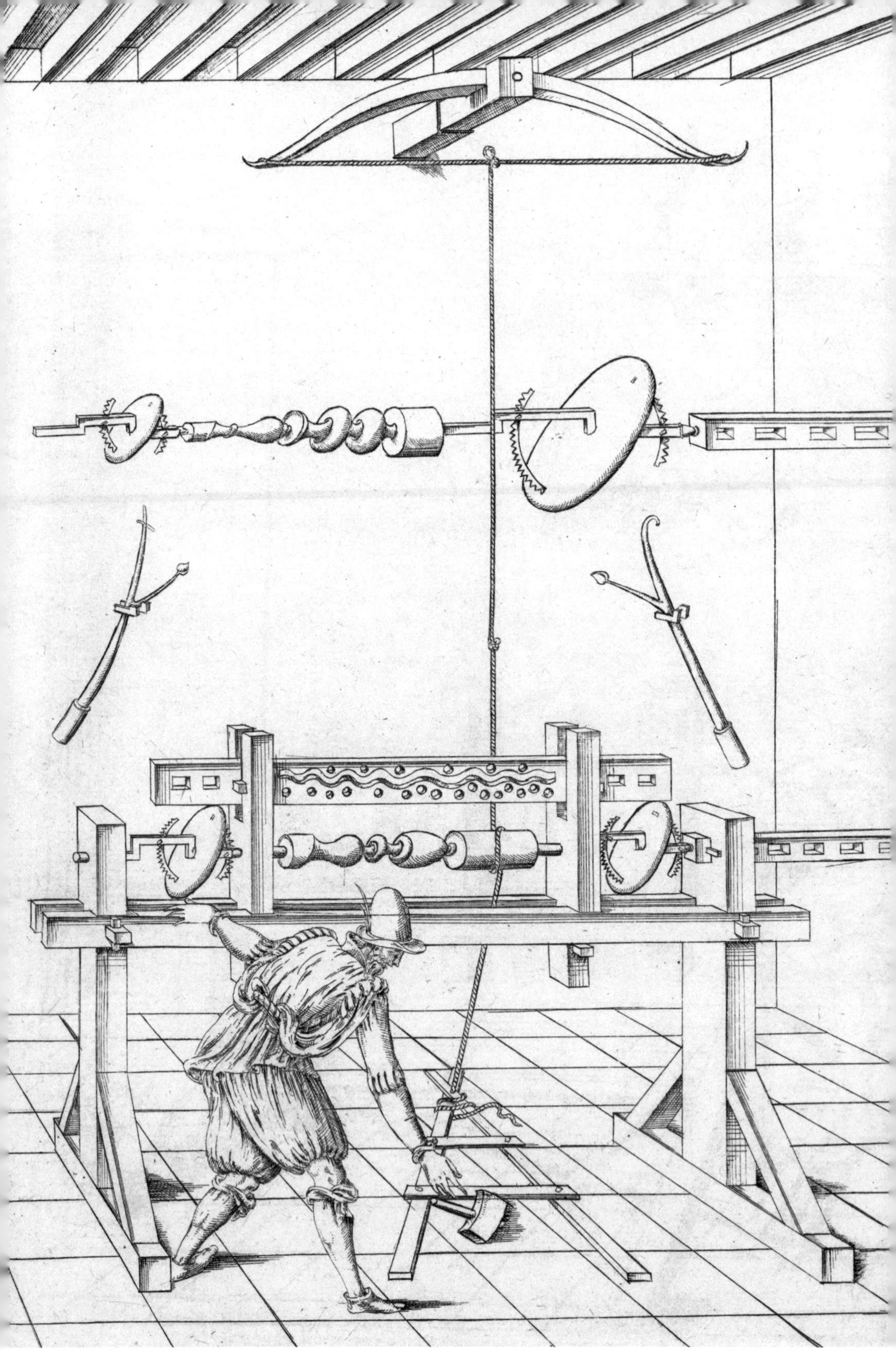

production with a rope rigged up to a bow protruding from the wall, providing the reciprocal motion required to cut into the material. In order to work this particular lathe, the turner would have been required to use both feet and hands while simultaneously guiding the material, a tiring prospect that would not have burnished the princely image or conveyed the restrained mastery of geometry to which sovereigns like Elector August aspired.

Improvements to the lathe sprang from the addition of new
rigging mechanisms that left the hands free, such as one that enabled
the lathe to be turned by pressing down on a treadle with one foot. 20
The central axis of the new "center lathes" spun by means of its attach-
ment to a flywheel—a wheel generating a continuous rotation in
one direction by means of an offset shank, crank, and treadle, itself
powered either via its connection to a treadle or by an assistant who
would turn the wheel himself. Programming tools called mandrils 21
or rosettes were also integrated, as were regulating mechanisms
known as rose-engines that could be mounted on the lathe spindle to
engrave ornamental curvilinear lines upon the surface of the worked
material. Made from metal disks, mandrils guided cutting tools to 22
make specific motions that would not have been possible to execute
by hand. The capacity of these lathes to combine continuous motion,
in both longitudinal and crosswise movements, while also rotating
their central spindle, meant that a material could now rotate ad
infinitum in three directions. At the same time, a cutting tool could
remain solidly fixed to the lathe, creating a reliably exact cut.

Taken together, these advances conspired to make the late sixteenth-century lathe more akin to a low-tech computer or a Jacquard loom, in which various mandrils and tools could be "plugged" into the machine with the turner knowing beforehand what the resulting forms would be. While master turners could attain dizzying heights of intricacy and fragility by combining mandrils, precision tools, and the newly continuous rotational capacities of the lathe, the amateur turner, initially under the guidance of one of his court artists, would have been able to turn elegant shapes without undue exertion, leave them on the lathe when his patience flagged, and return again to finish them at his leisure.

A turner at work on the lathe. The Latin inscription reads: "New geometrical [object], each cylinder realized in oval and conic form, with its decorations [*ornamentis*] turned from every kind of material." From Jacques Besson, *Theatrvm instrvmentorvm et machinarum* (Lyon, 1578). Ghent University Library, Belgium (BHSL.RES.1365, image 7).

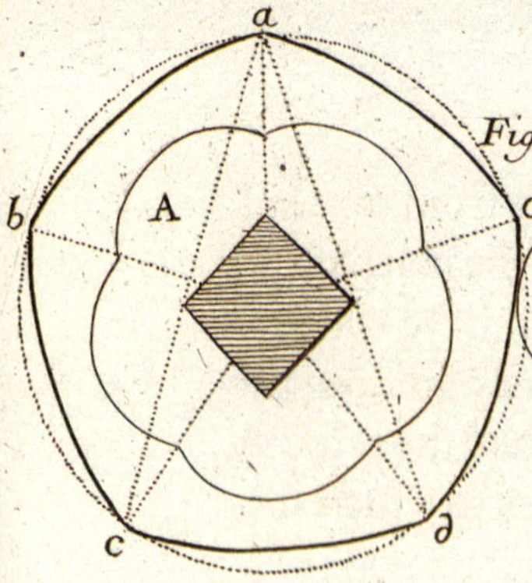

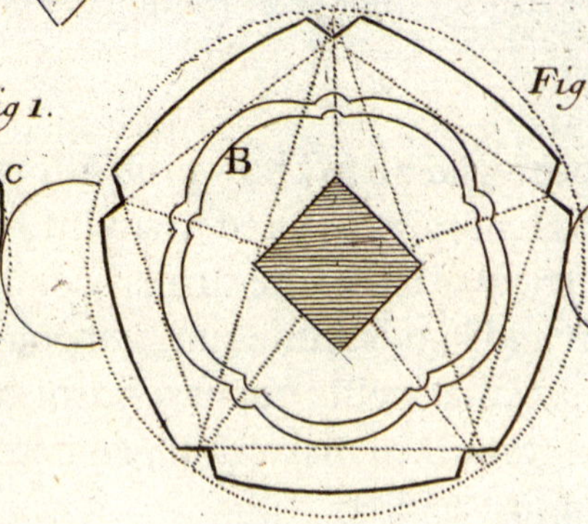

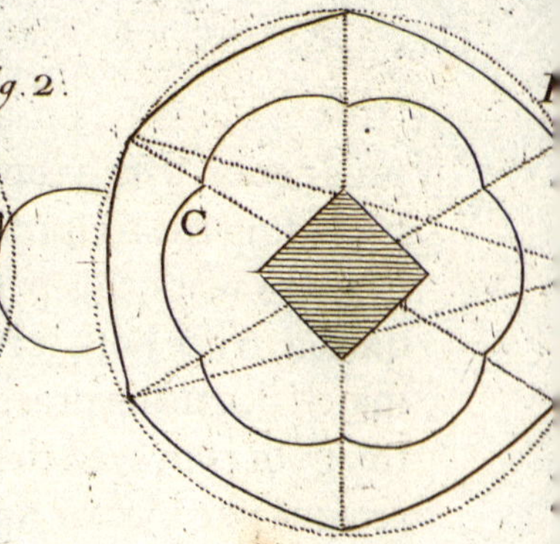

Fig. 4.

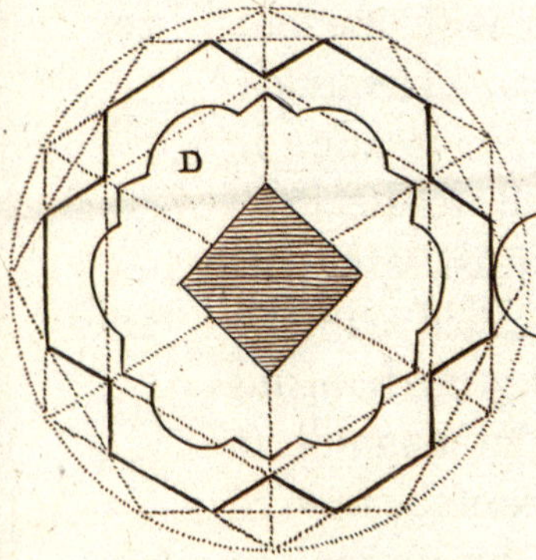

Fig. 5.

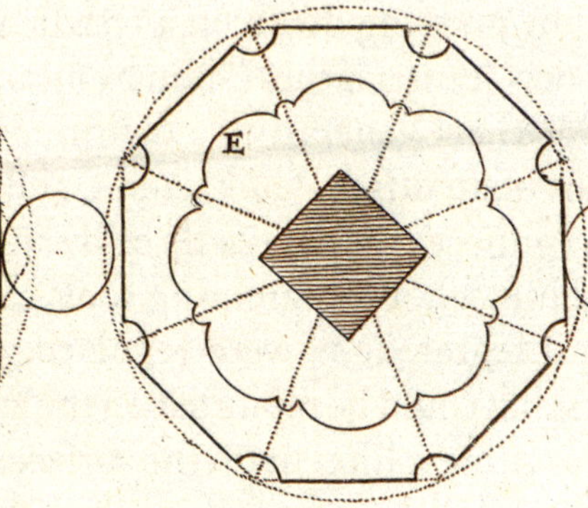

Fig. 6.

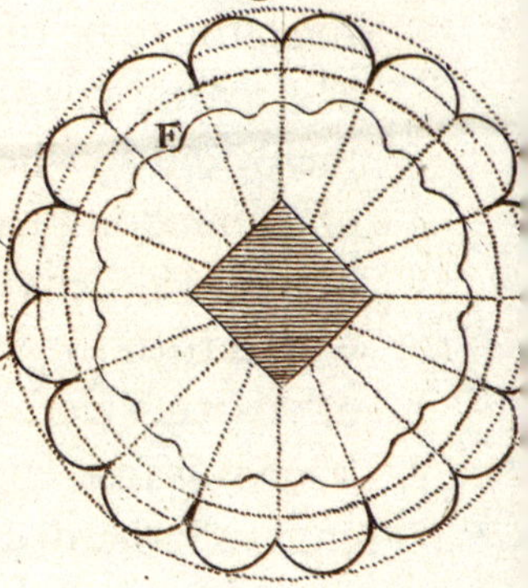

Fig. 7.

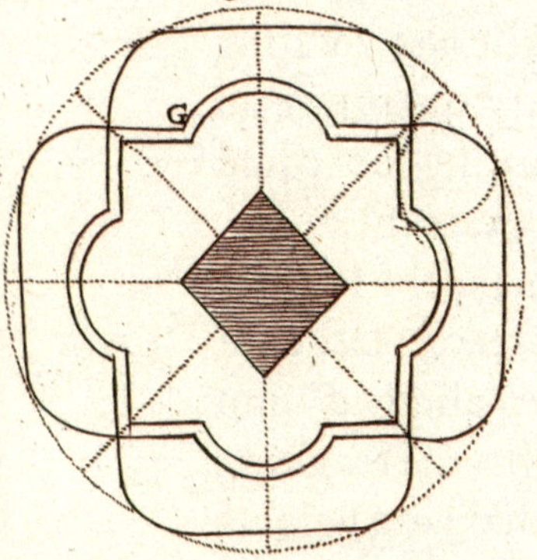

Fig. 8.

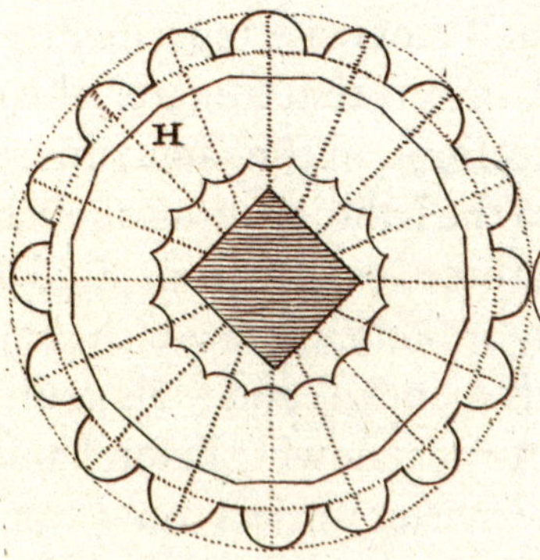

Fig. 9.

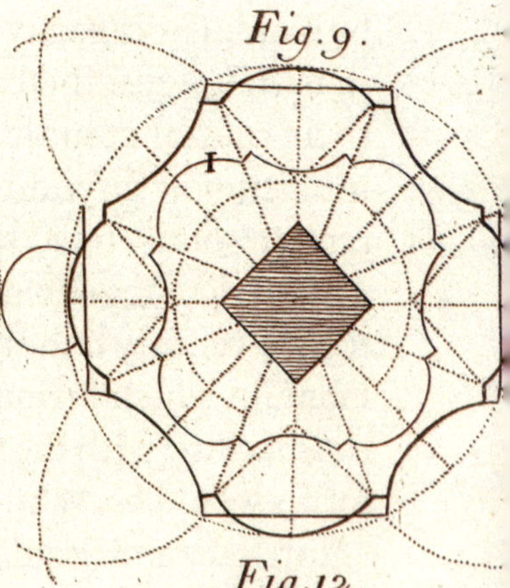

Fig. 10.

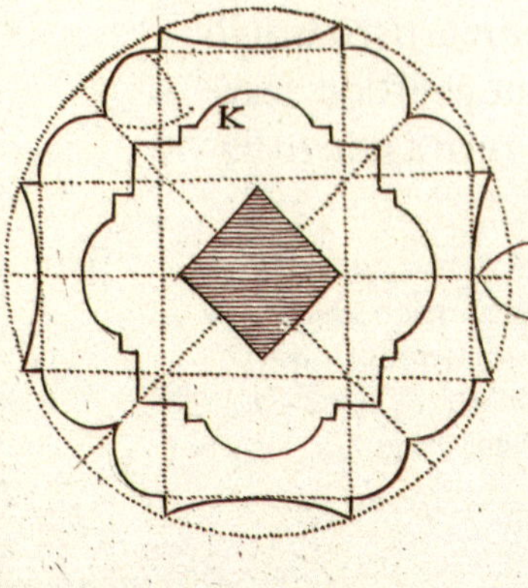

Fig. 11.

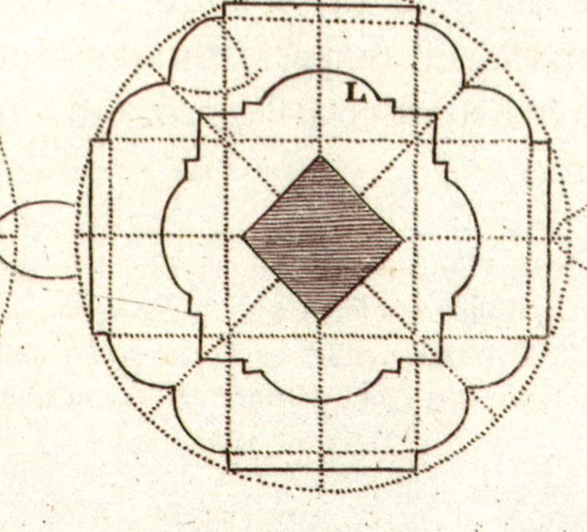

Fig. 12.

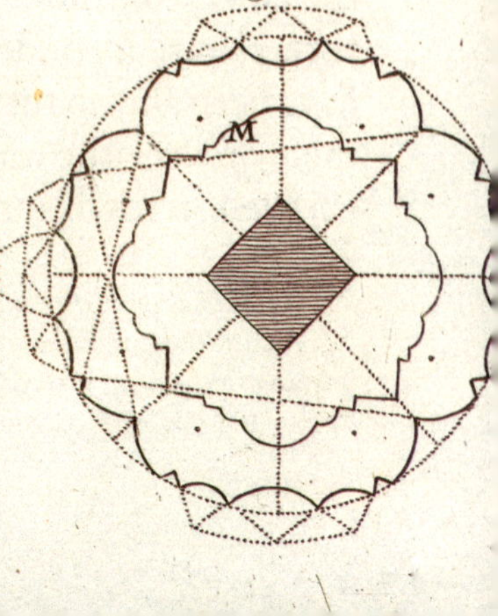

The attraction of working a costly material through a relatively
automatized process without years of technical training proved im-
possible to resist, and lathe turning came to be widely considered a
suitable pastime for European nobility. Like August, Emperor
Maximilian I (1459–1519) had a lathe installed in the Habsburg impe-
rial palace in Innsbruck and obtained a second machine in 1505 from
Tramin. Hans Wecker, the son of Georg Wecker, taught Emperor
Rudolf II, as did the Nuremberg turner Peter Zick the Elder. His son,
Lorenz Zick, went on to become court turner for Emperor Ferdinand
III (1608–1657). King Frederik III of Denmark (1609–1670), founder 23
of the *Kunstkammer* in Copenhagen, was an avid turner and collector
of decorative ivory pieces from southern Germany. The Wittelsbach 24
dukes were enamored with turning, and Duke Wilhelm V of Bavaria
(1548–1626) sent the Florentine Duke Francesco I de' Medici (1541–
1587) a hollow sphere by the master turner Giovanni Ambrogio
Maggiore (b. after 1550, d. after 1598), who had traveled to Bavaria
from Italy in 1574 on advice of the Milanese art dealer Prospero
Visconti and was the first turner to produce a *contrefait* sphere, or a
sphere enclosing multiple other spheres and carved items inside
itself. Francesco himself was known to carve decorative art objects 25
from rock crystal, jasper, and agate at workshops in his palace, the
Casino di San Marco, and the inventories of the ducal study list "un
castelletto da lavorar'… che se dice serviva al Gran' Duca Francesco." 26
Imbued with divine overtones, the practice of ivory turning had come
to be understood as a rarified craft exceptionally suitable for receiving
the sustainedly intense reworking and treatment required to trans-
form *exotica* into *artificialia*. Much as God, "the first lathe turner," had
created the miraculous sphere of the earth, so too did turning mimic
the creation of the globe through the force of perpetual rotation. 27

It was under the guidance of Elector August that the turn-
ing workshop at the Dresden court arguably become *the* center of
European turning innovation. Hainhofer noted pointedly that the
Kunstkammer possessed "a beautiful lathe used by August to turn,"
though this was only one of several ivory-turning facilities estab-
lished by the elector, who installed additional *Drechselkammern* at
his castles in Torgau, Augustusburg, and Annaburg. The illustri- 28
ous court turner Georg Wecker was employed on a permanent con-
tract in Dresden from 1578 and taught August's son and successor

Mandrils, from Charles Plumier, *L'art de tourner*… (Paris, 1749). Staatsbibliothek Bamberg (49 C 13, pl. XLV). Photo: G. Raab.

29 Elector Christian I to turn ivory. Egidius Lobenigk (d. 1595), the
first German to master the production of the *Contrafaitkugel* (*contre-*
30 *fait* sphere), was appointed court turner in 1584. He was succeeded
by Jakob Zeller (1581–1620), court turner from 1610, and Markus
Heiden (born 1597/98), who had previously worked for Duke Johann
31 Casimir of Saxony-Coburg (1564–1633). Scattered invoices in the
Finanzarchiv of the Sächsisches Hauptstaatsarchiv attest to pay-
ments provided to turners for their services and the agreeable terms
of their employment. The contract for Georg Wecker reveals that the
court was exceedingly excited to have him in residence and sought
to be unflaggingly generous. Wecker was to be paid two hundred *gul-*
den per year in quarterly installments; and "as a sign of appreciation
for his services, we have given and will pay every year 50 *gulden* to his
mother in Munich as long as she will live, so that he will be even more
32 inclined to stay with us in these lands."

More often than not, the delicate summits of the miniature ivory towers (*Saüle*) produced by the Dresden court turners were topped with open-worked *corpora* and other nested variants of the Platonic solids, frequently joined within layered *contrefait* spheres. A direct consequence of a long century of fascination with the limitless permutations of the solids, the *Saüle*'s mind-boggling precision had somewhat more in common with Jamnitzer's displays of graphic daring than they did with Pacioli's or Dürer's earnest disputations on geometry, though they differed in overall composition from many of the polyhedra that had preceded them. A pair of sumptuous confections by Lobenigk, for instance, incorporate continuous spiral incisions around each shaft and a capstone nested polyhedron (respectively a pyramidal tetrahedron and an icosahedron made of twenty equilateral triangles) at the summit. Hitherto restricted to two-dimensional media in drawing, print, and intarsia, or ephemera like the folded-up paper models used for perspectival and measurement studies, master craftsmen specializing in ivory turning had come as close as possible to fulfilling the geometrical fantasy of realizing *corpora* in three dimensions.

Numerous turned-ivory *corpora* remain in Dresden, though they also appear in many royal courts employing turners, markedly the *Kunstkammer* of Emperor Rudolf II, an inventory of which

Egidius Lobenigk (d. 1595). Ivory column with icosahedron (1591) and ivory column with tetrahedron (1588). Mentioned in *Inventar der Kunstkammer*, 1595, fols. 118 verso–199 recto. Bpk/Staatliche Kunstsammlungen Dresden (II 99 and II 77). Photo: J. Karpinski.

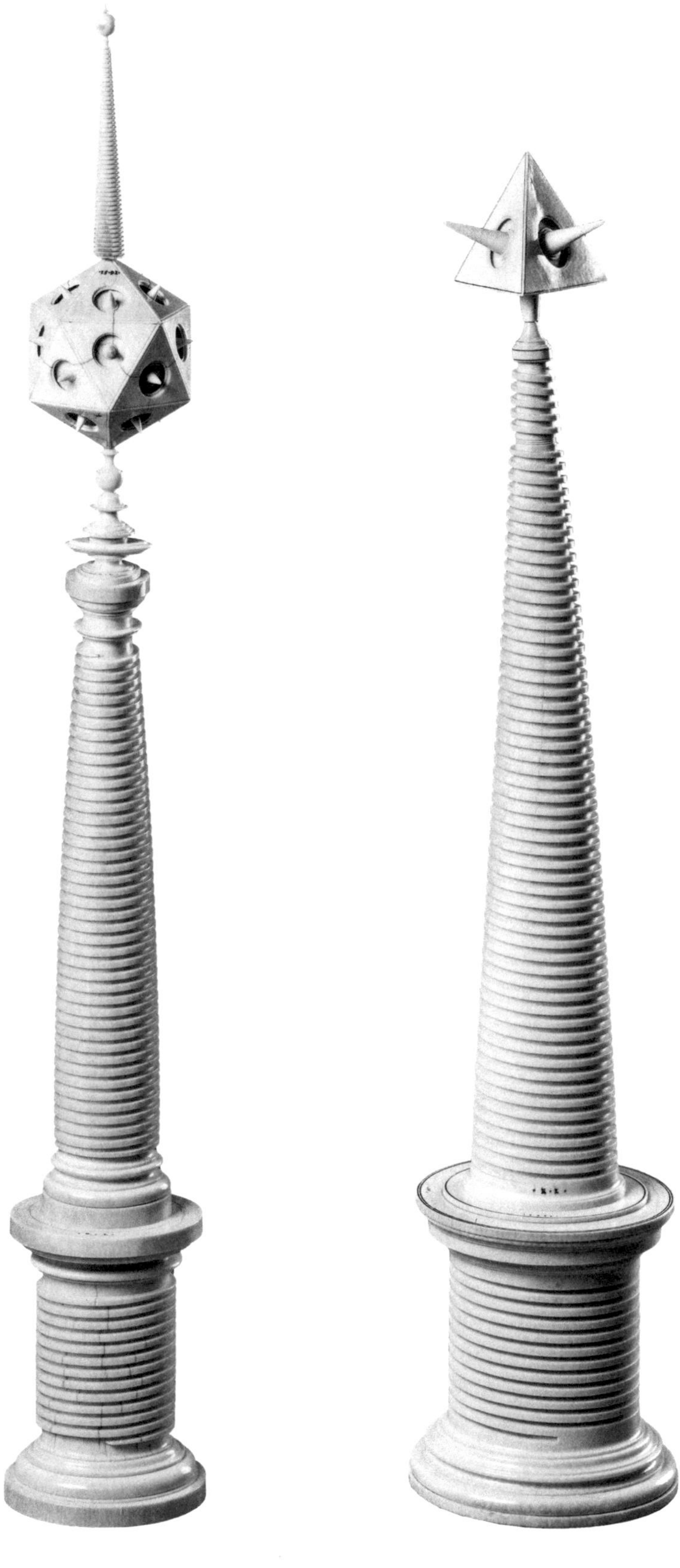

33 describes a number of turned *corpora*: "A spiral (screw-shaped) ovular
column of turned ivory on a base of three staggered levels. On it is
a round, hollow sphere which has been turned, within which is an
octahedron. On top of the sphere a screwed-on ivory pyramid has
34 been added"; "A triple-enclosed, one inside the other, ebony dodeca-
35 hedron, atop an ivory ornament: and small flowers"; "A long gilded
futral covered in leather and divided in quarters standing upon a
spiral ivory column. Upon it is a round hollow ball with holes with
more bodies inside. Above it is a three-cornered body, on whose
36 point stands a carved *Mercurius* from ivory"; "Two skeletal bodies
or turned ivory spheres, which have been enclosed five times inside
37 of the other"; and "A skeletal dodecahedron with a star and a small
38 surprise ornament turned inside."

The formal virtuosity of ivory *corpora*, having plainly outstripped the descriptive capacities of language, designated but one set of precious polyhedra scattered throughout courtly collections. In addition to being components of larger turned-ivory assemblages, polyhedra were also used as pedagogical aids for perspectival construction, as embellished, lathe-generated copies of workshop geometries rendered in ivory, and in Dresden as companion pieces to August's specialized research library. Appended to lids of ivory goblets that in other commissions might have been replaced by a carved figure of flowers, a royal crest, or a mythological scene, *corpora* functioned as signatories of technical knowledge gleaned from textbooks and the practical application of this knowledge in the production of new and experimental assemblages of geometrical forms. The boundary between geometrical research, as touted by practical authorities on the subject, and decorative art practice, as facilitated by the finest tools and equipment money could buy, made the Saxon court the site of a unique spatial and practical synthesis of textual and artisanal knowledge. Both tools and artworks, ivory polyhedra straddled the worlds of production and display, creating continuity between the geometry of the century's great polyhedrists and the performance of geometrical knowledge to astonished visitors in the *Kunstkammer*. Set against the backdrop of the evolving status of the *Kunstkammer* as a critical space for the demonstration of wealth, knowledge, and power, the transformation of raw ivory into turned geometry was fundamental to a courtly form of mathematics predicated on the synthesis of geometrical principles with the study of material performance—a practice that united texts on practical geometry with collections of prints for representing regular and irregular bodies as sources of inspiration. Here ivory polyhedra played a central role, delineating a matrix between the tools, texts, and drawings used to produce their component parts and the integral functions of the Platonic solids in the early modern workshop.

Crude and processed ivory had flowed into Europe since antiquity,
and not only as an unintended consequence of Hannibal's fateful
Alpine crossing. The victories of the Roman general and statesman
Lucius Scipio (d. after 183 BC) in Asia Minor were celebrated in Rome
with a triumphal procession that according to Livy, the classical his-
torian, included 1,231 ivory tusks among the overflowing spoils of
war. But during the fifteenth and sixteenth centuries, the numer- 39
ous Portuguese excursions into Africa and India opened up new and
lucrative opportunities for the procurement of ivory, which in turn
fueled the luxury markets for decorative art objects in the wealthy
urban centers of Renaissance Europe, as well as for ivory carvings
of non-European origin, which were prized for their mix of natu-
ralistic and intricate geometrical motifs. Often crafted with an eye 40
toward a western market, carved ivory salt cellars and bowls were vir-
tuosic pieces adorned with impressions of contemporary African life
as lived at the nexus linking European trade with the continent—
statuary of Portuguese explorers, ornament in relief derived from
animal patterns, and circular narratives that occasionally dramatized
the dangers of ivory's procurement.

The Munich *Kunstkammer* of Albrecht V of Bavaria and the
Hapsburg *Kunstkammern* swelled with the purchase of carved
African ivories, as did the *Wunderkammer* belonging to Ferdinand II,
Archduke of Tyrol in Ambras Castle at Innsbruck, which contained
twelve ivory items of African origin. Though relatively few examples 41
remain today, carved ivory goods were a common enough import
from Africa to be mentioned in the book of accounts of the Casa de
Guiné, the state-run commercial institution tasked with managing
the economic interests of the Portuguese empire and succeeded by the
Casa de Índia in 1503. Dürer's diaries from his trip to the Netherlands 42
in 1520–1521 regale the reader with exotica he encountered there,

including goods he claimed originated in Calicut, India—though
Calicut was a generic misnomer used for many non-European
43 regions, including West Africa. In Antwerp, Dürer recalled purchas-
ing "two ivory salt-cellars from Calicut" for three florins, which were
very likely examples of so-called "Afro-Portuguese" ivories created
44 for the European market by craftsmen working in West Africa.

European artisans from the mid-sixteenth century onward pio-
neered a competing method of reworking raw ivory into forms of
comparable virtuosity, relying less on skill in traditional sculpting
and chiseling techniques and more on the untapped potential of the
mechanical lathe (a machine that had hitherto been utilized primarily
for decorative architectural elements or furniture). Ivory, it had been
discovered, was exceptionally suitable for turning. Formed by spe-
cialized cells called odontoblasts, the dentine in ivory is organized in
a columnar pattern, one cone inside the other, and radiates out from
the center of the tusk to its edge. Together the layers of dentine create
an interwoven grain with structural strength in all directions, while
its intrinsic oiliness reduces brittleness. The coincidence between
the columnar pattern of the cells and their concentric arrangement
as cones is responsible for the strength and elasticity of the tusk,
structural properties that made it ideal for withstanding the cen-
45 trifugal stresses of the lathe. To give a sense, albeit unverified, of the
amount of ivory flowing into Europe by the early seventeenth cen-
tury, the ivory turner Markus Heiden states that there were over
300,000 tusks available when he selected the large piece of ivory in
Amsterdam for his 115-centimeter-tall ivory tower (1639), currently in
46 the Kunsthistorisches Museum in Vienna.

While ivory had not initially been the most profitable of Renais-
sance economic ventures, the slave trade, Portugal's most lucrative
business, established well-trodden mercantile routes, fortified trad-
ing outposts, and systems of exchange that were able to accommo-
date ivory until the explosion of its popularity in the seventeenth
century, by which point the commercial aspirations of other nations
and corporations—most famously the Dutch East India Company or
VOC (Vereenigde Oost-Indische Compagnie)—had overtaken that of
the Portuguese. But trading was a fraught and risky business for ele-
phant and human alike. Economic disparities with the local popula-
tions of the African port cities, the extremes of the climate, competition

from rival trading companies, and numerous informal microeconomies were a constant threat to the commercial monopolies of the Portuguese crown. And ivory was not just held in perpetuity, waiting to be sold to European buyers within purpose-built storehouses. It had to be obtained on demand, given the danger and expense involved in its procurement. As enumerated in a memorandum from 1715 by the Director-General of Elmina, a trading outpost run by the Dutch, European traders were obliged "to give out a great quantity of goods on credit, and these were sent off inland for the purchase of ivory
which could not be had otherwise." After the exchange was made, the 47
intermediary merchants would return to the coastal trading posts with the ivory loaded up on donkeys, where the ivory was deposited
in temporary storage facilities to await transfer onto European ships. 48
The ivory was only brought on board once the number of tusks tallied with the merchandise that had been provided by the European merchants on credit. If bound for Europe in the sixteenth century, the ships would likely have belonged to the Portuguese and also been carrying slaves, gold, and spices back to Lisbon. From there, the ivory would have been further transported by sea to multiple locales in Europe, namely Antwerp where the Portuguese had extensive trade connections, then Amsterdam or Hamburg.

Every time ivory changed hands, its value increased, and by the time it made its way to auctions and markets across Europe it had truly become a luxury item. A significant portion of the raw ivory used by August and the Dresden court's resident turners would have been purchased in Hamburg and subsequently pulled on horse-drawn boats up the Elbe River to Saxony, as this was more efficient
than transporting the ivory by wagon. An early eighteenth-century 49
design for an extension to the Zwinger palace complex in Dresden shows how the Elbe was used as just this kind of thoroughfare for delivering goods, which could be unloaded on the far shore and brought across the Elbe bridge or unpacked at a large staging area on the docks at the side of the palace. On the right side of the image, abutting the incongruously blank and ornamentless face of the unrealized extension, several boats are being pulled against the flow of the river into position at the foot of the castle, in much the way that
the African ivory would have arrived. 50

overleaf Anonymous. Unrealized design for the extension of the Zwinger, Dresden. Ca. 1730–1750. Kupferstich-Kabinett, Staatliche Kunstsammlungen Dresden (A1995-3152). Photo: H. Boswank.

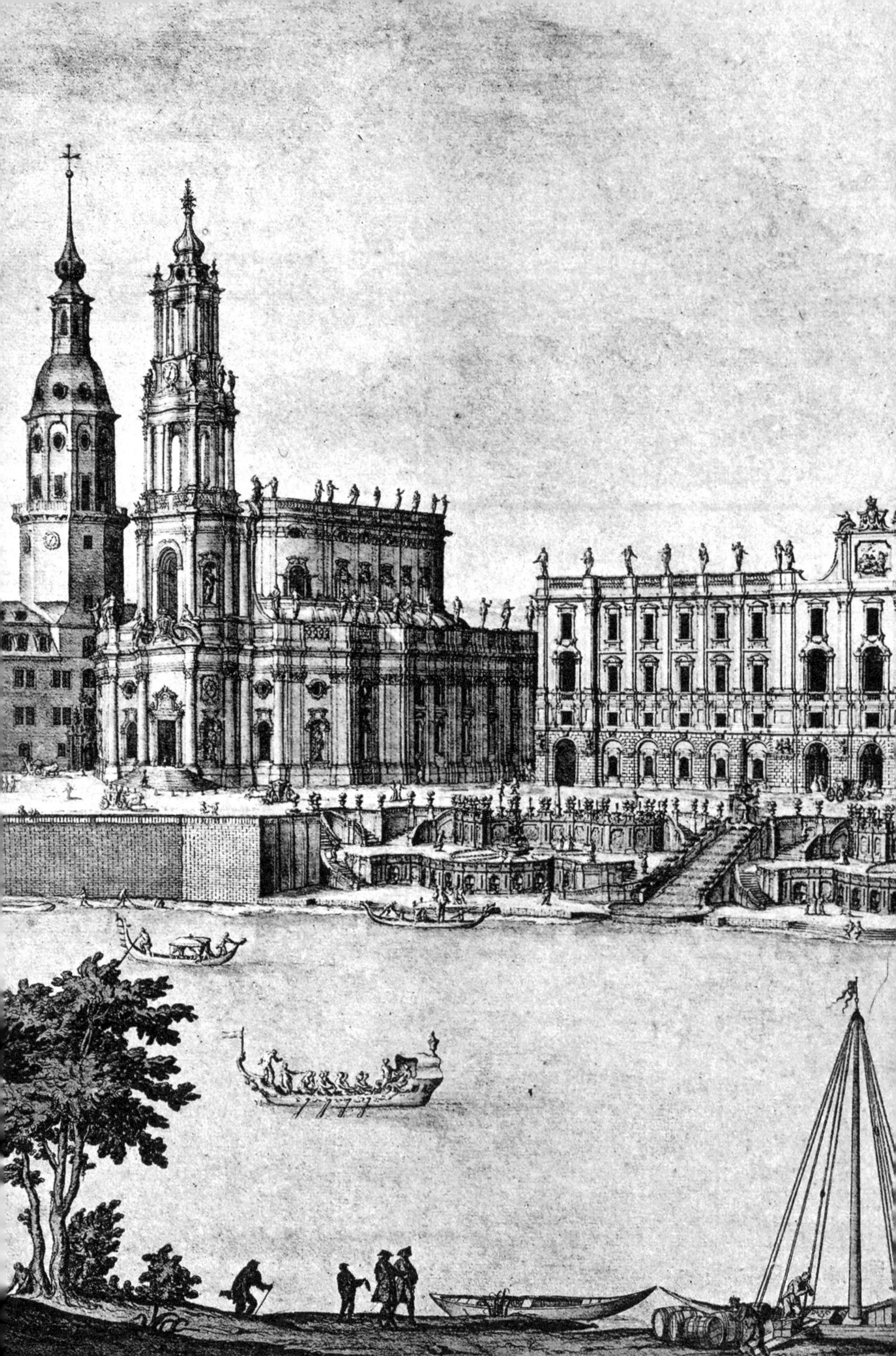

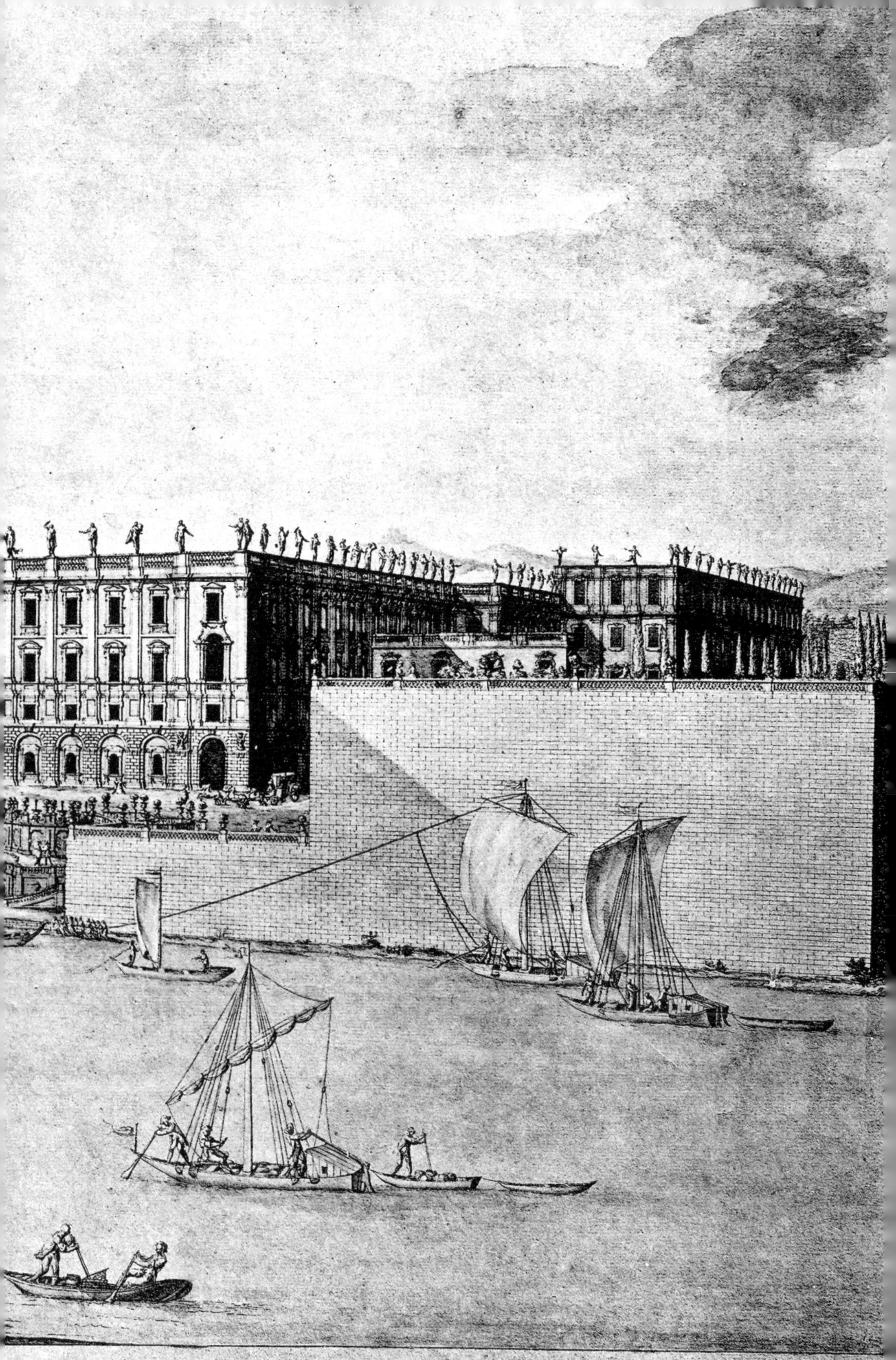

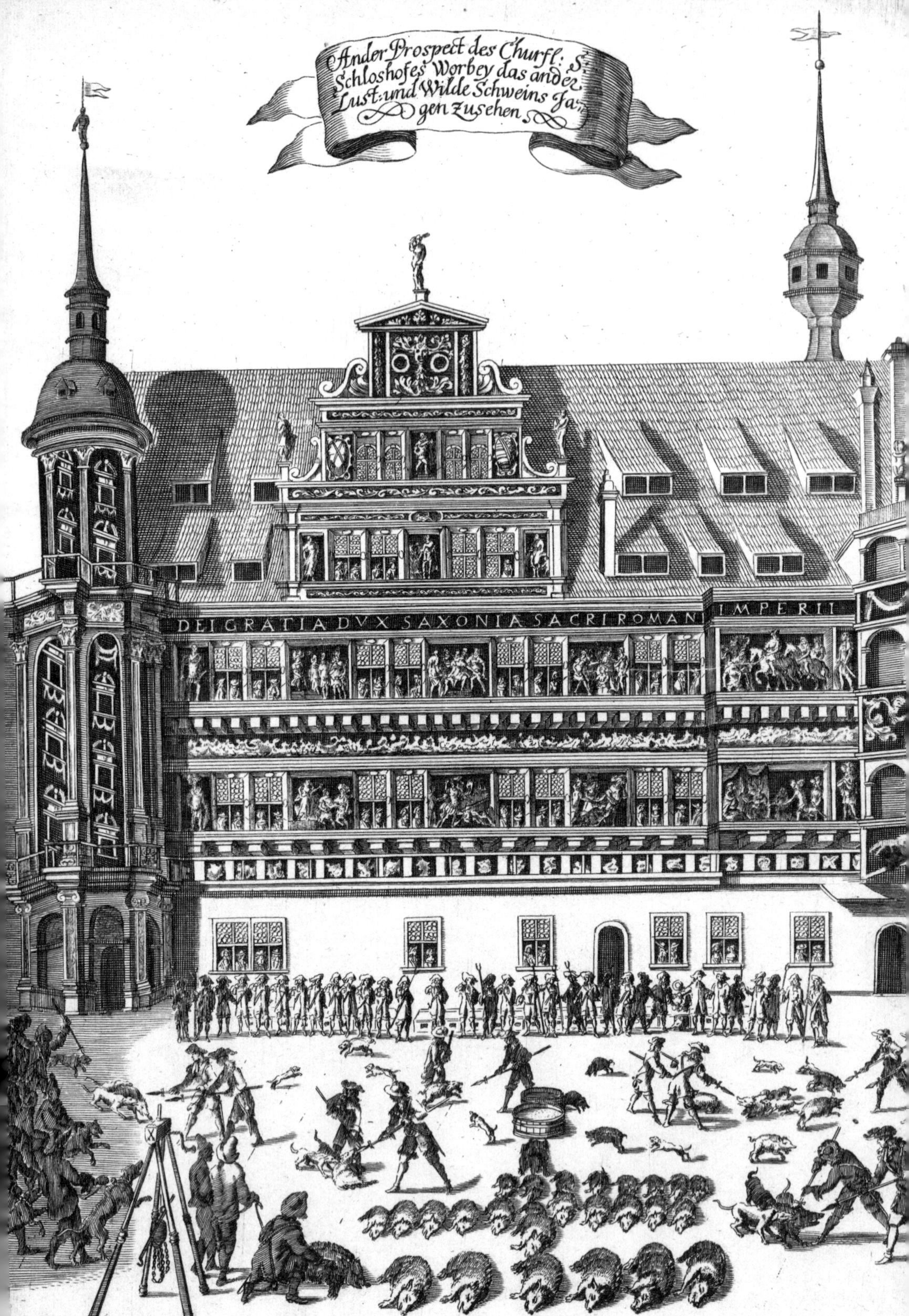
Ander Prospect des Churfl: S.
Schloshofes Worbey das ander
Lust: und Wilde Schweins Ja=
gen zusehen
DEI GRATIA DVX SAXONIA SACRI ROMANI IMPERII

After the ivory was unloaded at the river's edge, it would have been transported either by servants or on horseback up the external stairs that ran down to the water, carted around the near perimeter of the palace and into the large side courtyard. Here the ivory was unpacked from its travel casements and carried into the spiral stair tower at the courtyard's northwest corner. Precious goods flowed up and down these stairs, making the stairwell a secure space in which ivory in varying degrees of preparedness passed on its way to the *Kunstkammer* or to the storeroom above. Though there were stairs at each of the courtyard's four vertices, only the northwest stair led directly from the courtyard past the *Schatzkammer* (treasury) on the second floor and the *Kunstkammer* on the third floor to the fourth floor of the castle, which held the Elector's lathe machines and was used both as a storage space and workshop. Evidently the courtyard's large size made it amenable to all kind of macabre performances. Gabriel Tzschimmer (1629–1694), the Dresden statesman and chronicler, presents several contemporary tableaus of the courtyard and the activity occurring within it. In one scene, an audience watches as men with bayonets and dogs killed wild boars for sport. The stairwell that leads up to the *Drechselkammer* is located at the left in this image. In another scene, donkeys and cows are set upon by predatory animals, possibly wild dogs or wolves, while an appreciative audience enjoys the view from the safety of the castle's interior. The men carrying unwieldly deliveries up to the *Kunstkammer*, such as the one hundred-and-twenty-*pfund* tusk recorded in the 1619 *Kunstkammer* inventory, would have gazed down at the ritualized violence of these hunting reenactments through the distinctively slanted windows in the staircase that faced into the interior courtyard, their vertical movement traversing zones for the processing, preserving, storing, and display of art objects made from ostrich eggs, rhinoceros horns, bezoar stones, and nautilus shells, namely all the kinds of highly worked animal parts treasured by the wealthy.

The contents of the *Kunstkammer* continued to expand in the years after August's death. An additional large room on the same floor was annexed by the *Kunstkammer* and became responsible for housing masterpieces, including paintings by Lucas Cranach the Elder and

Pleasure hunt for wild boars in the large courtyard of the Dresden Castle, from Gabriel Tzschimmer, *Durchtigsten Zusammenkunfft* (Nuremberg, 1680). Bpk/Staatliche Kunstsammlungen Dresden (A 131503, p. 207, no. 23). Photo: A. Diesend.

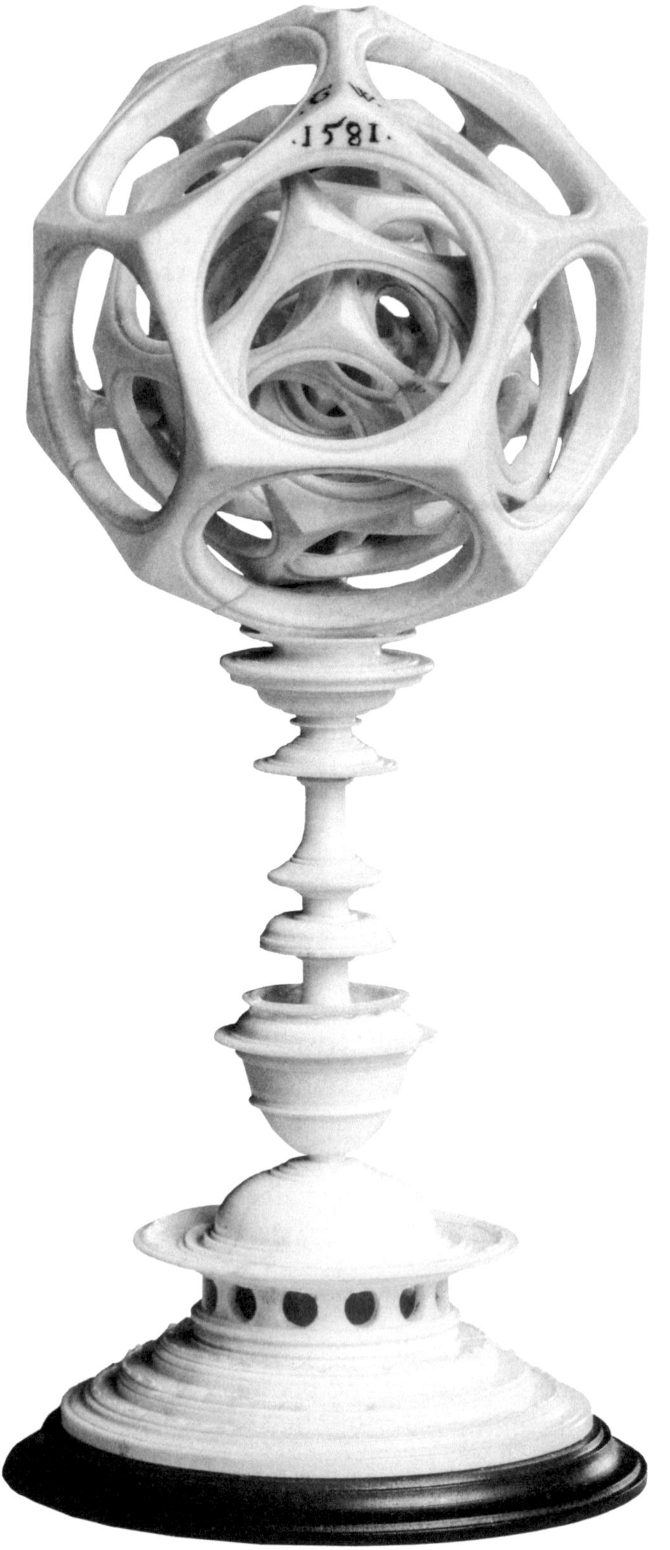
·1581·

Giuseppe Arcimboldo, magnificent prints by Dürer, lavish works in metal, *pietre dure* and elaborate carvings in wood, ivory, ebony, alongside maps, military busts, arms and armor, and collections of ceramics and glassware. As befitted their workmanship and status, turned ivory works by August and Wecker, such as a "large cup, turned three on top of the other, lid and base with little 'skeletal' [*durchsichtigen*] crowns," were transferred here after August's death
from other rooms of the *Kunstkammer*. This new space was also home 52
to an expanded repertoire of exquisite ivory *corpora*, characterized by facets filled with ornate, lacework-style ornament that bore only superficial relation to the models from the collection of mixed-mathematical instruments: "1 transparent ivory body, known as an octahedron. A transparent pyramid on all sides above a transparent, circular base made from bone. It was fastened by master Anton Örtel, carpenter, on June 21, 1620; 2 angled, transparent regular bodies from ivory, with double tetrahedra set one inside of the other, with a transpar-
ent base." The degree of embellishment here signified the elevation 53
of the commonplace study of perspective and geometry to a paragon of technical achievement.

Unlike the numerous images of polyhedral models circulating throughout Germany (including those in August's library), the ivory polyhedra found on *Säule* were aimed less at creating completely novel forms than at increasing degrees of embeddedness, the elegance of wrapping one body inside the other, the surprise of concealing seemingly irreconcilable polyhedra, and the degree of thinness to which ivory could be pushed. These polyhedra were typically symmetrical and nested one inside of the other, a constellation that according to Quiccheberg caused in viewers "a not trifling delight [to
be] experienced." Ingenuity was communicated not solely through 54
conjuring provocative new perspectival *irregulata* expertly rendered on paper, but, given the innate three-dimensionality of turned ivory, in the creation of objects that confounded the optical logic of the third dimension. While today's visitors to the Dresden palace or *Residenzschloss*—the jewel of the Staatliche Kunstsammlungen—may often be puzzled by such seemingly unique displays of geometrical virtuosity, the *Säule*, as a genre of craftsmanship, embodied the capacities of late sixteenth-century artisans to manufacture decorative art objects "straining the limits of tool credibility," their seeming

Georg Wecker (1566–1633). Skeletal polyhedron. 1581. Bpk/Staatliche Kunstsammlungen Dresden (II 290). Photo: J. Karpinski.

impossibility heightened further by the profound curiosity they
55 engendered. Working within extreme material constraints to produce maximal effects at a miniature scale, the industriousness, labor, and dexterity of the turner found expression through the production of "corpora impossibilia." Spatial in a way Stöer's drawings could never be, composed of a layered interiority Dürer's developed surfaces had conscientiously abandoned, the striking detail and relentless geometricality of ivory *corpora* were designed to flood the dispassionate gaze with insatiable curiosity.

Corpora in any material, precious or otherwise, had never been the ultimate aim of the German polyhedrists, for whom irregularity was primarily a graphic-perspectival pursuit and physical models most often reserved for the regular solids. Turning intricate follies in ivory, on the other hand, was an exclusive branch of artisanship in its own right, defined by the court turners' mastery of polyhedral geometry but also by the inimitable capacities of cutting at high rotational speeds: an ontology of the unattainable manifested in forms that strove to defy explication and concealing a plurality of worlds hidden in miniature. Their manufacture of variations on the theme of mutually enclosed bodies and assembly of parts into one fantastical whole ultimately relied on a matrix of highly specialized tools and drawings. It was also about learning how to perceive the act of turning in time and space—the inherent tactile pleasure, the almost magical appearance of emerging silhouettes, the body's attunement to the precise changes in pressure, and how the variety of tools at the turner's disposal required precise imaginative leaps in order to anticipate the effect they were designed to have upon material.

A section of ivory would first have been sandwiched between the lathe's two centers and fixed in place so that it could rotate at a high velocity. As the ivory would not have come into the *Drechselkammer* as perfectly symmetrical columns of material, the tusks would first have been split into segments that could be loaded into the lathe and the curvature of their profile made constant by applying pressure along its entire length, relative to the overall shape of the raw
56 material itself, until the piece was reconstituted as a smooth cylinder. The work of preparing the ivory for further articulation might well have been undertaken by apprentices, who would have stopped the lathe periodically to check on the progress of their smoothing process

Tools used in lathe turning. *Inventar der Drechselkammer*, 1622, fol. 11 recto. Staatliche Kunstsammlungen Dresden.

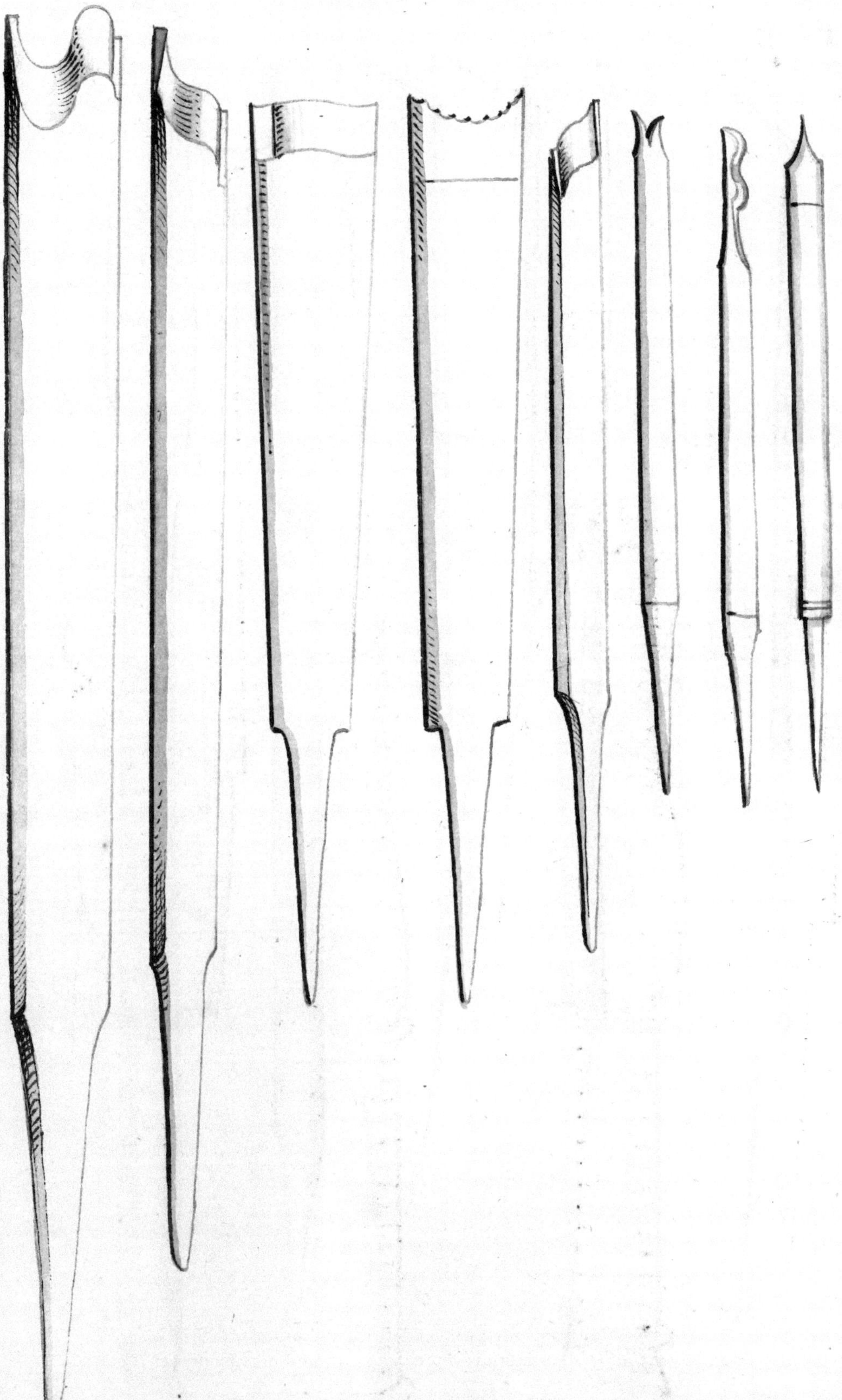

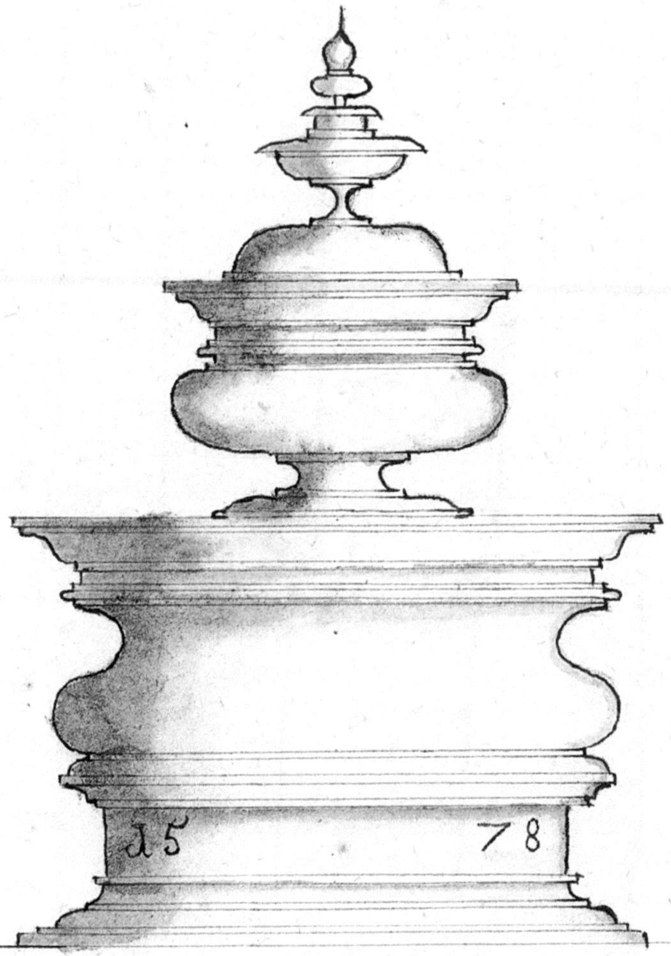

Drawing of a turned ivory vessel (1578), enclosed as a loose sheet with the *Inventar der Drechselkammer*, 1684, fol. 24 recto (drawing no. 4). Staatliche Kunstsammlungen Dresden.

since too much pressure would require them to even out the rest of the material to the same width. As slight changes in pressure created difference, a trained sensitivity of touch was important. Thus it was easier to proceed with caution from the outside—the original shape of the material—toward the creation of a stable core and rely on feel and the sound of the tools against the material to confirm how close one was to achieving a centered and constant base form.

The first *Inventar Drechselkammer*, compiled by the Dresden
court turner Johann Wecker on June 24, 1622, describes all the uten-
sils, accouterments, and machinery that were present in the work-
shop, including the number and contents of the workshops' many
drawers and also several *Drehe Bank* (lathes) and *klein Drehebänklein*
(miniature lathes). As techniques for turning nested polyhedra or 57
contrefait spheres were not described in early turning manuals and
there are no *Lehrbücher* to guide the learning of the required tech-
niques, these tools must have been guarded as proprietary secrets of
the turning trade. Incisors of noteworthy use were documented as 58
scaled drawings, preserving a repertoire of potential effects signified
by the edges of the tool themselves. As with the mandrils, a master
turner would have been intimately familiar with the effects wrought
by each tool and would have been able to imagine the shape it would
carve when impressed against swiftly rotating ivory.

Besides listing the accumulating contents of the *Drechselkam-
mer*, a later inventory from 1684 lists a series of hardline elevation
drawings of turned ivory vessels completed in a light-blue water-
color wash and dated 1578. Matching workshop tools dedicated to
cutting profiles with the highlighted profiles of the turned vessels
reveals the invisible choreography of tactile impressions used to
build up the impossibly articulated edge conditions of an ivory ar-
tifact. Though few if any working drawings exist, turners special- 59
ized in composing and mapping hardline designs for curved surfaces
in ivory. The drawings evidence the presence of pin pricks, which,
much as with Stöer's drawings, alludes to the use of precision instru-
ments to ensure that the contours of form would be absolutely sym-
metrical. Hardline elevation drawings composed at 1:1 scale such as
these would have been integral both as measurable visual references
for the turner used in the process of manufacturing *Säule* and as rec-
ords of distinctive works to be preserved for posterity. In contradis-

tinction to the *corpora irregulata* or Pacioli's skeletal polyhedra, the *Saüle*, having been created by turning and not by folding or perspective, directly translated into two-dimensional representations much more easily. There were no tricky vanishing points to compose, no relative diminishing of lengths in paper space. The frontal view of a regular, curvilinear object taken at an orthogonal angle had the capacity to fully capture an object's distinguishing marks and edge conditions, leaving the expression of three-dimensionality to the subjective art of shading at a gradient and not to the precise construction of geometry in perspective.

In order to make a nested geometrical showpiece, the raw ivory surface had to be made regular; tools had to be matched to drawn elevations; the lengths of the various components had to be decided upon and cut and an assembly strategy devised; templates made to measure the accuracy of spheres in order to ensure the regularity of their curvature, all while wasting as little of the precious material as possible. If the outermost appearance of the *Saul* was a sphere, a hemispherical wooden block was used to hold the material in place, whereas if a spiked star or stellated polyhedron was part of the
60 design, the block was square. Once a sphere was turned, the turner would have removed it from the lathe and used a compass to mark its surface with a field of equally spaced points (called constellation
61 points) corresponding to the requirements of the design. The ivory had become, temporarily, a drawing surface of regular curvature upon whose face geometrical calculations were plotted.

After the base geometry was graphically prepared, the ivory was reattached to the lathe. To turn a series of spheres trapped one inside the other, such as those depicted in the limp grasp of Manfredo Settala (ca. 1640) by Daniele Crespi, a turner used a specialized edge at a predetermined distance from the outermost surface of the outermost sphere, while each subsequent instrument, dentist-like, was inserted into the same opening hole in the outermost sphere. A different tool was deployed for each desired internal sphere, essentially grinding away smaller and smaller spheres that fit the curvature of the undercutting tool. To make a spiked star in a sphere, an essentially similar method was used. After drawing the opening holes on a prepared sphere, a curved undercutting tool and a specially made square-edged tool for forming the spikes were used to gently cut away the point of the star that protruded from each opening hole. The base of each spike terminated in a pentagonal base surface, with one surface corresponding to each spike, which was smoothed away using the square-edged undercutting tool. Then the outermost surface of the outer sphere was carefully ground down toward its center point in order to give the impression that the larger star had somehow become trapped within a smaller sphere.

Once these principles are apprehended—that to turn nested geometries specialized tools for each internal surface were deployed, that these tools had to directly correspond to the measurements drawn on an outermost sphere, that 1:1 scale drawings could be used to calculate the cutting angle, which would in turn define the angle of the undercutting tools, and that specialized jigs were used to hold tools at predetermined distances from the outermost sphere—it is perhaps easier to understand how this repertoire of techniques led to a prodigious, if morphologically related, output. To introduce lack of symmetry into the design of the column's capstone polyhedra would have been to fight the innate uniformity of motion of the lathe, which by design cuts toward stillness and regularity, the center of rotation. While in theory it was entirely possible to mark up the sphere with uneven intervals, unevenness may well have been perceived, within the imperative of turning a perfect sphere, as demonstrating a lack of rigor. After all, Euclid's definition of the Platonic solids had limited the set to those five polyhedra able to be inscribed into a sphere tangent to each face of the solid at the center of its face. Besides, objects with uneven curvatures or unequal sides would sit awkwardly within one another, making it exponentially more difficult to turn further interior shapes, which would have multiplied the lack of precision and inexactitude. By sticking primarily to the geometrically determined subdivisions of the regular and semiregular solids and their derivatives, including also *contrefait* spheres punctuated by star polyhedra, a turner gave himself the best chance of accurately anticipating where the material would be located when it came time to create further interior shapes.

Turning ivory at this, the highest of levels, demanded a sophisticated understanding of the relationship between drawing and modeling on curved surfaces that presaged the movement to non-Euclidean geometry. It was in interlinked courtly spaces,

self-conscious incubators of innovation, that ivory was crucially collected in its raw form and through the interplay of sophisticated sets of tools and drawings converted into exemplars of technical expertise. Held at the heart of the Dresden castle, the ivory-turning facilities attest to the weight given to the geometrical knowledge produced by artists and the authority of measurement as critical components in the substantiation of authority, the definition of ter-
62 ritory, and the extending of supremacy over nature. But to sit down at a lathe and dedicate oneself to the pastime of turning an artifact in ivory also called for a calculated dismissal of the ivory's innate and disturbing animality as well as the networks of trade, exploitation, and environmental disregard that stretched from the Indian subcontinent through Europe, down to Africa and back again. Reduced to exoticism, the residues of violence and ivory's animal origins so explicitly demarcated on the West African salt cellars or performed as spectacle in the courtyard of the Dresden palace ever linger, if repressed, behind the *Saüle*'s facade of fragility.

Formulaic, if combinable, the end result was a miraculous composition that strove to appear mysterious and inhuman, a class of art objects designed in hyperrealism as if for a mechanical gaze that could comprehend the vertiginous detail that the human eye could not. Much can be said of Crespi's portrait of Settala, which though it fails to capture the infinite interiorities of the *contrefait* effectively stages the ivory tower in the iconography of caprice; beyond function, beyond want, beyond its African context, an emblem of delicacy for the sake of delicacy. As the mathematician Joseph Moxon (1627–1691) noted in his *Mechanick Exercises* (1703), "some *Turners* … shew their Dexterity in *Turning*, and make others that know not
63 the way how it is done admire their Skill." The seeming impossibility of the nested geometries was their essential appeal. Not merely *Saüle*, these columns were also *Siegesaüle*, commemorating the culmination of highly refined artistic skills reified through the blinded, single-minded focus of a culture prizing precision calculation and manufacturing above all else.

Thus despite their allure, the intellectual and aesthetic framework established by the historical actors carries the risk of ethical recklessness. For inasmuch as the Dresden columns were paradigms of cutting-edge technology and had been conceptualized and

constructed according to the material and epistemological logics of turning geometry in three dimensions, they also carried, and carry today, a barely curbed subtext of cruelty. Months or years before they would be transformed into delicate polyhedral masterworks, the ivory which served as the raw material would have been nothing more than the highly developed upper incisors of an African elephant, grazing in the forests and plains off the coast of West Africa. The tusks that eventually made their way to Europe were likely not castoffs and would still have been attached to the animals' great heads at the moment of their poaching, as the more intact the ivory was when it reached Dresden and the greater its overall girth and circumference, the more creative options were potentially available to the ivory turners at the Dresden court. And though it was only from the eighteenth century onward that accounts of the killing of elephants for ivory appeared in Western travelogues, the basic approaches to cornering an elephant, and its frenzied response to a mortal threat, were similar to methods used two hundred years earlier by African poachers working on commission: traps that drove a lengthy beam-mounted spear between an elephant's shoulder blades; poisoned arrows shot from a heavy bow with "a pull of over a hundred pounds"; pitfalls with sharpened stakes at the base; tree platforms where hunters would lie in waiting, ready to plunge on the unsuspecting animal; assailing the elephant en masse with numerous spears at once; hamstringing it by the severing of its tendon with a sharp blade. Any of these violent strategies, most of which were 64
designed to initially weaken or cripple the animal, would cause it to run frantically through the woods or fields, trampling the vegetation in its wake until finally succumbing to fear, exhaustion, pain, and the killing blow. After the elephant was dead, the hunter needed to use a saw or ax to remove the tusks from the jaw, sometimes chipping away at the jaw's bony socket to extract the entire tusk from its base, which could add as much as twenty inches of the precious ivory to the overall length of the tusk. 65

Removed from the elephant, both physically and conceptually, its body discarded and forgotten on the killing field, the tusks were treated as one contiguous stream of material resources flowing into Europe—a plastic mass of elephant dentine that could be added, subtracted, and recombined. As imprints of sovereigns' rapacious taste

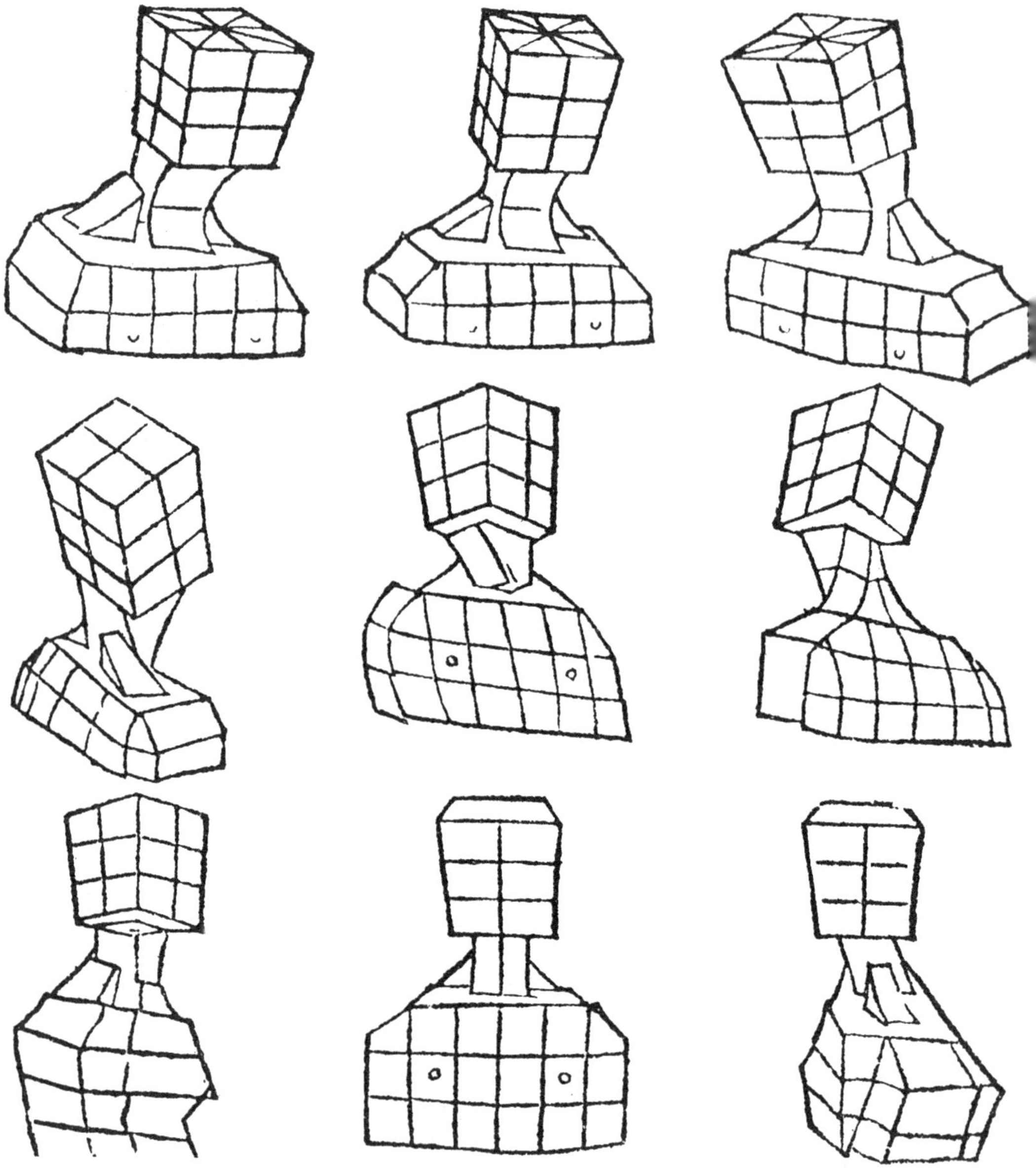

Nine stereometric torsos, from Erhard Schön, *Unterweisung der Proportion und Stellung der Possen …*, second edition (Nuremberg, 1542; [Frankfurt am Main, 1920]). Universitätsbibliothek Heidelberg (M 7134/1, fol. A III recto).

for the ingenuity afforded by hyperintricacy, the luxurious whiteness of the *Saüle*'s surfaces functioned as a new kind of technological ornamentation that, in its aim for a machined grace, sought the classical perfection embodied by the Platonic solids. To use ivory was to preserve a perceptual continuity between the prominent "paper projects" of the century's geometers and their further expression and elaboration as priceless physical models, through skill, precision, and technology to broach likenesses of what had heretofore been an unbuildable graphic reality, in seeming transcendence, through formalism, of nature.

Thus polished, ivory seemed to have uniquely lent itself to the plastic amateriality and acontextuality pursued by perspectival work on geometry. And yet once wrought in ivory, *corpora*, whispering of a nameless anxiety lingering at the edges of the century's polyhedral products, disturb this fantasy of guilt-free plasticity. The irregularity of the *irregulata*, they say, did not and could not evolve without giving off an ambiguous resistance to their own deformed (*contrefait*) and meticulous misshapenness. One has only to look at the stereometric bodies of the Nuremberger Erhard Schön to understand that dislodging geometry from a static, perspectival position had the potential to yield a foreboding affect, that in the gaining and privileging of geometric mastery something human was irretrievably lost, the figures on the page relegated to a state of irresolvable tension in which the eternal returning gives way to signs of strain, distress, and a sense of entrapment. The nine stereometric torsos from Schön's *Unterweisung der Proportion und Stellung der Possen ...* (1538) (Treatise on proportion and the position of poses) hint at this future, our future, one populated with representational and philosophical abstractions flexible enough to accommodate potential shifts in the priorities and targets of power. Analog prefigurations of the dehumanizing, digital gaze, Schön's geometric bodies (*corpora*) remain too abstract for complete identification and yet just human enough for objectification.

The noise of the lathe dies away; the visual interference of the material's inconsistencies gives way to the appearance of a stable object, which would also have gleamed a brilliant white as the wear and tear of the outer surface of dentine was ground down to its gleaming, virgin interior layers. Here, in front of August, once the numerous ivory shavings were swept away, the failures, fragments, the broken pieces, the very messy and physical process of preparation completed; after their transport from Africa to Lisbon, Antwerp, Hamburg and then up the Elbe to Dresden had been erased; the weight of carrying tusks through the *Grosser Schlosshof* and up the stairs to the *Drechselkammer* forgotten; after all the sweat and labor, the exchange of funds, the heat and poaching, the mosquitoes and the bribing, the bartering and the auctions, there would have lain a cylindrical segment, pristine in its simplicity, contextualless, an object so smooth and perfect that it could never have been created by human hands alone. Here would have lain an archetype of a new beginning, free from the aggression of imperfection and its animal origins, a whitewashed surface prepared to receive the knowledge and imprint of geometry.

Daniele Crespi (1600–1630), attrib. *Portrait of the Collector Manfredo Settala with a Hollow Sphere*. 1625. Veneranda Biblioteca Ambrosiana, Milan. Photo: ©Veneranda Biblioteca Ambrosiana/Mondadori Portfolio.

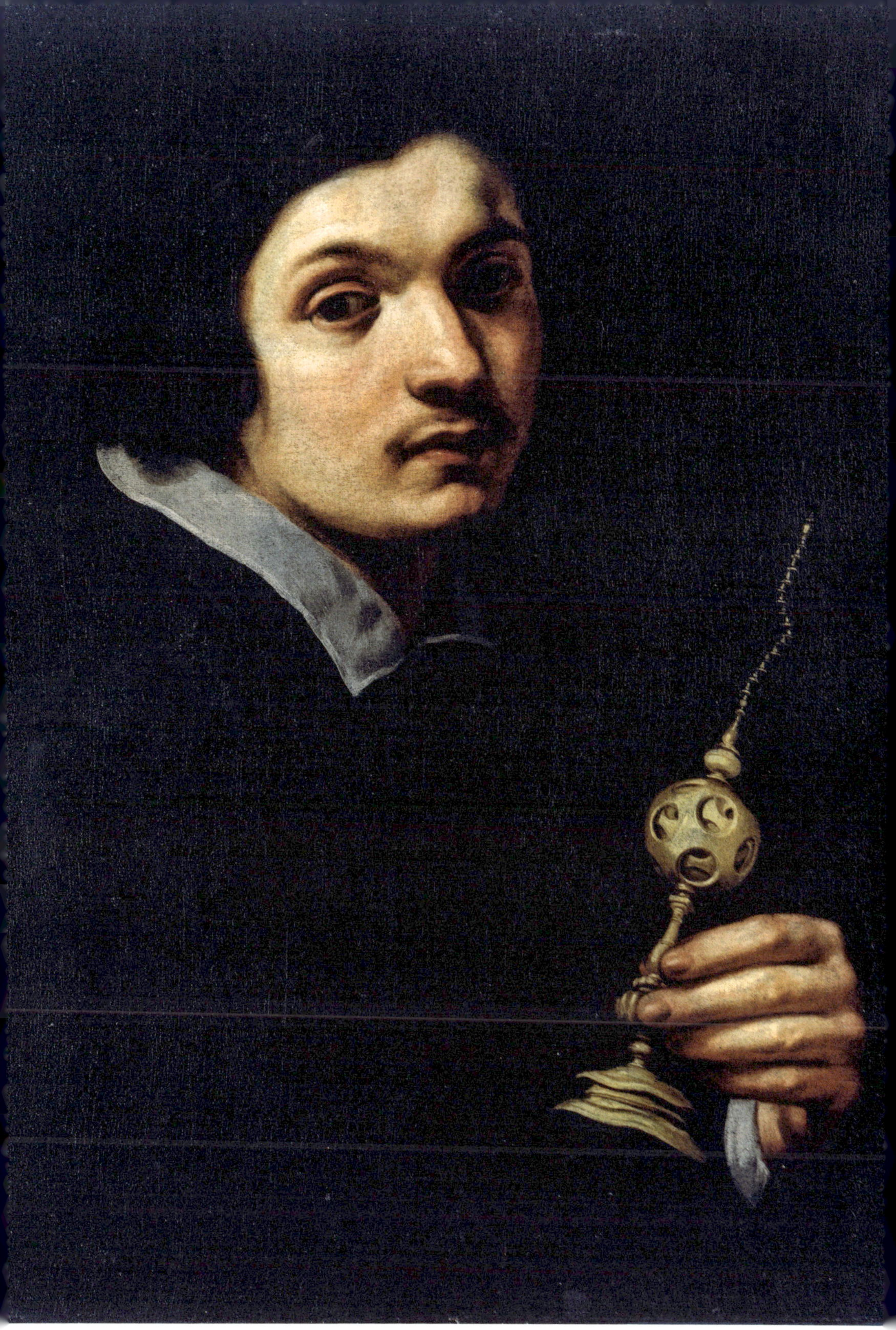

EPIL

COR

IRREG

ET REC

OGVE

ORA

VLATA

VLATA

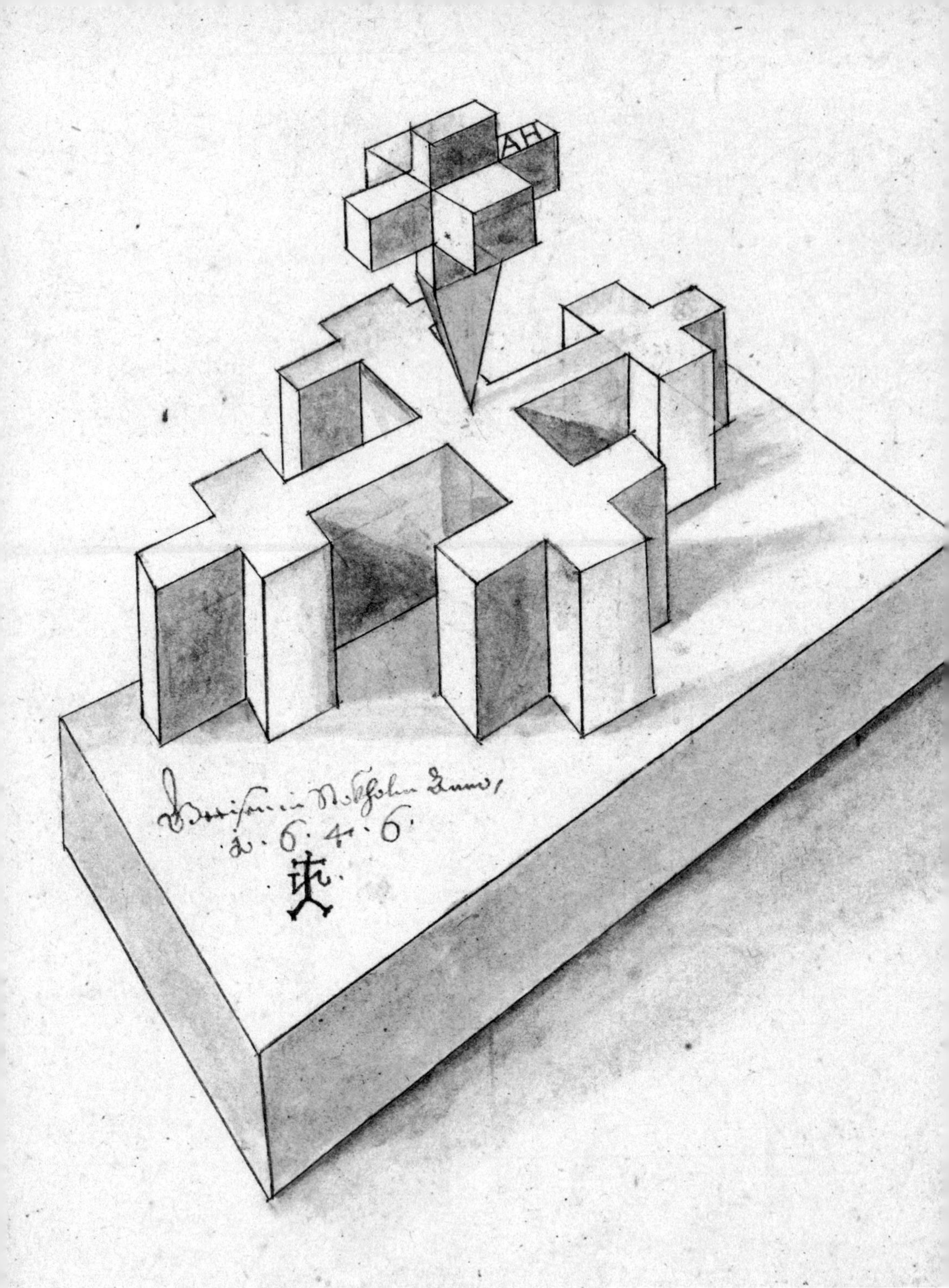

Adam Hörislamb (active 1645–1649). Cruciform geometrical structure. 1646. Statens Museum for Kunst, Copenhagen (KKSgb29469). Photo: J. Skou-Hansen.

So often deemed harmonic or divine and celebrated, like mathematical knowledge generally, for their "insularity from the mundane and messy facts of the physical world," the Platonic solids, as they were experienced and experimented with across the sixteenth century, strained at the boundaries separating "pure geometry" from an aesthetics of collecting and consumption, thus tracking the stirrings of an alternative history of geometrical abstraction that far predates
the modernist art movements of the twentieth century. The objects 1
themselves invite comparisons to Max Bill and Buckminster Fuller, to M. C. Escher and Donald Judd. They resemble the pedagogical models of synthetic or descriptive geometry produced by Gaspard Monge (1746–1818) at the École Polytechnique and Felix Klein (1849–1925) at the University of Göttingen, as well as the mathematical models of Joseph Caron (1849–1925) and Karl Rohn (1855–1920) at the Institut Henri Poincaré, which would eventually be transposed back into the second dimension in the photography of Man Ray and his painting
series *Shakespearean Equations* (1947–1954). The bone-white, polycar- 2
bonate models digitally printed in contemporary architecture practice are readily prefigured by the turned ivory works of Lobenigk, Wecker, and Maggiore.

The prolonged abandonment of the parallel postulate, the caesura that yielded the non-Euclidean geometry of Carl Friedrich Gauss (1777–1855), Bernhard Riemann (1826–1866), Felix Klein (1849–1925), and other nineteenth- and early twentieth-century mathematicians, would pave the way for a new hyperbolic, elliptic, and kinematic geometry. Einstein's towering achievements in twentieth-century physics loom on the horizon, including his 1915 general theory of relativity and its reframing of gravity as a curved space par excellence. In spite of their differences, Renaissance geometry was no straw man of Euclidean stability, trapped in the transcendent, proportional prison of Vitruvius. To believe so is to elide the vibrancy of geometry as it was understood and practiced by early modern artisans. A continuum of fluid forms, unstable, in flux, mapped onto and off of curvilinear surfaces and unfolding in a perpetual state of transformation, the *corpora* were the premier experimental subjects for an evolving geometricality pioneered by those with the hard-won skills to excel with pen, on copperplate, in ivory, and through the piecing together of tesserae from valuable woods. Their struggle to invent and explore

new geometrical forms foreshadows the pains to visualize a non-Euclidean geometry in the nineteenth century. Representing equations by the singularities they could produce was by no means self-evident, and despite the fact that non-Euclidean geometry would eventually open up dimensions beyond our experiential capacity, it owed its initial manifestation to visualizations conceived as models in a Euclidean space. It would take the Italian mathematician Eugenio Beltrami (1835–1900) to successfully conceptualize a surface of constant negative curvature as the rotation of a tractrix curve about its
3 axis before non-Euclidean geometry could begin to be legitimized. And Klein's highly influential theories on geometrical classification were predicated on *Anschaulichkeit* (spatial visualization), namely that the apprehension of surfaces in physical form was fundamental to
4 "knowing" and thus to the evaluation and doing of mathematics.

Despite the fact that the event of polyhedral geometry would only arrive as proto-topology (or "geometria situs") via Euler, by the last third of the sixteenth century mathematicians had already started to reincorporate irregular bodies, some of which had first been visualized by artists like Dürer or Jamnitzer, into the established format and logic of the Euclidean proof. In the third book of his *Problemata geometrica* (1583), Simon Stevin both truncated the Platonic solids and explicitly relied on Dürer's folding methods and patterns in *Underweysung* to derive a theory of the regular solids, and of several irregular solids which Stevin called "augmented regular solids" (*auctores regulares*) and
5 "truncated regular solids" (*truncata regularia corpora*). Stevin, like the French bishop François de Foix-Candale (1512–1594), René Descartes, and Euler, attempted to communicate the construction of these bodies in the languages of geometry and later algebra, outlining the proportional rules that described Dürer's (convex, semiregular) poly-
6 hedra by truncating regular solids at each of their planar vertices. Stevin concluded by further detailing how to form models of his creations. Deeming it superfluous to describe the process of folding the solids themselves, since "this matter is sufficiently known," Stevin related how to construct his augmented solids by joining or gluing folded-up pyramids in paper, wood, or metal to the face of each
7 regular polyhedron made from the same material.

Foix-Candale, for his part, added a new, sixteenth book to Euclid's *Elements* as well as a short tract on two of the semiregular

solids. Published in Paris in 1566 and translated in the famous
English edition by Billingsley introduced by the polymath John Dee's
Mathematicall Praeface, it covered mutually inscribed regular solids, as
well as two of the Archimedean (semiregular) solids, and may have
been known to Descartes, who himself produced a short essay of
algebraic solutions for the semiregular solids in *De solidorum elemen-
tis*, preserved only in a copy made by Leibniz and published posthu-
mously in the mid-nineteenth century. Descartes had also worked 8
with Johann Faulhaber in Ulm around 1620, the mathematician
whose *Newe geometrische vnd perspektiuische Inuentiones* (1610) detailed a
view of the regular solids as models hanging in his workshop. 9

Though it would not be until Euler's polyhedral formula,
proved by Adrien-Marie Legendre (1752–1833) and subsequently en-
larged to account for star-convex polyhedra, that a general study of
all polyhedra would be successfully accomplished, thus introducing
an algebraic ordering into what had become a mass of diverse and
vagariously classified solids by the late eighteenth century, the in-
stitutional impulse to classify and to separate what was and wasn't
polyhedral in nature had already emerged two centuries prior when
the multipronged engagement with and consumption of polyhe-
dral geometry by artists, artisans, and mixed-mathematicians was
at its zenith. Hirschvogel, Lencker, Stöer, Lobenigk, and many oth- 10
ers persisted in their use of the solids to generate knowledge in the
form of form, concerned neither with transcribing repeatable rules
of transformation that connected regularity to irregularity nor with
the impulse to order their creations by anything other than the most
tenuous of visual connections to their Platonic foundations. It was
this apparent disregard for mathematical description that enabled
later mathematicians to disregard the body of knowledge as a whole,
reducing the output to singularities and quibbles over which artist,
astronomer, or natural philosopher had "discovered" which solid
first. Thus, when Henri Lebesgue (1875–1941) claimed that "the first
important notions in topology [*géométrie de situation*] had been ac-
quired in the course of the study of polyhedra," he was not referring
to the early modern polyhedrists or courtly turners but to a truncated
lineage running from Descartes through Leibniz to Euler and on to
early topologists like Listing, Möbius, and Jordan, all of whom aided
in the proof and extension of topology as a branch of mathematics. 11

But the preponderance of emphasis on the "modelness" of geometry in the nineteenth and early twentieth century, indeed the notion of preserving the properties of objects under stress and deformation that would become the central tenet of topology, ties directly back to the strides made three hundred year earlier, when Renaissance artists first dared to inject *techne* into Euclid. They did this for different reasons, not because it was easier to communicate equations to students as outlandish shapes that would stick in the mind or to precisely map topological equivalences, but because they were object makers whose livelihood depended on their ability to make their wares and skills seductive. Even if they had had the capacity there was no reason for accompanying formulae, no reason to map *corpora* back into in the abstract language of mathematics once they had been extracted from the sanctity of *Elements* and made visible as bodies in their own right. Forcing the unreal to appear as real as possible, so real that it could almost be picked up and held, or indeed making the real utterly incomprehensible and incompatible with spatial perception was a calculated announcement of artistic daring that pulled the focus back to the objects and their creators, almost presaging the phenomenological manifesto articulated by Husserl, "to the things themselves."

If the early modern investment in polyhedral bodies were to be extrapolated to its utmost extent, a project might have been envisioned like the fledgling cosmology of Johannes Kepler, one of the founders of modern physics and the astronomer responsible for discovering the three laws governing planetary motion. Raised in obscurity and without the professional opportunities brought by noble birth, Kepler's feverish letters reveal in painstaking detail his belief that his career, if not his very survival, was fully dependent upon finding favor with a sympathetic patron, certainly in his early postgraduate years when his reputation had not yet been established. To rise above the morass of scientific theories, tractates, and pamphlets; to strike out on a career outside of the university and make a name for himself in real time, it was, in short order, essential to get noticed. Before Kepler became court astronomer to Rudolf II; before his collaboration with Tycho Brahe, his correspondence with Galileo and feud with Robert Fludd, before he successfully defended his mother in her trial against charges of witchcraft, before he published science fiction populated with extraterrestrial beings, before the premature death of his children, before, in other words, Kepler's life attained its extraordinary and tragic dimensions, lay only the yawning uncertainty of his future. And it was to secure as much as possible a route through the uncertainty that the young Kepler, only twenty-five at the time, published with the help of his former professor at Tübingen, Michael Mästlin (1550–1631), a bold, if eccentric, first work, *Prodromus dissertationonvm cosmographicarvm, continens mysterivm cosmographicvm* ... (1596) (hereafter *Mysterivm cosmographicvm*).

No mere heliocentric diagram composed of schematic lines that advertised a hypothetical nature, Kepler's cosmological system was depicted on a foldout page of *Mysterivm cosmographicvm* as a polyhedral model in the manner popularized by the century's artisans. The immediate aim of the book was to establish his reputation as an astronomer by impressing his local sovereign, Duke Friedrich I of Württemberg, who would also receive from Ludwig Hohenfeld that same year the distinctive polyhedral sundial embellished with Ovidian themes. But the physicality of the representation, the cross-hatching and shading, and the inclusion of an ornamental base in the style of Stöer, signaled most of all that the assertion of the theory's veracity was predicated upon an appeal to a phantom object lying beyond the image. The use of objecthood as a shorthand for realness while relying on realness to communicate the veracity and worth of geometrical knowledge was a rhetorical tactic and philosophical position, diametrical to the Cartesian *cogito*, borrowed directly from the polyhedrists. Namely, if a thing could be represented as a convincing object then it existed, and if it existed, it could speak to the truth of its own existence. 12

Kepler's youthful epiphany in *Mysterivm cosmographicvm*—one that he would return to throughout his life—conjectured that the intervals between the six known planets could be accounted for by nesting the Platonic solids one inside the other, or as he would later enumerate, "Cubus, inter et Pyramis, inter et Dodecaedron, inter et Icosahedron, inter et Octaedrum." Evocatively representing the space 13
between the planets as skeletal Platonic solids, Kepler's polyhedral model portrays the fuzziness of the existing observational data on the planetary orbits as a series of embedded half-spheres, the thickness of whose rims was supposed to account for the data's variability.

Hence the model's purpose was not to represent what *could* be seen, but rather to represent what could *not* have been seen, namcly the proportional distance between the planets' positions (the planetary intervals) while accommodating the relative indeterminacy of late sixteenth-century orbital positioning data, which Kepler in one of his great triumphs would later come to formulate, accurately, as being elliptical in constitution.

But before he could come to this understanding, it was compulsory for Kepler to pass through what had been a classical paradigm of harmonic proportion and geometry, ingeniously reinterpreted in his polyhedral cosmos albeit one already superseded by a century of polyhedral experimentation. In an early letter addressed to Mästlin on October 3, 1595, Kepler unequivocally rejected the possibility of structuring the intervals between the planets with any *corpora* other than the five regular solids. "Remove the irregular bodies [*irregularia corpora*], namely on the foundation that creation is of the most
14 well-ordered nature." This schism between a divinely centered and symmetric geometrical cosmology and the abstract legacy of more fluid artistic inventions plays out in the diagram of the model itself, a transalpine and temporal hybrid divided as it is between nested Platonic solids in the mode of Pacioli's *De divina proportione* resting upon a plinth made from the kind of *Rollwerk* that could easily have complemented any of Germany's irregular, post-Dürerian *corpora*. Moreover, despite the enduring effectiveness of the image in communicating the essence of his new vision of the universe, Kepler contended that works of art incorporating polyhedra were too commonly produced to be special enough for the *Kunstkammer* object and courtly gift he had envisioned being produced for Duke Friedrich to accompany the publication of *Mysterivm cosmographicvm*. In initially rejecting the graphic model as an insufficient blueprint for the ambitious object, an astronomical exhibit/table fountain/drinking cup he called a "Credentzbecher," Kepler dismissively posed the rhetorical question of whether there was anyone "who had never seen those five mathematical bodies [the Platonic solids] set in large or small
15 half-divided globes?"

The answer of course was "no." By the late sixteenth century, it would have been almost impossible to visit a self-respecting *Kunstkammer* in Europe without encountering one of the solids in any of

their myriad expressions. Inventory records for Duke Friedrich alone
reveal that he collected many and "various turned ivory things" (*aller-
lej gedrehte Helfenbainene Sachen*) in addition to a robust repository of
perspectival and polyhedral literature. Had they been aware of it, the 16
princely accession of these books would have been all the more satis-
fying for their authors, whose belief in the untapped potential of the
knowledge gleaned from rigorous craftsmanship had never been
truly embraced by the university. Yet though they were forced to
clamor from the outside for the value of their experiments, to self-
consciously inflate their work with humanist rhetoric, or to ignore the
academy altogether, the seeds of change that would lead to Monge,
Beltrami, Klein, as well as to Descartes and Euler, had been planted.
While it would not be until later in the seventeenth century that
approaches lifted straight from the workshop environment and the
Lehrbücher it had produced would become regularized in institutional
pedagogy, a vista had been cleared toward reforms in mathematical
instruction that would stress the importance of learning through
experience and experiment, rather than from rote memorization. 17
Professional academies would progressively enforce the acquisition
of knowledge by tracing an arch from the concrete to the abstract in
order to connect the world, with all its variation, to the mapping of
its underlying principles. 18

The geometrical restrictions upon Kepler's "most well-ordered
nature" introduced an ambiguous scalar tension absent from the ear-
lier polyhedral models though already present in Plato's *Phaedo*, which
described the earth "when looked at from above ... [as] in appearance
streaked like one of those balls which have leather coverings in twelve
pieces." Plato knew full well, as did the nineteenth-century pioneers 19
of non-Euclidean geometry, that in broaching ideas or beliefs that
could not be apprehended because they lay beyond the dominion of
human perception, it was indispensable to work through the mate-
rial and spatial, to appropriate tactics from familiar and observable
environments in order to develop new directions for thought. Thus,
in magnifying the existential scope of the solids to encompass the
structure of the known universe, Kepler also collapsed the universe
in on itself, turning it into a covetable and precious paper object ap-
propriate for display within or upon the ornate surface of a Renais-
sance cabinet. This dual act of compression and expansion was not

without a palpable sense of the vertiginous claustrophobia that came from being on the edge of the unknowable. Kepler's polyhedra, as complete as they purported to be in exposing the geometrical logic underpinning reality, were also the very leitmotif that would render the cosmological theory so imminently discardable.

Indebted to a fiction of stability stretching back to antiquity, the spatial regularity that defined the Platonic solids in Euclid and would define Kepler's opening foray into the public eye was essential to preserving the harmony of cosmic order if not also a measure of order in Kepler's own life, a life that would continue to unfold in a state of perpetual anxiety and professional uncertainty viscerally preserved on the margins of his monumental accomplishments and in his correspondence with family, colleagues, and patrons. Opening up the possibility of enmeshing the cosmos in the limitless expanse of untamable multiplicities and abstract geometrical experimentation pioneered by European polyhedrists threatened to undermine the very organization of existence that Kepler had sought to establish. What Duke Friedrich needed to establish his notoriety and political influence, indeed what Kepler himself needed in that moment, was the promise of stability that had become ever more fleeting with Dürer's unfolding of the Platonic solids. Irregularity did not only apply to the countenance of shapes, regardless of formal symmetry, but to the assumptions embedded in their unregulated and perpetually shifting formations. Kepler's insistence on removing the irregular bodies was an offer to preserve the fiction of cosmic order in a rapidly expanding and unbridled world, and to keep disorder at bay, if only for a little while longer.

RHOMBICOSIDO-DECAHEDRA

1 Varying mathematical definitions of polyhedra, sometimes conflicting or ambiguous, have been offered across different contexts and timeframes. Grünbaum, "Polyhedra," 43. For the sake of simplicity, I have used the words polyhedra/solids/bodies/*corpora* interchangeably to refer to the general class of regular and irregular polyhedral geometries as they were understood and explored by Renaissance artists.

2 See as further examples the portrait of the engineer Agostino Ramelli (1531–ca. 1600) from *Le diverse et artificiose machine* (1588), and the frontispieces to Jean Cousin's *Livre de perspective* (1560) and Peter Apian's *Instrument Buch* (1533).

3 On the Codex Forster and polyhedral geometry, see Kemp, *Leonardo da Vinci*, 109–112.

4 Weisstein, "Regular Polygon."

5 *Mathematicae mixtae*, often set in opposition to *mathematicae purae*, encompassed a set of practices produced from a field of diverse interactions between scientific, technological, and social areas, each with its own tangled history. Epple, Kjeldsen, and Siegmund-Schultze, "From 'Mixed' to 'Applied' Mathematics," 658. For the sake of simplicity, I refer to this broad epistemological category as "mixed mathematics," though an equally valid term might be "practical mathematics." For further exposition, see Brown, "Evolution of the Term 'Mixed Mathematics"; and Bennett, "Category of 'Mixed Mathematics.'"

6 See Laue, *Perspectiva*, for discussion of the allegorical themes and the historical illustrations on the Newbattle Abbey Casket.

7 Andreas Schöner's *Gnomonice* (Nuremberg, 1562) contains a sundial in the shape of a rhombicuboctahedron net that indicates the locations for gnomons and is decorated with rudimentary hour lines on each face. See the foldout image after page 30 in *Gnomonice* and Welper, *Neu-vermehrte Welperische Gnomonica*, 22. See also the gilt polyhedral sundial attributed to Nicolaus Kratzer (ca. 1487–1550), royal astronomer to King Henry VIII of England (1491–1547): Museum of the History of Science at Oxford, Lewis Evan Collection (54054). Three wooden polyhedral sundials (one dodecahedron [1587] and two truncated octahedra) by Stefano Buonsignori (d. 1589), cartographer and court astrologer to Francesco I de' Medici, are richly colored affairs, accented in gold leaf and each containing a depression to hold a magnetic compass for orienting the dial toward the local magnetic meridian. For the Buonsignori sundials, see Museo Galileo, Florence (2456, 2458, and 2459). The museum also holds an anonymous, seventeenth-century polyhedral sundial in the shape of a truncated tetrahedron (3261).

8 On the Hohenfeld sundial, see Maes, "Entzerrung der Sonnenuhr"; and Eichholz-Bochum, "Polyeder-Sonnenuhr."

9 Bennett and Johnston, *Geometry of War*, 6.

10 Dürer was also the first to print basic instructions for constructing paper polyhedral dials, which he did in *Underweysung der Messung*, and in all likelihood the fashion for sundials shaped like polyhedra was directly related to the prominence the solids enjoyed.

11 "Vnd will nun fort faren / vnnd zu dem andern theyl greyffen / darinn du mancherley schöner Corpora vnd Kugel auß ligenden grundt zusamen flechten vnnd schliessen (Recht als ein Spinnire Netz / von einem langwirigen faden / jhre behausungen jneinander zeucht vñ flicht)." Hirschvogel, *Grundtliche anweysung*, last page of first chapter (unpaginated).

12 Among the many examples are *Der Rhein (Melancholia)* (1982–2013) by Anselm Kiefer and *Untitled (Divine Proportion)* (2006) by Ai Weiwei.

CORPORA REGULATA ET IRREGULATA

1 On the pre-Euclidean history of the regular polyhedra, see Malkevitch, "Milestones," 53–55.

Polyhedral model of the universe, from Johannes Kepler, *Prodromus dissertationvm cosmographicarvm, continens mysterivm cosmographicvm* (Tübingen, 1596).
ETH-Bibliothek Zürich (Rar 1367: 1).

2 That is, there are only five regular, convex polyhedra whose faces are made from equal, regular polygons meeting in the same way at every vertex of the solid so that the entire polyhedron is inscribable within a sphere.

3 Vasari, *Vite* (1997), 73. Translation by author. For the Neoplatonic elements in Vasari's concept of *disegno*, see Rubin, *Giorgio Vasari*, 241–242. On theories of *disegno*, see Barzman, "Perception."

4 For further reference on the spread of editions of Euclid, see Andersen and Bos, "Pure Mathematics," 710–711, n. 24.

5 Zamberti worked directly from the Greek into Latin without any of the Arabic intermediaries. His poor grasp of mathematics prompted Luca Pacioli to revise the Campanus edition and republish it in 1509. The 1572 translation of *Elements* by Federico Commandino (1509–1575) became the definitive edition and remained in use until the nineteenth century. Cromwell, *Polyhedra*, 108. On the centrality of Euclid in the Middle Ages, see Mackinnon, "Portrait," 147; Folkerts, "Euclid."

6 On Gessner, see Blair, "Humanism and Printing."

7 The small stellated dodecahedron was the first of two regular concave polyhedra—also known as "Kepler-Poinsot solids"—to be discovered by Renaissance-era artists. The second is the great stellated dodecahedron from Jamnitzer's *Perspectiva corporum regularium*. Burt et al., "Beyond Archimedian Solids," 682. On the four Kepler-Poinsot solids including the great and small stellated dodecahedron, see Field, "Star Polyhedra"; P. Gruber, "History of Convexity"; and Brigaglia, Palladino, and Vaccaro, "Historical Notes."

8 For Vasari's comments on Pacioli copying Piero della Francesca, see Vasari, *Le vite* (1966–1987), 3:258 and 264.

9 Field, *Piero della Francesca*, 120.

10 *Libellus de quinque corporibus regularibus* details six Archimedean solids created through truncating the Platonic solids. Piero della Francesca, *De corporibus regularibus*, 556, and Davis, *Mathematical Treatises*, 58.

11 Two quantities are in "golden ratio" if their ratio to each other is the same as the ratio of their sum to the larger of the two quantities.

12 Their modern monikers are the truncated cube, truncated tetrahedron, truncated dodecahedron, truncated icosahedron, truncated octahedron, truncated cuboctahedron, truncated icosidodecahedron, cuboctahedron, icosidodecahedron, rhombicuboctahedron, rhombicosidodecahedron, snub cube, and snub dodecahedron. Each Archimedean solid may be generated by evenly manipulating the faces of Platonic solids, through truncation, expansion, and cantellation.

13 Barbaro, *Pratica*, 56. It was not until the astronomer Johannes Kepler's *Harmonices mundi libri V* (1619) that all of the Archimedean solids were published and illustrated together for the first time. Field, "Archimedean Polyhedra," 242, n. 1.

14 Kemp, *Science of Art*, 62.

15 The manuscript versions of *De divina proportione* contain one extra image not present in the printed version, making the count of geometrical bodies sixty in total. In the Geneva codex it is titled "superflua ex errore" and in the Biblioteca Ambrosiana codex "pyramis laterata exagona vacua." Biggiogero, "Life and Works," 12. For a comparison of the various manifestations of, and precursors to, *De divina proportione*, see Pacioli et al., *Antologia* (2010). See Pacioli, *De divina proportione* (2009), for reproductions of the Milan and Geneva manuscripts.

16 Gnann, *Parmigianino*, 1:187–188. The Archimedes was printed with five colors and was probably based on a design by Raphael. Gnann, *Chiaroscuro*, 64. For further references on Ugo da Carpi's chiaroscuro prints, see chapter 4 in Gnann, *Chiaroscuro*.

17 Tormey and Tormey, "Renaissance Intarsia"; and Grasselli, "Piero della Francesca," 196.

18 Biggiogero, "Life and Works," 8.

19 Kotrč, "Dodecahedron," 217–219. See also Duhem, "Plato's Theory of Space."

20 Plato, "Timaeus," *Dialogues*, 3:475.

21 Plato, "Timaeus," *Dialogues*, 3:478, commentary by Jowett.

22 Ficino, *All Things Natural*, 94. On the relation between the Platonic solids and atomism, see Lucretius, *De rerum natura*, first printed in Lombardy in 1473. For further reading, see Palmer, *Reading Lucretius*.

23 On the composition of the triangles which make up the elements in the *Timaeus*, see also Artmann and Schäfer, "'Fairest Triangles'"; and the commentary on the *Timaeus* in Cornford, *Plato's Cosmology*.

24 Plato, "Timaeus," *Dialogues*, 3:474–475.

25 Allen, "Ficinian *Timaeus*," 238–239. The first edition of *Compendium in Timaeum* was published in Ficino's *Platonis Opera omnia* (1484), while the final version appeared in his *Commentaria in Platonem* (1496). This latter version was republished in the subsequent editions of *Platonis Opera omnia* of 1561, 1576, and 1641. Allen, "Ficinian *Timaeus*," 247, n. 2. Manuscripts of Calcidius's translation and commentary on Plato's *Timaeus* have been documented in most important libraries of Renaissance Italy. See appendix I in Hankins, "Study of the *Timaeus*," for a complete list of these manuscripts.

26 Ficino, *All Things Natural*, 93.

27 For the widespread influence of Platonism on the arts of the Italian Renaissance, see Ames-Lewis, "Neoplatonism"; Cheney and Hendrix, *Neoplatonic Aesthetics*; *Hendrix, Platonic Architectonics*; and Cheney, "Giorgio Vasari's Fine Arts."

28 Aristotle, *De caelo*, 30.

29 On knots in Leonardo and Dürer, see Bambach, "Leonardo, Tagliente, and Dürer."

30 Kemp, *Science of Art*, 77.

31 "Als ich in meiner Jugend durch wunderliche mittel / vñ fast von mir selbsten / ohne einig Geometrisch Fundament in wissenschaft der fünff corporum regularium (derer halben Euclides seine Elementa Geometrica vnd Arithmetica geschriben) gelanget / dieselbe lusts halbẽ / auß Papier / Holtz / vñ Stein schnidte." Schwenter, *Geometriae*, "Vorrede an den Leser."

32 Syndram and Minning, *Die Inventare der kurfürstlich-sächsischen Kunstkammer*, [vol. 3], *Das Inventar von 1640*, fol. 319 recto. The high-quality and heavyweight paper conducive to making court-quality models may have been by Samarkand or Abadi, though there would have been a wide range of less-precious papers available for internal use in workshops.

33 Quiccheberg, *Inscriptiones*, 68, Fourth Class, Inscription 5.

34 Quiccheberg, *Inscriptiones*, 67, Fourth Class, Inscription 2.

35 The caption to the Faulhaber image reads: "Cubus NB. wann der ligende Grund vff das ligende Taffelin gemacht wirdt / so kompt das Corpus vff den Tisch darunder / vñ hinwiderumb / u. Gerbrauch zur Perspectiv." Faulhaber, *Inuentiones*, 32.

36 Gabbarelli, "Solid Mistake," 237. See Gabbarelli also more generally on the iconographic evolution of *Diogenes*.

37 See Popham, *Catalogue*, 2, pl. 109, no. 107, for sketches at the Uffizi relating to the Diogenes print. The studies on the same sheet have been identified as including Diogenes' right hand and arm and left leg. Ekserdjian, *Parmigianino*, 213. The word *giometria* is scrawled on the blank page of the pictured book and stands in as a placeholder for a future dodecahedron that will be added later. For discussion of the relationship between Parmigianino, Caraglio, and Ugo da Carpi, see Ekserdjian, *Parmigianino*, 219–220.

38 Mackinnon, "Portrait," 134.

39 There are numerous examples of polyhedra by Leonardo. See fols. 518 recto, 849 recto, 930 recto, and 1040 recto in the Codex Atlanticus in the Biblioteca Ambrosiana in Milan and MsE 56r in the Leonardo Manuscripts, located in the Bibliothèque de l'Institut de France, Paris. Fol. 930 recto in the Codex shows an icosahedron drawn by Leonardo with a hole drilled into its upper triangular facet through which a string to hang it from has been threaded.

40 See Hart, "Fra Giovanni's Intarsia Polyhedra."

41 Quote as reprinted from a press release for the exhibition "Medici Splendour"

(Florence, 2013). Vasari also relates that installed in the sacristy is "a ball with seventy-two faces made by the goldsmith Piloto, which is very beautiful." Vasari, *Lives*, 9:43.

42 Vasari, *Lives*, 2:131, with reference to Paolo Uccello.

43 The secondary literature on *Melencolia I* is voluminous. In addition to Panofsky, *Life and Art*, and Klibansky, Panofsky, and Saxl, *Saturn and Melancholy*, see more recently Doorly, "Dürer's *Melencolia I*"; Büchsel, *Zeichen und Emotion*; Hoffmann, *Im Zwielicht*; Merback, *Perfection's Therapy*; and Bubenik, "Shape of Things."

44 See Panofsky, *Life and Art*, 161, on the relation of geometrical tools in *Melencolia I* to the geometrical paraphernalia that surround allegories of geometry, such as Gregor Reisch's *Margarita philosophica*. On allegories of geometry generally, see K. Park, "Allegories of Knowledge." The melancholic influence of Saturn may also account for the presence of a globe, or celestial globe, in allegories of melancholy.

45 There has been much debate about the exact definition and form of Dürer's polyhedron. As per Klibansky, Panofsky, and Saxl, the truncated rhomboid in *Melencolia I* refers to "a figure formed of six rhomboids which, owing to the cutting off of two opposite points (those in which the sharp angles of the rhomboid meet), has been transformed into an octahedron." Klibansky, Panofsky, and Saxl, *Saturn and Melancholy*, 400. A summary of some of the disagreements over the polyhedron's precise shape are summarized on 400–402. For further reading, see Lynch, "Geometric Body"; and Schreiber, "New Hypothesis." For a full description of the tools in *Melencolia I*, see Klibansky, Panofsky, and Saxl, *Saturn and Melancholy*, 327–329. The Sächsische Landesbibliothek in Dresden possesses the preparatory drawing of the polyhedron in *Melencolia I*, which is depicted with a floating eye to signify the point of sight used to generate the solid's perspectival angle.

46 "Pour arpenter & a mesurer terre, Voire tous corps & toute Arquitecture, Suiure me doit, l'homme qui veut aquerre Le sçauoir mien, de diuine nature."

47 On the fragility of allegorical structures, see Reitman, "Allegory's Lament."

48 See Gnann, *Parmigianino und sein Kreis*, 125–128; and Ekserdjian, "Two Drawings," 369–372.

INSTRUCTION IN MEASUREMENT

1 Goldthwaite, *Economy of Renaissance Florence*, 198.

2 "Nichts bessers ist, dann kunst auff Erdn, Nichts nutzlichers kan gfunden werdn, Als Kunst: Kunst ist ein trewer Gfehrt, Drumb seind Künstler allr Ehren wehrt."

3 Petz, "Urkundliche Nachrichten," 238.

4 Rupprich, *Schriftlicher Nachlass*, 2:9.

5 Petz, "Urkundliche Nachrichten," 239. Johann Schöner's introduction to *Joannis de Regiomonta de triangulis* (Nuremberg, 1533) also confirms that Pirckheimer had purchased many manuscripts pertaining to ancient mathematics from the libraries of Regiomontanus and Walther; Dürer, *Painter's Manual*, 14–15.

6 For instance, "the foot is one seventh of the entire height. Its height equals one third of its length. The ankle is at the midpoint of the height of the foot. The toes measure a third part of the length of the foot." Dürer, *Human Figure*, 60.

7 "Doch nam ich mein eÿgen ding für mÿch vnd las den Fitrufium, der beschreibt ein wenig van der glidmas eines mans." Rupprich, *Schriftlicher Nachlass*, 1:102.

8 Panofsky, *Life and Art*, 251.

9 Rupprich, *Schriftlicher Nachlass*, 2:10.

10 '(4r: 1523 Januar), ad 13 ditto verkauft wir dem Albrecht Dürer 10 pücher von des Ber enharts Walthers püchern, so den malleren dienstlich sein vnd durch Wilbolt Pirchamer geschetz worden vnd zalt an münz fl. 10'; Abrechnung des Gemeinen Almosen (Nr. 42) für die Zeit vom 2. IX. 1522 bis 14. VI. 1523 im

Nürnberger Stadtarchiv; Rupprich, *Schriftlicher Nachlass*, 1:221.

11 Dürer may have understood the titular *messung* (measurement) to be more akin to geometry, or *geometria*—literally "earth measurement"—given that in 1538 the translation of *Underweysung* into Latin by Camerarius substituted the word *messung* for *geometria*. Dürer, *Painter's Manual*, 10; Rupprich, *Schriftlicher Nachlass*, 1:310.

12 Kavaler, *Renaissance Gothic*, 43; Dürer, *Painter's Manual*, 16. Dürer directly copied certain diagrams from these fifteenth-century books. For instance, figure 16 in Book II of *Underweysung der Messung* is a copy of figure 27 in Roriczer's *Geometria deutsch*. Shelby, *Gothic Design Techniques*, 117. Strauss notes that Dürer adopted the construction of a right angle and the method of locating the center of a circle from Roriczer and the construction of a pentagon from Ptolemy. Dürer, *Painter's Manual*, 16–17.

13 J. Smith, "Dürer as Collector," 17.

14 Rupprich, *Schriftlicher Nachlass*, 1:222, n. 1; and Rupprich, "Kunsttheoretischen Schriften," 236. The 1522 inventory is reprinted in Petz, "Urkundliche Nachrichten," 247–262. A later 1563 inventory is published in Zinner, *Regiomontanus*, 161–168.

15 In his dedication to Pirckheimer in the foreword to *Vier Bücher*, Dürer declares: "In order that these instructions may be better understood, I have issued a book [*Underweysung der Messung*] about measurements which describes lines, surfaces, solid bodies, etc. Without this book my instructions may not be fully understood. It is therefore necessary for anyone who wishes to engage in this art to first be well acquainted with measurement. He should know how to draw the ground plan and the elevation of an object in the matter employed by skillful stonemasons for daily use. Lacking this, he might not comprehend my teachings in every respect." Translation in Strauss, Complete Drawings of Albrecht Dürer, 4:2384.

16 The earliest precursor is the French craftsman and architect Villard de Honnecourt (born ca. 1175), who utilized basic stereometric geometries overlaid on natural forms. The work of Villard survives on thirty-three sheets of parchment in the Bibliothèque nationale de France. Villard draws several stereometric figures composed of triangles as well as simple linear geometry superimposed on top of animals and human heads. See Bucher, *Architector*, 15–39, which also reprints Villard's drawings, as well as the multiple introductory essays in Erlande-Brandenburg et al., *Villard de Honnecourt*.

17 See Rupprich, *Schriftlicher Nachlass*, 3:261, for the original German. Translated in Dürer, *Human Figure*, 208.

18 Rupprich, *Schriftlicher Nachlass*, 1:222.

19 Rupprich, *Schriftlicher Nachlass*, 2:374–377; Strauss, *Complete Drawings* of Albrecht Dürer, 6:2817–2823. Dürer also translated portions of a text on perspectival theory copied from Piero della Francesca's celebrated treatise *De prospectiva pingendi*. "Item prospectiua ist ein lateinisch wort, pedewt ein durchsehung." Rupprich, *Schriftlicher Nachlass*, 2:373. See Panofsky, *Dürers Kunsttheorie*, 42–43; and papers at the British Museum, Sloane 5228/202.

20 "Jtem als jr mir zw sagett, so jr weill möcht haben, wolt jr den Ewklide jn tewczsch bringen, wolt jch geren wissen, ob jr etwas doran gemacht het." Rupprich, *Schriftlicher Nachlass*, 1:113.

21 Dürer's copy of *Elements*, which he bought in 1507, is in the Herzog August Bibliothek, Wolfenbüttel. The inscription in Dürer's hand reads: "Daz puch hab ich zw Venedich vm ein Dugatn kawft im 1507 jor. [I purchased the book in Venice for one ducat in the year 1507.] Albrecht Dürer."

22 "Litterarum quidem studia non attigerat, sed quae illis tamen traduntur, maxime naturalium et mathematicarum rerum scientiae, fere didicerat. Aeque is praecipua ut intelligebat et re explicare noverat, ita et oratione sciebat declarare. Testantur hoc scripta eius geometrica, in quibus quid de illa scientia possit requiri, quatenus quidem tractandam sibi iudicavit, non video." Rupprich, *Schriftlicher Nachlass*, 1:307–308. On Dürer's Latin literacy, see Price, *Albrecht Dürer's Renaissance*, 10.

23 Rupprich, *Schriftlicher Nachlass*, 3:328.

24 Price, *Albrecht Dürer's Renaissance*, 14. For a survey of Dürer's projects for Maximilian I, see Rebel, *Albrecht Dürer*, 305–323.

25 Rostenberg, "Three Nuremberg Patricians," 23.

26 On Dürer's illuminations for Pirckheimer, see Rosenthal, "Dürers Buchmalereien"; and Eckert and Imhoff, *Willibald Pirckheimer*, 84–88.

27 Dürer, *Painter's Manual*, 24.

28 "vnnd auch nicht alleyn den maleren / sonder Goldschmiden Bildhaweren Steynmeßen Schreyneren vnd allen den so sich das maß gebrauchen dienstlich seyn mag."

29 The Delic or Delian problem, also known as "doubling the cube," describes a situation in which the edge of one cube is given and it is necessary, using only a compass and straightedge, to construct a second cube of double the volume of the first cube.

30 Dürer, "Aesthetic Excursus," *Vier Bücher*, Book III, fols. T recto and T ii recto. Translated in Ashcroft, *Documentary Biography*, 873 and 875.

31 "Aber eyn punckt ist ein solch ding / das weder Grôß Leng Breyt oder Dicken hat / Vnd ist doch ein anfang vnd ende / aller leiblichen ding / die man machen mag / oder die wir in vnsern synnen erdencken mûgen."

32 "Wie dañ das die hochuerstendigen / diser kunst woll wyssen / vnd darumb erfûllt keyn punckt keyn stat / dann er ist vntzerteylich / vnnd er mag doch auß vnnsern synnen oder gedancken / an alle end oder ort gesetzt werden / Dañ ich mag mit dem synn ein punckten hoch in lufft werffen / oder in die tyffen fellen / da hyn ich doch mit dem leib nit reichen kan."

33 "Darumb will ich alle ding / die ich in diesem bûchlin beschreib / auch darneben auffreissen / auff das meyn darthon / die iungè zû einer einbildung vor augen sehen / Vnnd dest baß begreiffen." Dürer, *Underweysung*, Book I, fol. A ii recto.

34 On the capacity of *Einbildungen* to be generated from the imaginative faculty as well as from the five "external" senses—vision, hearing, smell, taste, and touch—see K. Park, "Organic Soul," 465–473.

35 "So dañ die bauleût auch maler vnd ander etwan schrift an die hohen gemeûer pflegen zû machen / so thût not das sie recht bustaben leren machen." Dürer, *Underweysung*, Book III, fol. K ii recto.

36 See Bucher, *Architector*.

37 S. Schmidt, "Georg Hartmann," 271. On Hartmann's unfolded paper sundials, see S. Schmidt, *Altered and Adorned*, 77. On the relations between Dürer and Hartmann, see Friedman, *History of Folding*, 55–57.

38 McNamee, "Imagining Solidity," 96–104.

39 Sanders, "Regular Polyhedra," 513. Bovelles's *irregularia corpora* are not discussed in an artistic context but rather in exploring how solid angles might fill the space around an imaginary point. Bovelles, *De geometricis corporibus*, fol. 191 recto, proposition 7. For translation, see Sanders, "Regular Polyhedra," 550–551.

40 Zorach, "Meditation," 329. On Bovelles and the ambiguity of visual representation, see Zorach, "Meditation," 329–339; and Oosterhoff, *Mathematical Culture*, 173–175.

41 Bovelles, *De geometricis corporibus*, fol. 191 verso, proposition 10. Translated in Sanders, "Regular Polyhedra," 554.

42 Panofsky, *Life and Art*, 259. There are nets of all five Platonic solids as well as seven Archimedean solids in the first edition of the *Underweysung* from 1525, and nine in the second 1538 edition.

43 Dürer's use of the word *ganz* is itself a loaded proposition. *Ganz* can also mean exposed; undone, and accessible, implying that the act of opening up the polyhedra makes them more accessible as geometrical forms.

44 "Auch sind noch vill hûbscher corpora zûmachen / die auch in einer holen kugel mit all jren ecken an rûren / aber sie haben vngleyche felder / der selben wil ich eins teyls hernach auf reyssen / vnd gantz aufgethan / auf das sie ein yetlicher selbs zamen mûg legen / welicher sie aber machen will der reyß sie grôsser auf ein zwifach gepabt papier / vñ schneyd mit einem scharpfen messer auf der einen seyten all ryß durch den einen pogen papiers / vnd so dan all ding außdem vbrigè

papier geledigt wirt / als dañ lege man das corpus zůsamen / so lest es sich geren in den rissen piegen / darumb nym des nachfolgeten auf reyssens acht / dan soliche ding sind zů vill sachen nůtz." Dürer, *Underweysung*, Book IV, fig. 34[a].

45 Barbaro was taught perspectival geometry by Giovanni Zamberti, brother of Bartolomeo Zamberti—who published the retranslated Euclid in 1505. Cromwell, *Polyhedra*, 132. For further reading on the influence of Dürer's *Underweysung* in Italy see Fara, "Ricezione italiana," as well as Aikema, *Dürer e il Rinascimento*. Steck's appendix has a small reproduction of a drawing of an unfolded net of a dodecahedron with ornamental polygons, which he attributes to Leonardo da Vinci's work for Pacioli's *De divina proportione*. Steck, *Dürers Gestaltlehre*, pl. XVIII, fig. 67. But no substantiating evidence or identifying information is provided as to the location of this drawing. If it exists at all, the drawing was more likely drawn or inserted into the margins of a copy of *De divina proortione* sometime after the publishing of *Underweysung*.

46 "So ich daforen manicherley corpora wie man die mach anzeigt hab / wil ich auch leren so man soliche gemecht ansicht wie man die in ein gemel můg pringè / zů solichem wil ich das schlechtest corpus fůrnemen / also den wůrffel / darpey anzeygè das man mit allen cȯpern also handelen mag." Dürer, *Underweysung*, Book IV, fig. 51a, fol. O iii verso.

47 Panofsky, *Life and Art*, 245.

48 For Dürer's hands-on role in editing *Underweysung* in anticipation of a second edition, see Andrews, "Personal *Underweysung*," 412–416.

49 Textbooks after Dürer that featured polyhedral nets tended to use perspectival solids as cross-reference images. Andrews, "Personal *Underweysung*," 420.

50 Although there is no specific provenance information on the Albertina woodblocks, they may have come from Ambras Castle in Innsbruck, which still has in its possession the woodblocks belonging to Maximilian for Dürer's *Ehrenporte*. Three of the woodblocks have Andreae's monogram. Fischer, and Sternath, "Archimedean Solids," 467.

51 Woodblock HO 2006/718 is one of the only completely finished woodblocks in the set. It matches figure 37 (Book IV, fol. N ii recto) from *Underweysung*. HO 2006/701 has been copied from figure 38 (Book IV, fol. N ii verso), and HO 2006/722 from figure 40 (Book IV, fol. N iii verso). In this last example, the image on the block has been rotated in comparison to the same image in *Underweysung*. If a block such as this one had been used on the *Underweysung*, it would have needed to be positioned at an ungainly angle on the printed paper to generate the orientation of Dürer's net. If the block cutter was working backward from *Underweysung*, it is likely that he did not heed the original orientation of the Dürer geometry and situated the net as he pleased, a decision which has no impact on the ability to subsequently cut out the printed net.

52 Schreiber, Fischer, and Sternath, "Archimedean Solids," 458.

53 "Das aber solche maler wolgefallen in iren yrthumben gehabt / ist alleyn vrsach gewest / das sie die kunst der messung nit gelernet haben / an die keyn rechter werckmañ werdè oder seyn kan." Dürer, *Underweysung*, Introduction.

54 "vnd dise corpora magst du von einer grȯß zwifach durch einander reyssen / also das alweg das ein eck durch des anderen planum sticht. ... Du magst auch auf dise cȯrper auf einen yeden plano einen spitzen punct setzen / nider oder hoch erhaben so von vill ecken als der planus hat darauf er stet." Dürer, *Underweysung*, Book IV, fig. 34a.

55 Brigaglia, Palladino, and Vaccaro, "Historical Notes," 201 and 204.

56 Warburg, *Der Bilderatlas Mnemosyne*, 38, pl. 23a.

57 Warburg, *Bilderatlas Mnemosyne: The Original*, 54.

58 "Ceste figure a esté estimée par les anciens la plus excellente & parfaicte de toute la geometrie, tellemet qu'ilz l'ont accomparée

à ceste grande machine du ciel." de Meung [?], fol. A ii recto. From plate 23a of the *Mnemosyne Atlas*, reprinted in Warburg, *Bilderatlas Mnemosyne: The Original.*

GEOMETRIA AND THE LEHRBUCH

1 Rutkin, "Astrology," 549; and Frank, "Mathematics," 307.
2 Cosgrove, "Ptolemy and Vitruvius," 22.
3 Ramus, *Via regia ad geometriam*, 1–2. See John Dee's analysis of geometry's virtues and application in his *Mathematicall Praeface*.
4 This was not a uniquely early modern phenomenon. Medieval mathematics had already shown signs of expanding into new branches of mathematical knowledge that sought to apply theory to practice—such as in the science of vision (*perspectiva*), the science of weights, and the science of "engines" (*ingeniis*). K. Park, "Allegories of Knowledge," 361.
5 Johnston, "Mathematical Practitioner," 111.
6 The war is known in German as the *Deutscher Bauernkrieg* or *Revolution des gemeinen Mannes.* The defeat of the peasants at the Battle of Frankenhausen, a bloody skirmish fought on May 15, 1525, signified the end of the war.
7 Melanchthon was appointed a Greek professorship in 1518 and would later come to be known as the *Praeceptor Germaniae*. For further reading on the "radical reformers," see Baylor, *Radical Reformation*, and the writings of Thomas Müntzer (ca. 1489–1525), Andreas Karlstadt (1486–1541), and Felix Manz (ca. 1498–1527).
8 Melanchthon, *Orations*, xiv.
9 Skinner, *Foundations*, 199.
10 Melanchthon, *Orations*, xxii.
11 On stipends in the German university, see Wright, "Educational Policy," 418–419.
12 Grendler, "Universities," 9–10.
13 Wright, "Educational Policy," 414.
14 Melanchthon was personally involved only in establishing three schools: Magdeburg (1524), Eisleben (1525), and Nuremberg (1526). Scheible, *Philipp Melanchthon*, 36.
15 Hannam, "Teaching Natural Philosophy," 117. On mathematical instruction in England, see also Johnston, "Mathematical Practitioner," 99–103.
16 From Tunstall, *De arte supputandi*, 178. Quoted in Johnston, "Mathematical Practitioner," 118.
17 Grendler, "Universities," 6–7.
18 Pedersen, "Tradition and Innovation,"466.
19 Folkerts, "Mathematik," 2.
20 Brockliss, "Curricula," 590. The university did on occasion experiment with craft practice and its intersection with mathematics. Jacques Lefèvre d'Étaples (ca. 1455–1536) and his students at the University of Paris engaged in extensive trials in order to bring his manuscripts into print, in so doing creating a place for the acquisition and honing of craft skills within the university classroom. See Oosterhoff, *Mathematical Culture*, 86–121.
21 Brockliss, "Curricula," 589.
22 Thorndike, *Sphere of Sacrobosco*, 42–43.
23 Hanschmidt, "Elementarbildung," 39.
24 Wesoly, "Berufsausbildung," 110–111; and Schmitt, "Education and Learning," 73.
25 Prass, "Ausbildung,"149.
26 Seidenfuß, *Geometrische Perspektiv*, 135. First editions of *Lehrbücher* remain today in many German libraries, including in Augsburg, Bamberg, Berlin, Dresden, Freiberg, Gotha, Halle, Hannover, Heidelberg, Hesse, Mainz, Munich, Nuremberg, Rostock, Tübingen, and Wolfenbüttel. See for instance the preface to Heinrich Lautensack's *Des Circkels vnnd Richtscheyts* (1564) and Hieronymus Rodler (d. 1539), who references the difficulty that young artists had in understanding Dürer's *Underweysung der Messung* and *Vier Bücher der menschlicher Proportion* in the preface to his *Eyn schön nützlich büchlin vnd vnderweisung der kunst des Messens* ... (Simmern, 1531).
27 For overviews of the perspectival and geometrical texts of Jamnitzer, Lencker, Hirschvogel, and others, see Eimer, "Abstrakte Figuren"; Richter, *Geometrischer Körper*, 54–92; and Seidenfuß, *Geometrische Perspektiv*, 125–207. On the function of play and playfulness in the pursuit of early

modern mathematical knowledge, Rothstein, "Making Trouble."

28 Moran, "German Prince-Practitioners," 260.

29 Scheible, *Philipp Melanchthon*, 37.

30 Methuen, "Role of the Heavens," 387.

31 "Mir scheint der Schritt von der Geometrie zu der Perspetiv, und zwar zu einer stereometrischen, der natürlichste zu seyn, indem Geometrie und Perspectiv sehr genau miteinander verbunden sind …" Nilson, *Anleitung*, vi.

32 Doppelmayr, *Historische Nachricht*, 199. For further reading on Hirschvogel, see Schwarz, *Lebensbeschreibung und Zeichnungen*; Schwarz, *Ein deutscher Meister der Renaissance*; and Schaper, *Hirschvogel von Nürnberg*.

33 Doppelmayr, *Historische Nachricht*, 156. See Hirschvogel's *Moscouia Sigmunds Freyherrns zu Herberstain Neyperg vnd Guetenhag &c verteütscht* (1557) and his maps of Austria—*Das in dem Ertzhertzogtumb vnter Osterreich*.

34 *Katalog der Gedenkschau*, 6.

35 "Hierauff ist mir von einem gutem Bekandten / Augustin Hirschvogels / weyland Burgers in Nürnberg / Geometria geliehen wordẽ / hinter dise hab ich mich mit grossem eyser gemacht / vnd weil sie sein leicht / schlecht vnd gerecht / sie mit lust durch studiert / biß mir vnter deß auch Wolff Schmids von Bamberg Geometria Anno 1539 zu Nürnberg gedruckt / vngefehr vnter die Hand kom̃en / ein sonderlich sein vnd wolgegründet Büchlein für die anfahenden / darauß ich dann gelernet / was proportio vnd proportionalitas, wie etliche zureden pflegen / seye / vnd kan mit warheit sagen / daß ich auß grund gedachter beeder Wercklein / hernach auch Albertum Dürerum, Vitruvium vnd andere Auctores mit nutz lessen kȯñen." Schwenter, *Geometriae*, "Vorrede an den Leser."

36 Blackburn, "Eloquence of Silence," 222. The phrase *Spero Fortuna Regressum* was also the motto of the Dutch publisher Jasper Tournay and was accompanied by an image of Fortune on a globe. With thanks to Lorraine Daston for this reference.

37 "Gvnstiger Herr / vnd fürderer / Nach dem Bißher / durch vnsere vorfordern / ein langezeit dise edle / vñ nützliche kunst des messens (Perspectiua in Latein genant) in Deutscher sprach ganz verborgen gehalten / vnd den gemeinen man / zulernen schwerlich zubekom̃en / Auch den meren thail in Griekischer vnd Lateinischer sprach verfast." Hirschvogel, *Grundtliche anweysung*, Introduction.

38 "Solche magst du von dickem zůsamen gebapten Papir auß schneyden vnnd zůsamenlegen / das dir ein materlich Corpus daraauß wirt." Hirschvogel, *Grundtliche anweysung*, fol. c recto.

39 "Its use [the drawing instrument] is clear from number 8, where it is seen how on the hinge [*Regel*], the Octahedron, which is lying on one of its sides, can be pushed back and forth and moved onto the drawn line of the affixed paper." "Sein Gebrauch ist auß No. 8 klar / da gesehen wirt / wie auf der Regel / das Octöedrō auf seiner fl̊achen einer ligend / hin vnd wider geruckt / vnd auf die gerissene lini deß aufgekleistersten Papirs / kan geführt werden." Halt, *Perspektivische Reiß Kunst*, 18.

40 For further information on Jamnitzer and Dürer's perspectival setups, see Bedini, "Perspective Machine"; and Hauschke, "Dürers Perspecktivtische." On the Jamnitzer house, see Mulzer, "Jamnitzerhaus."

41 "Wie er dann ein Buch außgehen lassen von den fůnff Regulierten Corporibus, vnnd derselben vnauffhȯrlichen darauß volgenden Corporen / Die er also durch diese Art der Perspectiu außbringt. Er hat seinen Perspectiu Tisch inn einer sondern Stuben inn seinem Hauß / so angeschraubt vnnd sich nicht bewegen kȯnnen / stehendt gehabt / das er die Saiden dargegen an einer schrauben der Stubenwendt anlagen können / vnd also an solchem Ort seines Hauses / seine Perspectiu zu werken verbunden gewesen ist." Pfinzing, *Kurtzer Extract*, 9.

42 It has been argued that the images may be attributed to Lorentz Stöer, who had a preexisting relationship with Pfinzing having worked together with him on the Pfinzing-Atlas (ca. 1594). Hauschke, "Dürers Perspecktivtische," 182.

43 As an example, the page describing the dodecahedron in Ramus's *Arithmeticae libri duo: Geometriae septem et viginti* (1569) shows three ways of visualizing the polyhedron, including an unfolded net, and reads: "1. Dodecahedra have 30 sides, 60 angled planes, and 20 solids. 2. If twelve equal pentagons with solid angles are orderly combined, they will create [comprehend] a dodecahedron. As you see here." "[1. Dodecaedi latera sunt 30, anguli plani 60, solidi 20. 2. Si duodecim quinquangula ordinata æqualia solidis angulis componantur, comprehendent dodecaedrum. Ut hic vides." Ramus, *Arithmeticae libri duo*, 178.

44 Employing labeled diagrams in multiple viewpoints (plan, elevation, and the unfolded *corps developpé*), Cousin collapses graphic information, linked by construction lines, together on the same page, for easier cross-referencing of different representations of the same solid.

45 Wolff, *Elementa*, 187–188.

46 Doppelmayr, *Historische Nachricht*, 161. The full title is *Des Circkels vnnd Richtscheyts auch der Perspectiua, vnd Proportion der Menschen und Rosse, kurtze, doch gründtliche vnderweisung deß rechten gebrauchs: Mit vil schönen Figuren, aller anfahenden Jugent, vnd andern liebhabern dieser Kunst, als Goldschmiden, Malern, Bildhauwern, Steinmetzen, Schreinern.* On Lautensack, see Heuer, "Kinesis and Death."

47 Bryan, *Dictionary of Painters*, 186.

48 "Zum Vierdten / wenn du die vierung zum sechs eck wilt machen / so mustu ein siebendtheil an der ein seiten daruon thun / wie am dreyeck / so ist sie recht / darnach mach ein Creutz vber ort darinn / vnd auch ein gerads Creutz / so gibt das gerad Creutz der leng nach an beyden enden zwey eck / so du denn der breyten nach vber zwerg ein jeglichs halb theil der vierung in zwey theil theilest / so hastu vnden vnd oben an beyden seiten zwey eck / wie es hie ist auffgerissen." Lautensack, *Circkels vnnd Richtscheyts*, 20.

49 "Hie hab ich es auff ein schlechte art auß dem grundt auffgezogen in die Perspectiff." Lautensack, *Circkels vnnd Richtscheyts*, 20.

50 P. Smith, *Body of the Artisan*, 66.

51 Goulding, *Defending Hypatia*, 36. Nevertheless, the geometrical definitions in Ramus's textbook *Arithmeticae libri duo* (1569) remain lengthy, overwrought, and without accompanying explanatory images.

52 P. Smith, *Body of the Artisan*, 67. See also Oosterhoff, *Mathematical Culture*, 299.

53 See also the entry on Jamnitzer in Doppelmayr, *Historische Nachricht*, 160–161. On the urban proximity of numerous Nuremberg artisans, see Mulzer, "Jamnitzerhaus," 63.

54 Waddington, *Ramus*, 210. Further promoted in Germany through the writings of Henning Rennemannus and Frédéric Beurhusius and sometimes combined with the logic of Melanchthon into what is known as "Philippo-Ramism," Ramus's writings also began to exert an influence in Germany and on Germans' self-perception of the worth of their mixed-mathematical practices. Hooykaas, *Humanisme*, 113.

55 Bennett, "Mechanical Arts," 684. See also the genre of paper scientific instruments for which Nuremberg was renowned. Kremer, "Experimenting with Paper"; and S. Schmidt, *Altered and Adorned*.

56 Daston, "Physicalist Tradition," 269.

57 See Rowe, "Mathematical Models"; and Sattelmacher, "Geordnete Verhältnisse."

58 Marr, "Walther Ryff," 136. On the audience for textbooks of practical knowledge and their role in the shaping of "mathematical communities," see Marr, "Oddi's *Dello squadro*," 166.

LOVER OF THE ART OF PERSPECTIVE

1 Hauschke, "Mathematical Instruments," 2–3.

2 Hayward, *Virtuoso Goldsmiths*, 38–39. In order to become a master goldsmith in Nuremberg, an applicant needed to produce a columbine cup, a gold ring set with a precious stone, and a steel seal die.

3 Doppelmayr, *Historische Nachricht*, 160. For standard descriptions of *Perspectiva*, see Richter, *Geometrischer Körper*, 80–82;

and Seidenfuß, *Geometrische Perspektiv*, 214–224.

4 "eine sehr liebliche vnd holdselige Kunst (welch von den Gelerten Optica vnd sonst gemainlich die Perspectiua genent wird)." Jamnitzer, *Perspectiva* (1568), fol. a ii recto.

5 Jamnitzer succumbed to the plague in 1585. The inscription on the grave reads "WENTZEL JAMNITZER ALT 78 IM 1585," and "CHRISTUS IST MEIN LEBEN—STERBEN IST MEIN GEWI[NN]." See Hauschke, "Soziale Aufstieg," for a wider discussion of Jamnitzer's grave plate in the context of the graves of other goldsmiths in the Johannisfriedhof. Taking posthumous self-fashioning in polyhedra to another level of investment, the Gorges Monument (1635) in the Salisbury cathedral covering the tomb of Helena, Marchioness of Northampton (1548/49–1635) and her husband Sir Thomas Gorges (1536–1610) is adorned with two cuboctahedra, with a celestial globe with a dodecahedron perched on top.

6 Farmer, *Virtuoso Craftsman*, 95.

7 The number of vertices (*V*), edges (*E*), and faces (*F*) in a polyhedron always satisfy the equality of Euler's formula. On Euler and topology, see Richeson, *Euler's Gem*. For a selection of mathematical expressions chosen by mathematicians specifically for their "beauty," see *Concinnitas Portfolio*.

8 Goldthwaite, *Wealth and the Demand for Art*, 212.

9 Evidently intense local competition and the high levels of skill required of commissions exerted pressure on makers to corner niche markets of craft production in Nuremberg. This atomization of expertise is clearly documented in *Die Hausbücher der Nürnberger Zwölfbrüderstiftungen*, a portrait series of retired Nuremberg craftsmen belonging to the Twelve Brothers Foundation, a Nuremberg-based charity organization for retired craftsmen. For instance, one Thomas Erngast, depicted in 1535 with his profession listed as a *Ringkelschmid* (buckle maker), is shown producing reams of buckles with a hammer and chisel. The image of Frycz Zorn from 1482 identifies him as a *Nagler* (nail maker), while Ving ... N.N. (full name unknown), from as early as 1414, was already known as a *Vingerhuter* (thimble maker). See *Hausbuch*, Amb. 279.2° fol. 24 recto (Landauer I); Amb. 317.2° fol. 101 recto (Mendel I); Amb. 317.2° fol. 5 verso (Mendel I).

10 On scientific instruments in Nuremberg and central Europe generally, see *Schätze der Astronomie*; Stephenson, Bolt, and Friedman, *Universe Unveiled*; and Strano et al., *European Collections*.

11 In his letters, Philipp Hainhofer details the immense, multiyear coordination required to produce the *Pommerscher Kunstschrank* (Pomeranian cabinet of curiosities) for Duke Philip II of Pomerania. Jamnitzer's correspondence with Archduke Ferdinand of Austria, between 1556 and 1562, similarly elucidates the collaborative nature of artisanal practice. Smith, "Matter and Meaning," 47. For further examples, see Neudörffer, *Nachrichten*, 125; and Brandl, "Kunst und Handwerk," 59.

12 On the implications of the shifts in the social and cultural landscape of Reformation-era Germany for artists and other craft practitioners, see Morrall, "Defining the Beautiful," 83 and 86; and Koerner, *Reformation of the Image*, 13 and 27.

13 For a general survey of goldsmith drawings, see Farmer, *Virtuoso Craftsman*; and Warncke, *Ornamentale Groteske*, vol. 2.

14 The contents of the Amerbach-Kabinett are held jointly at the Kunstmuseum and the Historisches Museum in Basel.

15 Söll-Tauchert, Beuing, and Roda, *Kunstkammer*, 249. While it is likely that the Kabinett contains more items from Jamnitzer's workshop, the Kabinett's various inventories only mention Jamnitzer by name in regard to one other item—a mold of an ornamental saddlebow also made for Maximilian II. Inventar D, 1585–1587, says "Item, a saddlebow with a caritas and the back part of the saddle cast with plaster (of Emperor Maximilian's saddle, so made in silver by Gamützer of Nuremberg)." "Item ein Sattebogen mit einer Caritas vnd hinder theil eines satels

mit gÿbs (von keiser Maximilian sattel, so Gamützer Ze Nornberg in silber gemacht) abgossen." Landolt, *Amerbach-Kabinett*, 148. In Inventory G (July 1662, corrected in 1664), the same saddlebow is referred to again as "Item two, a plaster piece of Emperor Maximilian's riding saddle." "Item Zweÿ stukh von Gÿps von Kaÿs. Maximiliani reitsattel." Landolt, *Amerbach-Kabinett*, 187.

16 Maximilian II is known to have given the coffer to his son, Philip II, who passed it on to his sister who lived in the Monasterio de las Descalzas Reales in Madrid. The coffer remains there today as a reliquary.

17 Landolt, *Amerbach-Kabinett*, 59–60.

18 Landolt, *Amerbach-Kabinett*, 53.

19 Warncke, *Ornamentale Groteske*, 2:57.

20 On the art of the goldsmith in medieval practice, see *De diversis artibus* (ca. 1122) by Theophilus, a comprehensive manual of twelfth-century craft knowledge produced by a working artist with an extensive section on metalworking.

21 "Petern Kuster, dem goldschmid, auflegen, die warheit anzuzeigen, woher ime der Wenzls Gamitzers mödel und kunsteisen komen und wers ime geben hab. Im fal dan, dass ers sich widersetzen wirt, sol er ins loch geschafft werden." From the *Ratserlässe der Reichsstadt Nürnberg*; in Frankenburger, *Beiträge*, 6, no. 22.

22 For further information on the *Basler Risse*, see Shelby, *Gothic Design Techniques*, 58–59; Ueberwasser, "Spätgotische Bau-Geometrie"; and Mager, "Technik."

23 On the Basel Holbein drawings, see Müller, "Holbein oder Holbein-Werkstatt."

24 Translated by author from Schönherr's paraphrasing of Strada's letter to Ferdinand, the original quote not having been reprinted. Schönherr, "Wenzel Jamnitzers Arbeiten," 291.

25 Anonymous, British Museum (BM 1978, 1216.21). On the history of the *Pocket Book*, see the appendix in Rowlands, *Drawings by German Artists*.

26 These drawing typologies have a number of interesting precedents in ornamental drawings produced for Gothic architecture. Kavaler, *Renaissance Gothic*, 43. For further discussion of the wimperg and baldachin drawings in the *Basler Risse*, see Seeliger-Zeiss, *Lorenz Lechler*, 49–57.

27 Kayser, "Intellectual and Artisan," 49.

28 "dass ich die löbliche, nützliche und synreiche kunst der perspectif also erfunden, dass dergleichen zuvor nie gesehen worden." Schönherr, "Wenzel Jamnitzers Arbeiten," 294.

29 "Perspectiva corporum regularium. Das ist / ein fleyssige Fůrweysung / Wie die Fůnff Regulirten Côrper / daruon Plato inn Timaeo / Vnnd Euclides inn sein Elementis schreibt. [&]c. Durch einen sonderlichen / newen / behenden vnd gerechten weg / der vor nie in gebrauch ist gesehen worden / gar kůnstlich inn die Perspectiua gebracht / Vnd darzu ein schône Anleytung / wie auß denselbigen Fůnff Côrpern one Endt / gar viel andere Côrper / mancherley Art vnd gestalt / gemacht / vnnd gefunden werden můgen." Translated in Veltman, "Geometric Games," 13.

30 Brigaglia, Palladino, and Vaccaro, "Historical Notes," 204–207.

31 "Der erst vnter den Fůnf regulirten Côrpern Ist ein Corpus gemacht / von vier gleichseitigen triangeln / Flechen oder Pôden gerader Linien / darauff es gestelt werden mag / hat sechs seiten oder gerader Linien Zwôlff flache Winckel vnnd vier Côrperlicher Eck. Auß disem drianglichten Corpore oder Kegel / sind verner drey vnd Zwainzig ander Côrpora geursacht / vnd vf mancherley vnterschidliche Art zu werckh gezogen. Wie hernach gesehen wirt." Jamnitzer, *Perspectiva* (1568), fol. A 1.

32 Luminet, "Geometrical Imagination," 255.

33 Lucretius, *De rerum natura*, II.1013–1022.

34 The Lencker quotation is from Veltman, "Geometric Games," 55, n. 94, in turn quoted from the introduction by Flocon in Jamnitzer, *Perspectiva* (1568/1981).

35 "Das Gott ... ist gôtig vñ trew / das es aber alles schon vnd artlich geordnet /.../ Himel vnd erden vnd die herrlichen liechter Soñ Mohn / vnd Stern / damit er den Himel gezieret hat." Jamnitzer, *Perspectiva (1568)*, "Vorrede," fol. a iii recto.

36 "How miraculously that with the love of God alone, the four elements and the companion five universal essences are ordered under heaven. All earthly bodies and precisely we men and everyone else are measured according to his complexion and qualities. So do fire and water prefer to be composed and maintained, both likewise in body as to the eye. And still fire is not extinguished by water, and likewise water is not completely consumed through fire." "Wie wunderlich hat nur der liebe Gott die vier Elementa / vnd derselbigen fünffte wesenheyt vnter dem himel geordnet / darauß alle jrdische Côrper / vnd wir menschen selbs / genaturet vnd gemessigt werden / vnd ein jeder sein Complexion vnd eigenschafft hat / also / das das Fewer vnd Wasser / bedes zugleich in einem Côrper als im auge verfasset vnd erhalten werden mûgen / vnd doch also / das das fewr vom wasser nit erlescht / desgleichē das wasser durchs fewr nit gentzlich verzert / vnd aus gedruckent wird." Jamnitzer, *Perspectiva* "Vorrede."

37 "vnd daneben anzeigt / so wenig mûglich sey / das vber die vier natûrliche Element / vnd derselben funfte wesenheyt / noch andere mehr Elementa in der natur sindt / so wenig kônnen auch andere mehr Corpora regularia / das sind andere mehr Côrpor von gleich fôrmigen / vnd gleich grossen Bôden gemacht werden / dazu so vergleicht er noch dieselben fûnff Côrper den Elementen der natur selbst / dañ gleicher weyß / wie der Pyramis Trilaterata, nemblich / der kegel oben zu gespitzt ist / also hat auch das fewer vnd flammen oben seine spitzen / .../ vnd wie alle andere irrdische Côrper / von disen vier Elementen der Natur zusamen gesetzt vnd miscirt werden / in den lebendigen Creaturen / als menschen vnd sich / sowol / als in den Vegitatilibus, Nemblich laub vnd graß / sowol werden auch ander Geometrica Corpora auß disen fûnf Corporibus, ohn endtlich miscirt vnd zusamen gesetzt / wie in disem meinem vorhabenden werckh / in die 140. vnterschiedliche Côrper gesehen werden / gantz vnnd durchsichtig / Ich hab aber sonderlich dise Côrper in meiner newen Perspectif gebrauchen wôllen / dieweyl sie mancherley vnd viel vnterschiedliche Eckh / seythen / winckel vnd spitzen / einwartz vnd außwartz gekehrt haben." Jamnitzer, *Perspectiva* "Vorrede."

38 See Newman, *Promethean Ambitions*; and P. Smith, *Body of the Artisan*.

39 Newman, *Promethean Ambitions*, 115.

40 Haug, "Artificial Interventions," 83. On art and alchemy, see Newman, *Promethean Ambitions*, 115–163; Battistini, *Astrology, Magic, and Alchemy*; Principe, *Secrets of Alchemy*; and Dupré, Kerssenbrock-Krosigk, and Wismer, *Art and Alchemy*.

41 "Opvs Paramirvm," in Paracelsus, *Essential Theoretical Writings*, [cf. H 1:69], 305.

42 See Paracelsus, "Opvs Paramirvm," 1 Cor 3:12–15 [cf. H 1:71], 305, n. 2. Florentine treatises from the period award Vulcan, or Hephaestus, god of fire and all craftsmen, the status of being a quintessentially modern artist, one capable of making his imagination tangible with his *fervor ingenii*, or enthusiastic spirit of invention. Göttler, "Allegories of Fire," 136.

43 Keller, "Science of Salvation," 490; and Haug, "Artificial Interventions, 83.

44 Battistini, *Astrology, Magic, and Alchemy*, 309.

45 Quoted from Haug, "Artificial Interventions," 82, n. 7.

46 Jeanneret, *Perpetual Motion*, 1.

47 Thanks to Lorraine Daston for mentioning the shield of Achilles in this context.

48 Daston and Park, *Wonders*, 277.

CUT APART IN MANY WAYS

1 The art of intarsia extends back to Asia Minor and found popularity in the crafts of the Roman Empire, though the techniques faded after the empire's decline. See "Der Werkstoff und die Technik" in Flade, *Intarsia*, 379–409; Raggio and Wilmering, *Gubbio Studiolo*; and Ramond, *Masterpieces*.

2 On the discovery of the *Bom Jesus*, see Hauptmann, Schneider, and Bartels, "Shipwreck of Bom Jesus."

3 See "Five Things to Know about the Richest Person in History," Medievalists.net.
4 Stetten, *Kunst-, Gewerb- und Handwerks-Geschichte*, 1:114.
5 "Es ist aine lange Zeit her diese Statt Augspurg hoch berüemt gewesen und (ist es) noch von wegen ihrer Kistler-Arbait, daß man in der Frembde und auch in dieser Statt uns das Lob gibt, wie man so scharpff (künstliche) Dinge allhie einlegt, welches kainem Maler möglich ist, von (mit) Farben nachzukommen." Quoted in Hellwag, *Deutschen Tischlerhandwerks*, 457; translation in Alfter, "Marked Augsburg Furniture," 52. On the tension between intarsia craftsmen and cabinetmakers in early modern Augsburg, see Hellwag, *Deutschen Tischlerhandwerks*, 470–471.
6 Vasari, *On Technique*, 262–263.
7 Vasari, *Lives*, 2:132–133.
8 The chest is recorded in the 1596 Ambras inventory, PA 21, 318. See Boeheim, "Urkunden und Regesten."
9 Distelberger, "The Castrucci and the Miseroni," 32. For further reading on Prince Karl's tabletop, see Vincent, "Prince Karl I."
10 See Laue, *Perspectiva*, 18–19. On the cabinet belonging to the Metropolitan Museum, see Andrews and Donefer-Hickie, "Miniature Collector's Cabinet." The wood used on the interior is Hungarian ash, whose grain is similar to marble and thus used to create the illusion of stone. (With thanks to Wolfram Koeppe.) There are eleven pieces of furniture, in public and private collections, originating in the Nuremberg workshop of the Master of Perspective. See Laue, *Perspectiva*, 6–29.
11 See Seipel, *Spielwelten der Kunst*, 160–162; and Andrews, "Writing and Reading Box."
12 Koeppe, *Art of the Royal Court*, 150–151. The finials are located at the Museo dell' Opificio delle Pietre Dure, Florence (1905: 2705a–b).
13 See Colsman, *Möbel*, 132–139.
14 For the alternative cover, see Graphische Sammlung, Munich (21268 Z [1566]).
15 Jervis, *Printed Furniture Designs*, 4.
16 Vredeman de Vries, quoted in Heuer, *City Rehearsed*, 104. The two books are entitled *Den Eersten Boeck Ghemaect opde Twee Columnen Dorica en Ionica* (the first book on the subject of the Doric and Ionic columns) and *Das ander buech Gemacht auff die zway Colonnen, Corinthia vnd Composita* (the other book treating the two Corinthian and Composite columns).
17 "Geometria et Perspectiva Hier Inn Etliche zerbrochne Gebew den Schreiner in eingelegter Arbait dienstlich auch vil andern Liebhabern zu sonder gefallen geordnet vnnd gestalt Durch Lorenz Stöer Maller Burger Inn Augspurg. LS. Mit Rö: Käy: Maÿ: u aller genedigste Privilegio nit nach zetruckhen, 1567."
18 Attributed to Bartholomew Weisshaupt (active between 1560 and 1580), Royal Monastery of San Lorenzo de El Escorial, Madrid (10014373). See also A. Gruber, *History of Decorative Arts*, 401.
19 With thanks to Virginie Spenlé for noting this relationship.
20 Victoria & Albert Museum, London, Museum number 4250:1 to 33 -1858. See Möller, *Wrangelschrank*, no. 70, fig. 179; and Aguiló-Alonso, "In Search of Wisdom."
21 "Lorenz Stör ist weniger bekannt als er verdient." Stetten, *Kunst-, Gewerb- und Handwerks-Geschichte*, 1:283. Stöer's fortunes may be changing. Belgian-based publisher Zones Sensibles reissued *Geometria et perspectiva* in 2018.
22 "Entwürfe, die einen kühnen und lebhaften Geist verrathen." Stetten, *Kunst-, Gewerb- und Handwerks-Geschichte*, 1:283. Stetten, however, admits that "two stone tablets by him of considerable size, on which biblical and moralistic phrases have been inscribed in a masterly manner, in the most delicate way embossed, etched, and gilded." "Ich weis von ihm zwo steinerne Tafeln von ahsehnlicher Größe, auf welche biblische und moralische Sprüche mit schreibmeisterischen Zügen eingefaßt, auf das zierlichste erhaben, geäzet und vergoldet sind." Stetten, *Kunst-, Gewerb- und Handwerks-Geschichte*, 1:284.
23 Thieme, Becker, *Allgemeines Lexikon*, 91.
24 Voltelini, "Urkunden und Regesten," p. lxvii, no. 6475.

25 The later 1617 edition was published in Augsburg by Steffan Michelspacher.

26 "Lorennz Stör, mahler, hat sein bürgerrecht auffgesagt, gewöndlich verschreibung geben unnd ist damit inn die losungstuben gewiesen worden" [1557, II, 8b], May 24, 1557. Hampe, *Nürnberger Ratsverlässe* 1:527, no. 3667. The notice from August 13, 1595 reads "Lorentzen Störs, mahlers von Augspurg, forderung der gemachten abriß des eingenommenen augenscheins halben, das klein waidtwerckh betreffendt, weiln er von dem, so er dem kays. commißario alberait zugestelt, 50 f. und von dem, so er solchem original nachmachen soll, 40. f. und von dem gemachten vorwerckh 8f. forderen thuet, soll man auf die verordtnete herren stellen, wie sie mit ihme abkommen mugen." Hampe, *Nürnberger Ratsverlässe* 2:246, no. 1386.

27 In his autobiography, Kraus recounts how the prince, desperate to move his books from Vienna, which was still occupied by the Soviet Union after the end of World War II, smuggled them to a schoolhouse in the village of Schaan in Lichtenstein and after some negotiations with Kraus and under pressure of discovery, sold the books on the condition that they be removed in three days. Kraus, *Rare Book Saga*, 153–154.

28 There is an additional record of the sale to Kraus on the actual reference card (165.2.1, where it is referred to as *Geometria et perspectiva corpora regulata et irregulata*) within the card catalogue of the Lichtenstein Museum in Vienna—it says "verkauft an Krauss 1948." The Harvard manuscript, referred to as *Geometria practica um 1600*, is described in a letter dated July 26, 1948 in the archives of the Lichtenstein Museum as one of the books that Kraus wished to purchase. In the section on "Mathematik und Astronomie" in the handwritten catalogue of Hauslab's donations to the museum, "System der Hauslab-Sammlung," reference is made to a book of geometrical drawings—"Geometr. Zeichnen, allg. darstellende Geometrie"—that probably refers to the Stöer volume.

29 The watercolor wash appears to allude gesturally to wood in a few drawings.

30 Morrall, "Art, Geometry, and the Landscape," 210.

31 The shadows cast by the models are used as notional, or gestural, proof of the models' three-dimensionality and have not been subjected to any kind of geometrical rigor.

32 "die der Euclides corpora regularia nennet." Dürer, *Underweysung*, Introduction to Book IV.

33 Winckelmann, *Writings on Art*, 61.

34 Several of Halt's unfolded polyhedra are copied from Dürer's *Underweysung*, and there are comparable geometrical forms to those found in Barbaro's *Perspettiva* (1569), Lorenzo Sirigatti's *La practica de prospettiva* (1596), as well as Jamnitzer's *Perspectiva*.

35 "Werden hernach 23. andere vnterschidliche Corpora formiret." Jamnitzer, *Perspectiva* (1568), fol. A VI.

36 "auß disem werden auch hernach dreyvndzweintzig vnterschiedliche Corpora mancherley art vnd form fůrgerissen vnd in die Perspectif gebracht." Jamnitzer, *Perspectiva* (1568), fol. D III.

37 Staats- und Stadtbibliothek Augsburg, 4 Cod. Aug. 247. The hexahedron and the cube are both six-sided polyhedra.

38 The page numbers which run continuously through the volumes are later interventions.

39 See Ebelmann, *Decken*, fol. 19.

40 The inventory number for the Wolfenbüttel manuscript is Cod Guelf. 74.1 Aug. 2°.

41 Cod Guelf. 74.1 Aug. 2°: page 10 = the upper right drawing on fol. C II in *Perspectiva*; page 9 = middle right drawing on fol. C II in *Perspectiva*; page 7 = upper right drawing on fol. D V of *Perspectiva*; page 12 = upper left on fol. C I in *Perspectiva*; page 25 is copied from fol. E IIII in *Perspectiva*. Pages 27 and 28, which state "Geometria Perspectiva Corporum Regularium" and "Geometria et Perspectiva," refer directly to Stöer's work, although they are not drawings from the Munich volume.

42 The full introductory inscription reads "Perspectief Buch Darinnen ordentlich zu befindenn / die stuck / welche der

Durchlauchtige Hochgeborne Fürst vnndt Herr / Herr CHRISTIANUS, Hertzogk zu Sachsenn, Landgraf in Thüringen, Markgraf zu Meissen auf Hansenn Lenckers Burgers zu Nürnbergk vnterthnige vnterweisung vonn dem letzten tagk Februarij dess 1576. Jars an vor sich mitt eigener handt gerissen hat."

43 Sächsische Hauptstaatsarchiv Dresden, 10037, Rentkammer-Rechnungen, Nr. 183, fol. 228r. See also Kappel, "Drechselnde Herrscher," 166–168. For further information on Lencker in Dresden, see Gluch, "Hans Lencker."

44 The title and attribution are scrawled on the fore edge of the book.

45 The main watermark of the manuscript's first section is a variation of the "arms of Nuremberg," an eagle inside a shield, found on paper produced in Nuremberg from 1585 to 1591, and also possibly in Augsburg. Reference image 919, Briquet, *Filigranes*. There are pages with the same watermark in the second section, though the latter is also composed of sheets with a watermark representing the coat of arms of Freising. Reference image 2255, Briquet, *Filigranes*: Hallstadt 1598–1599, Neumark 1598, Eichstadt 1604.

46 Payne, "Order and Disorder."

47 Benjamin, *Arcades Project*, 417.

48 On self-erasure and early modern drawing, see Andrews, "Architectural Gesture."

49 Universitätsbibliothek München, 2 Cod. Ms. 582a.

50 Witcombe, *Copyright in the Renaissance*, 53.

THE VIOLENCE OF WHIMSY

1 Württembergische Landesbibliothek Stuttgart, Cod. Hist. 2 1068.

2 "Astronomische, Geometrische, Geographische, Scenographische, oder perspectivische vnd typographische, insonderheit von schönen sphaeris armillaribus vnd artificialibus, globis coelestibus & terrestribus, von metall, holz vnd andern materien, Astrolabia, qvadranten, qvadrata, declinatoria, nocturnalia, mancherleÿ horaria solaria, cylindros, truncos, varia corpora regularia & irregularia." Hainhofer, *Reisen nach Innsbruck und Dresden*, 175.

3 "Ahn regulirten corporibus, von holz und gepappten papir gemacht." Syndram and Minning, *Die Inventare der kurfürstlich-sächsischen Kunstkammer*, [vol. 1], *Das Inventar von 1587*, fol. 52 recto.

4 The complete inventory list of polyhedra from the *Kunstkammer* is as follows: "I Pyramis ex regularibus primum. I Cubus ex regularibus corporibus secundum. I Octohedrus ex regularibus tertium. I Icosahedrus ex regularibus quartum. I Dodecahedrus ex regularibus quantum. I Conus ambligonius. I Conus oxigonius. I Conus orthogonius. I Holtzern gevirter cubus mit einem andern achteckigten cubo darinnen. I Rundt corpus von mößenen drath, inwendigk mit den regulirten corporibus von guldenen fäden gezogen. I Holtzern birnbeumen cubus in einem kleinen khestlein. Hat Abraham Riese gemacht." Syndram and Minning, *Die Inventare der kurfürstlich-sächsischen Kunstkammer*, [vol. 1], *Das Inventar von 1587*, fols. 52 recto–52 verso.

5 Syndram and Minning, *Die Inventare der kurfürstlich-sächsischen Kunstkammer*, [vol. 3], *Das Inventar von 1640*, fol. 318 recto.

6 "I Octahedrum, darein ein hexaedrum und in selbigen hinwiederumb ein tetraedrum geschloßen ist," and "I Hexaedrum, darein ein octaedrum geschloßen ist." Syndram and Minning, *Die Inventare der kurfürstlich-sächsischen Kunstkammer*, [vol. 3], *Das Inventar von 1640*, fol. 319 recto.

7 "I Kleiner Cubus," "I Durchbrochenen tetraedrum, inwendig ein octaedrum," and "I Prisma, so man in 5 stück zernehmen kan, welche herr Doctor Christof Pincker zu Leipzigk gemacht." Syndram and Minning, *Die Inventare der kurfürstlich-sächsischen Kunstkammer*, [vol. 3], *Das Inventar von 1640*, fol. 319 recto.

8 "zur demonstration radicis quadratae und cubicae dienlich." Syndram and Minning, *Die Inventare der kurfürstlich-sächsischen Kunstkammer*, [vol. 3], *Das Inventar von 1640*, fols. 318 recto–318 verso.

9 See the 1615 plan of the Royal Kunstkammer in Dresden by David Uslaub, its art chamberlain and sole keeper. Sächsisches Staatsarchiv, 100024, Geheimes Archiv, Loc. 7324/1, fol. 258 recto. For histories of the Dresden castle, see Syndram, *Schloss zu Dresden*.

10 See also Hainhofer, who describes: "Aine schöne Mathematische librej oder bibliotec zu der Arithmetica, Geometria, Astronomia, Architectur, perspectiv vnd andern künsten mehr. Schöne von helffenbain gedrehte coni vnd pyramides, darauf mancherleÿ corpora regularia & irregularia zu sehen." Hainhofer, *Reisen nach Innsbruck und Dresden*, 173. The first and second room described in the 1587 inventory are referred to as the fifth and fourth room respectively in a later 1619 inventory, though the physical location remains the same. See Syndram and Minning, *Die Inventare der kurfürstlich-sächsischen Kunstkammer*, [vol. 2], *Das Inventar von 1619*.

11 The list of titles in the Elector's possession included "Perspectiva corporum regularium Wenzel Gamitzers"; "Underweisung der meßung" by Dürer; "Perspectiva Danielis Barbari in lingua italiana"; "Hansen Lenckers von Nurnbergk perspectiva, illuminirt"; "Augustin Hirschvogels eigentliche und grundtliche anweisung in die geometria ..."; "Erhardt Schön von Nurnbergk underweisung der proportion und stellunge der poßen"; and "Perspectivische kunststucken, so mein gnedigster churfurst und herr, herzogk Christian zu Sachen etc., selbsten gerißen"—the set of Duke Christian's geometrical drawings. Syndram and Minning, *Die Inventare der kurfürstlich-sächsischen Kunstkammer*, [vol. 1], *Das Inventar von 1587*, fols. 165 recto / 175 recto–196 verso / 206 verso.

12 "kunstlichen dröhewergk von helfenbein ... welchs hogstgedachter hertzogk Augustus ... selbsten gedröhet." See the section on turned ivory objects made by Elector August in Syndram and Minning, *Die Inventare der kurfürstlich-sächsischen Kunstkammer*, [vol. 1], *Das Inventar von 1587*, fols. 131 verso / 141 verso–136 verso / 146 verso. For Lobenigk, see fols. 137 recto / 147 recto–137 verso / 147 verso. For Wecker, see fols. 138 verso / 148 verso–139 verso / 149 verso.

13 "Aif ainem selzam gedreheten Becher aine durchbrochene kugel, in welcher andere 24 kugeln alle beweglich zu befinden." Hainhofer, *Reisen nach Innsbruck und Dresden*, 171.

14 The entire quote reads "Aine dichte kugel auswendig anzusehen, inwendig aber erkennt man durch ain loch aine andere mit schönen stern vnd schnecklen durchbrochen: veber das in selbiger ain bix mit zweÿen liden, in welcher bix zweÿ contrafet, Churfürsten Christiani secundi samt desselben gemahl geschnitten, die lider aber mit gulden bändern angeschlagen, alles aus ain stuck inwendig gemacht, vnd die kugel noch aus wendig di basso rilievo geschnitten ist." Hainhofer, *Reisen nach Innsbruck und Dresden*, 171–172.

15 Quiccheberg, *Inscriptiones*, 71.

16 Moran, "Valuation of Precision," 211 and 222. See also Moran, "German Prince-Practitioners," 254; and Ash, "Introduction," 14–16. On the shift from mixed mathematics to a mechanical science, see Henninger-Voss, "Working Machines."

17 See Lefèvre, *Picturing Machines*.

18 Syndram, "Amassing Magnificence," 38.

19 Long, "Authorship of *Ars*," 21.

20 Though pole lathes were still restricted to one half of the motion—that of the lathe rotation toward the tool—they were not necessarily incapable of continuous motion. Holtzapffel, *Hand or Simple Turning*, 27–28.

21 In this case, the wheel would be called a "hand flywheel." Leonardo may have been the first European to design a flywheel. Connors, "*Ars Tornandi*," 218. For a history of the development of the center lathe in particular, see Holtzapffel, *Hand or Simple Turning*, 29–57. See also Maurice, *Sovereigns as Turners*, chap. 5.

22 Maurice, *Sovereigns as Turners*, 134.

23 Maurice, "Princely Art of Turning," 21.

24 See Gundestrup, *Royal Danish Kunstkammer*; Bencard, *Kongelige Kunstdrejere*; and Hein, *Treasure Collection*. The collection in Rosenborg is replete with evidence of

ivory polyhedra. For a catalogue of Rosenborg's ivory collection, see Hein and Kristiansen, *Rosenborg*, 92–107.

25 Maurice, *Sovereigns as Turners*, 38. Duke Francesco de' Medici was not the only Italian recipient of a turned object from Wilhelm. The 1585 Tribuna inventory at the Uffizi in Florence lists a present to Grand Duke Ferdinando I of a turned ebony ball enveloping a smaller ivory ball enclosing portraits of Duke Wilhelm, his wife, and son. Casazza, "Mattias de' Medici," 123. See also Schmidt and Sframeli, *Diafane passioni*.

26 Archivio di Stato di Firenze, Inventario della Guardaroba, p. 75, no. 509, quoted in Bedini, *Patrons*, 409.

27 See Friedrich Friese's *Der vornehmsten Künstler und Handwercker Ceremonial-Politica* ... (1708) and Johann Martin Teuber's *Dreh-Kunst* (1740).

28 "ain schöner drehzug zum Agstain [August] zu drehen." Hainhofer, *Reisen nach Innsbruck und Dresden*, 165. On the turning workshops, see Syndram, "Three Turned-Ivory Pieces," 142.

29 Syndram, Kappel, and Weinhold, *Baroque Treasury*, 54.

30 Kappel, "Turned Ivory Works," 178.

31 For a history of the carved ivory artworks in Dresden, see Kappel, *Elfenbeinkunst*.

32 "vnd zur erfözligkeit solcher seiner Dienste, wollen wir ime irhelich die zeit seiner Lebens, Vnd langen er vnnser besteltter Dreqler sein und bleiben wirdet, Zwei Hudert gulden Agünz, zur den Vier Quarteal zeitten ... Vnd das er desto williger Im dieser Landen bei vns bleiben möge, seiner Mutter zur München, so lange sie am leben sein wirdet, alle jahr Funfzigt gulden groshen zu geben vnd folgen zulassen bewilligett." Finanzarchiv, Sächsische Hauptstaatsarchiv Dresden, 10036, Rep. LII Gen 1925, Loc. 33342, fols. 506 recto–507 verso. For additional references to contracts and payment, see references and catalogue entries in Syndram and Scherner, *Princely Splendor*, 176–197.

33 All the turned ivory recorded in Rudolf's inventory falls under the category of *von Helffenbain, Ebenholtz und anderm gedrechselte Sachen* (of ivory, ebony, and other turned things), a list of one hundred and sixty items. Published in Bauer and Haupt, "Kunstkammerinventar Kaiser Rudolfs II."

34 "Ein *colonna a ovato* schraubenweis von helffb: gedret uff einem tristaffelten ebnin fueß, dabey ein runde holgedrehte kugel, darin das corpus *octoedron*, auf die kugel gehört beyligendt eingeschraubter *pyramus* von helffenb." Bauer and Haupt, "Kunstkammerinventar Kaiser Rudolfs II.," fol. 95, no. 891.

35 "Ein trifach ineinander geschlossen corpus *dodecaedron* von ebano, obenauf ein zierd von helffenb: und blümlein." Bauer and Haupt, "Kunstkammerinventar Kaiser Rudolfs II.," fol. 96, no. 905.

36 "Inn einem langen gefierten mit leder überzognen vergulten futral ein von helffenbain geschraufte *colonna*, oben darauf ein runde hol gelocherte cugel und mehr corpora darin, oben darauf ist ein triegget corpus, uff dessen spitz stehet ein von helffenb: geschnitzter *Mercurius*." Bauer and Haupt, "Kunstkammerinventar Kaiser Rudolfs II.," fol. 97, no. 953.

37 "Zwey durchbrochne corpora oder von helffenb: gedrehte kugeln, dern jedes fünffach ineinander geschlossen." Bauer and Haupt, "Kunstkammerinventar Kaiser Rudolfs II.," fol. 99, no. 990.

38 "Ein durchsichtig corpus dodicaedron, inwendig ein stern und ein kunststuckhle daran gedreht." Bauer and Haupt, "Kunstkammerinventar Kaiser Rudolfs II.," fol. 99, no. 1006.

39 Livy, *History of Rome*, Book 37.

40 Shalem, *Oliphant*, 18–37. See Vickers, *Ivory*.

41 Mark, "Towards a Reassessment," 198. For a list of the items in the Ambras *Wunderkammer*, based upon the 1596 inventory of the collection, see Bassani, *African Art*, 3–8. They include one saltcellar, various pieces of cutlery, and several oliphants. See also Fagg, *Afro-Portuguese Ivories*.

42 Ryder, "Note on the Afro-Portuguese Ivories," 363. Ninety percent of ivory carvings amassed in the sixteenth century have disappeared from German collections,

leaving approximately 150 known items in existence. Mark, "Towards a Reassessment," 201. See also Mark and Horta, *Forgotten Diaspora*, 135–136.

43 Massing, "Quest for the Exotic," 118.

44 Dürer, *Journeys to Venice and the Low Countries*, 73. Mark argues that the adjective "Afro-Portuguese" should be replaced with "Luso-African" to more accurately reflect the identity of the sculptors who produced the saltcellars. Mark, "Towards a Reassessment," 189–190.

45 Shalem, *Oliphant*, 14.

46 Maurice, *Sovereigns as Turners*, 56. On Heiden, see Haag, "Ivory Goblet." See also Heiden's treatise on ivory turning, *Beschreibung eines von Helffenbein gedrehten Kunststücks ...* (1640).

47 Ryder, *Benin and the Europeans*, 133. Contemporary findings based on nuclear DNA analysis of the ivory cargo aboard the shipwreck of the Portuguese trading vessel the *Bom Jesus* suggests that at least a proportion of ivory traded in the sixteenth century was sourced from the savanna and mixed habitats of West Africa closer to the Portuguese trading ports rather than from deep rainforests. Flamingh et al., "Sourcing Elephant Ivory," 623.

48 As the Scottish explorer Mungo Park (1771–1806) observed in *Travels into the Interior of Africa* (1799), "the ivory thus collected, is seldom brought down to the Coast by the hunters themselves. They dispose of it to the itinerant merchants, who come annually from the Coast with arms and ammunition, to purchase this valuable commodity." M. Park, *Travels*, 244. Merchants traded goods procured in India and brought to Africa via Portuguese and later Dutch and British traders expressly for the purpose of purchasing ivory. See Ryder, *Benin and the Europeans*, 40–41.

49 Early eighteenth-century records attest to the bringing of sculptures from Rome to Dresden via ship from Hamburg. Zimmermann, "Dresdener Skulpturensammlung," 13. Both male and female African elephants possess tusks, while the tusks of the Asiatic or Indian elephant grow only on males and were considered to be of lesser quality than the whiter and finer-grained African tusks. The ivory artifacts found in Dresden likely came from Africa based on their color and textural quality.

50 Another possible route to Dresden could have been through the Leipziger Messe, a material fair in the neighboring city of Leipzig selling exotica like coconuts, shells, and ivory to decorative art workshops across Europe. On the history of the Leipziger Messe, see Buschmann, *800 Jahre*.

51 Syndram and Minning, *Die Inventare der kurfürstlich-sächsischen Kunstkammer*, [vol. 2], *Das Inventar von 1619*, fol. 319 recto.

52 "I Großer becher, drey auf einander gedrehet, deckel und füßlein mit durchsichtigen crönlein ... No. 18." Syndram and Minning, *Die Inventare der kurfürstlich-sächsischen Kunstkammer*, [vol. 3], *Das Inventar von 1640*, fol. 439 verso. "I Helfenbeiner becher, so das letzte stück ist, welches weiland churfürst Augustus zu Sachßen hochlöblichster gedechtnus selbst mit eigener hand gedrehet hat. Ist von Georg Weckern, hofdrechslern, eingeantwortet worden den 16. octobris anno 1567. No. 100." Syndram and Minning, *Die Inventare der Kurfürstlichkurfürstlich-sächsischen Kunstkammer*, [vol. 3], *Das Inventar von 1640*, fol. 444 verso.

53 "I Durchbrochen corpus von helfenbein, so octaedrum genennt wird. Uf allen flechen mit auch durchbrochenen pyramidibus erhaben uf einen beinern durchbrochenen schnirckelfuß. Hat meinem gnädigsten herrn Anthonius Örtel, tischler, darmit angebunden 1620, am tage Johannis. 2 Eckigte durchbrochene helfenbeinerne corpora regularia, als duppelte tetraedra in einem der gesezt, mit durchbrochenen füßen." Syndram and Minning, *Die Inventare der kurfürstlich-sächsischen Kunstkammer*, [vol. 3], *Das Inventar von 1640*, fol. 455 verso.

54 Quiccheberg, *Inscriptiones*, 72.

55 Connors, "*Ars Tornandi*," 223.

56 In *Mechanick Exercises* (1703), Joseph Moxon concisely described the basic principles behind evening out a work piece on a

center lathe relative to a central axis. See Moxon, *Mechanick Exercises*, 167.

57 Staatliche Kunstsammlungen Dresden (68).

58 For eighteenth- and nineteenth-century textbooks on turning, see Plumier, *L'art de tourner*; and Bergeron, *Manuel du tourneur.*

59 See Sächsische Hauptstaatsarchiv Dresden, Loc. 9835/15, fol. 20. The 1591 inventory of Georg Wecker's workshop is also in the Sächsische Hauptstaatsarchiv Dresden. Drawings of some of his tools are contained within fols. 23 verso–37 recto, represented in a light-blue wash similar to the drawings in the 1684 inventory, while the tools of Egidius Lobenigk are listed in fols. 29 verso–30 verso.

60 Springett, *Woodturning Wizardry*, 24. In contemporary parlance, these wooden elements are called chucks and jigs.

61 On how to set out the surface of a sphere, see Springett, *Woodturning Wizardry*, 44–47. Drilling or turning holes on a clearance point allows waste material that would not otherwise be easily accessible from the main points to be removed. Springett, *Woodturning Wizardry*, 47.

62 Korey, *Geometrie der Macht*, 17.

63 Moxon, *Mechanick Exercises*, 214.

64 Moore, *Ivory*, 41–47.

65 Burack, *Ivory and Its Uses*, 41.

EPILOGUE: CORPORA IRREGULATA ET REGULATA

1 Phillips, *New Math*, 5.

2 See Grossman and Sebline, *Man Ray*.

3 Rowe, "Euclidean Geometry," 58.

4 On the range of geometrical models produced by nineteenth- and early twentieth-century mathematicians, see Rowe, "On Models."

5 Stevin, *Principal Works*, 2:124–125.

6 In order to obtain the last two of Dürer's polyhedral "inventions," Stevin applied a twofold truncation to a cube in which both vertices are removed as are sides by parallel planes. Dijksterhuis, *Simon Stevin*, 44.

7 Stevin, *Problemata geometrica*, 82, translated in Stevin, *Principal Works*, 2:295. Stevin ascribes his inspiration to one Frans Cophart at Leiden, who, Stevin says, had faceted a cube by cutting out twelve tetrahedra and erroneously believed he had found a sixth regular solid. Stevin, *Problemata geometrica*, 47, translated in Stevin, *Principal Works*, 2:225.

8 Sasaki, *Descartes's Mathematical Thought*, 132.

9 Manders, "Algebra," 186.

10 Federico, *Descartes on Polyhedra*, 65.

11 Pont, *Topologie algébrique*, 13.

12 On the printing of Kepler's polyhedral model, see Andrews, "Tabula III," particularly 294–301.

13 Kepler, *Gesammelte Werke*, 13:32.

14 "Removenda verò irregularia corpora, nempe in conditu ordinatissimae creaturae." Kepler, *Gesammelte Werke*, 13:35.

15 "Dan wer hatt nie grosse vnd kleine in halb abgetheiltte kugelln, samptt denen funff Corporibus mathematicis ... gesehen?" Kepler, *Gesammelte Werke*, 13:52. The term *Credentzbecher* refers to a large chalice intended to be displayed on a credenza or sideboard. On Kepler's failure to manufacture a model of his new planetary theory in gold, silver, and jewels, see Andrews, "Gilding Kepler's Cosmology."

16 Hainhofer, "Stutgarter Rayss," 306.

17 In Germany, professional training would become the domain of specialized schools (*Handwerksschule*), partially through the educational reforms of Wolfgang Ratke (1571–1635) and John Amos Comenius (1590–1670).

18 *Encyclopaedia Britannica*, "Wolfgang Ratke."

19 Plato, "Phaedo," *Dialogues*, 2:258.

overleaf Jost Amman, after Wenzel Jamnitzer. Spiral model, from *Perspectiva corporum regularium* (Nuremberg, 1568). The Metropolitan Museum of Art, New York, Harris Brisbane Dick Fund, 1924 (24.45.1, fol. H IIII).

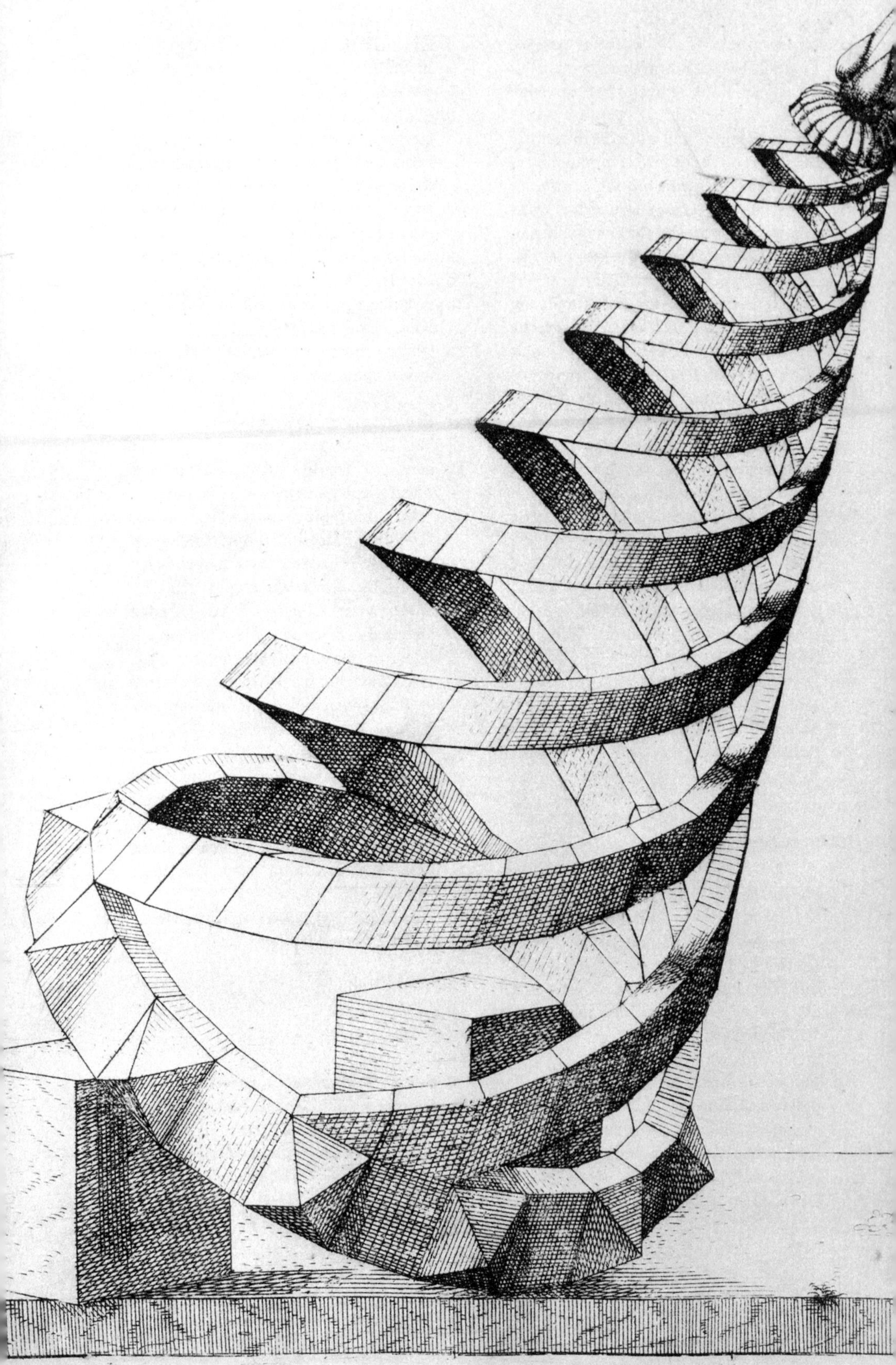

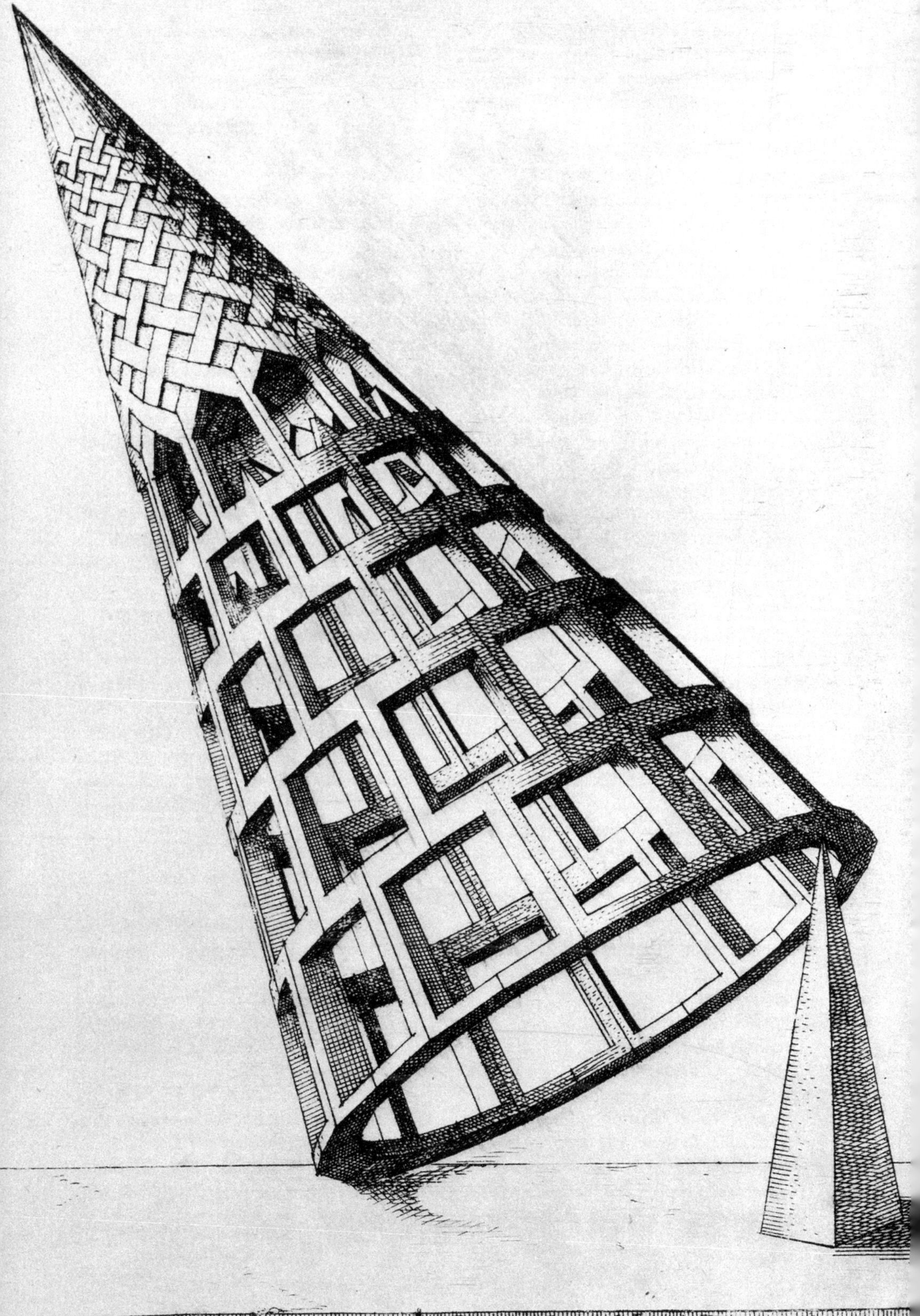

Aguiló-Alonso, María Paz. "In Search of Wisdom: An Iconographic Programme in a 16th-Century German Bureau with Marquetry." In *B'09: Bulletin* (Bilbao Fine Arts Museum), no. 5 (2010), 15–63.

Aikema, Bernard, ed. *Dürer e il Rinascimento tra Germania e Italia*. Milan: Palazzo Reale, 2018.

Alfter, Dieter. "Gemarkte Augsburger Möbel: Ein Gütesiegel für Herkunft und Qualität" / "Marked Augsburg Furniture: A Warranty Seal for Provenance and Quality." In *Möbel für die Kunstkammern Europas: Kabinettschränke und Prunkkassetten / Furniture for European Kunstkammer: Collector's Cabinets and Caskets*, 42–57. Munich: Kunstkammer Georg Laue, 2008.

Allen, Michael J. B. "The Ficinian *Timaeus* and Renaissance Science." In *Plato's Timaeus as Cultural Icon*, edited by Gretchen J. Reydams-Schils, 238–250. Notre Dame, IN: University of Notre Dame Press, 2003.

Ames-Lewis, Francis. "Neoplatonism and the Visual Arts at the Time of Marsilio Ficino." In *Marsilio Ficino: His Theology, His Philosophy, His Legacy*, edited by Michael J. B. Allen and Valery Rees, 327–338. Leiden: Brill, 2002.

Andersen, Kirsti, and Henk J. M. Bos. "Pure Mathematics." In *The Cambridge History of Science*, vol. 3, *Early Modern Science*, edited by Katharine Park and Lorraine Daston, 696–724. Cambridge: Cambridge University Press, 2006.

Andrews, Noam. "Albrecht Dürer's Personal *Underweysung der Messung*." *Word and Image* 32, no. 4 (2016): 409–429.

Andrews, Noam. "The Architectural Gesture." *Log*, no. 33 (2015): 137–155.

Andrews, Noam. "Gilding Kepler's Cosmology." *Journal for the History of Astronomy* 52, no. 1 (February 2021): 3–32.

Andrews, Noam. "Tabula III: Kepler's Mysterious Polyhedral Model." *Journal for the History of Astronomy* 48, no. 3 (August 2017): 281–311.

Andrews, Noam. "Writing and Reading Box (cat. 69)." In *Making Marvels: Science and Splendor at the Courts of Europe*, edited by Wolfram Koeppe, 154–155. New York: Metropolitan Museum of Art, 2019.

Andrews, Noam, and Ana Matisse Donefer-Hickie. "Miniature Collector's Cabinet (cat. 67)." In *Making Marvels: Science and Splendor at the Courts of Europe*, edited by Wolfram Koeppe, 152–154. New York: Metropolitan Museum of Art, 2019.

Aristotle. *Aristotle's "de Caelo" III: Introduction, Translation and Commentary*. Translated by Theokritos Kouremenos. Stuttgart: Franz Steiner, 2013.

Artmann, Benno, and Lothar Schäfer. "On Plato's 'Fairest Triangles' (*Timaeus* 54a)." *Historia Mathematica* 20, no. 3 (1993): 255–264.

Ash, Eric H. "Introduction: Expertise and the Early Modern State." *Osiris* 25, no. 1 (2010): 1–24.

Ashcroft, Jeffrey. *Albrecht Dürer: Documentary Biography*. 2 vols. New Haven: Yale University Press, 2017.

Bambach, Carmen [Carmen Bambach Cappel]. "Leonardo, Tagliente, and Dürer: 'La scienza del far di groppi.'" *Achademia Leonardi Vinci* 4 (1991): 72–95.

Barbaro, Daniele. *La pratica della perspettiva*. Venice: appresso Camillo, & Rutilio Borgominieri fratelli, al segno di S. Giorgio, 1569.

Barzman, Karen-edis. "Perception, Knowledge, and the Theory of *Disegno* in Sixteenth-Century Florence." In Larry J. Feinberg, *From Studio to Studiolo: Florentine Draftsmanship under the First Medici Grand Dukes*, 37–48. Oberlin: Allen Memorial Art Museum, Oberlin College, 1991.

Bassani, Ezio. *African Art and Artefacts in European Collections, 1400–1800*. London: British Museum Press, 2000.

Battistini, Matilde. *Astrology, Magic, and Alchemy in Art*. Los Angeles: J. Paul Getty Museum, 2007.

Bauer, Rotraud, and Herbert Haupt. "Das Kunstkammerinventar Kaiser Rudolfs II., 1607–1611." *Jahrbuch der Kunsthistorischen Sammlungen in Wien* 72 (1976).

Baylor, Michael G., ed. and trans. *The Radical Reformation*. Cambridge: Cambridge University Press, 1991.

Bedini, Silvio A. *Patrons, Artisans and Instruments of Science, 1600–1750*. Aldershot: Ashgate, 1999.

Bedini, Silvio A. "The Perspective Machine of Wenzel Jamnitzer." *Technology and Culture* 9, no. 2 (April 1968): 197–202.

Bencard, Mogens. *Kongelige Kunstdrejere.* Copenhagen: Rosenborg, 1995.

Benjamin, Walter. *The Arcades Project.* Translated by Howard Eiland and Kevin McLaughlin. Cambridge, MA: Belknap Press of Harvard University Press, 1999.

Bennett, Jim. "How Relevant Is the Category of 'Mixed Mathematics' to the Sixteenth Century?" *Oberwolfach Reports* 10, no. 1 (2013):677–680.

Bennett, Jim. "The Mechanical Arts." In *The Cambridge History of Science*, vol. 3, *Early Modern Science*, edited by Katharine Park and Lorraine Daston, 673–695. Cambridge: Cambridge University Press, 2006.

Bennett, Jim, and Stephen Johnston. *The Geometry of War, 1500–1750.* Oxford: Museum of the History of Science, 1996.

Bergeron, Louis-Éloi. *Manuel du tourneur.* 2nd ed. Paris: Chez Hamelin-Bergeron, 1816.

Besson, Jacques. *Theatrvm instrvmentorvm et machinarum.* Lyon: apud Barth. Vincentium, 1578.

Biggiogero, Giuseppina Masotti. "On the Life and Works of Luca Pacioli." In *De divina proportione*, 3–14. Biblioteca Ambrosiana and Officina Bodoni. Milan: Mediobanca di Milano, 1956.

Billingsley, Henry, trans. *The Elements of Geometrie of the Most Auncient Philosopher Evclide of Megara.* London: Iohn Daye, 1570.

Blackburn, Bonnie J. "The Eloquence of Silence: Tacet Inscriptions in the Alamire Manuscripts." In *Citation and Authority in Medieval and Renaissance Musical Culture: Learning from the Learned*, edited by Suzannah Clark and Elizabeth Eva Leach, 206–223. Rochester: Boydell Press, 2005.

Blair, Ann. "Humanism and Printing in the Work of Conrad Gessner." *Renaissance Quarterly* 70, no. 1 (2017): 1–43.

Boeheim, Wendelin, ed. "Urkunden und Regesten aus der K. K. Hofbibliothek." *Jahrbuch der kunsthistorischen Sammlungen des Allerhöchsten Kaiserhauses* 7 (1888): XCI–CCCXIII.

Bovelles, Charles de. *Quoe in hoc volumine continētur .... De geometricis corporibus.* Paris, 1510.

Brandl, Rainer. "Zwischen Kunst und Handwerk: Kunst und Künstler im mittelalterlichen Nürnberg." In *Nürnberg 1300–1550: Kunst der Gotik und Renaissance*, edited by Gerhard Bott, 51–60. Munich: Prestel, 1986.

Brigaglia, Aldo, Nicla Palladino, and Maria Alessandra Vaccaro. "Historical Notes on Star Geometry in Mathematics, Art and Nature." In *Imagine Math 6: Between Culture and Mathematics*, edited by Michele Emmer and Marco Abate, 197–211. Cham: Springer, 2018.

Briquet, Charles-Moïse. *Les filigranes: Dictionnaire historique des marques du papier dès leur apparition vers 1282 jusqu'en 1600.* 4 vols. Amsterdam: Paper Publications Society, 1968.

Brockliss, Laurence. "Curricula." In *Universities in Early Modern Europe (1500–1800)*, edited by Hilde de Ridder-Symoens, 565–620. Vol. 2 of *A History of the University in Europe.* Cambridge: Cambridge University Press, 1996.

Brown, Gary I. "The Evolution of the Term 'Mixed Mathematics.'" *Journal of the History of Ideas* 52, no. 1 (1991): 81–102.

Bryan, Michael. *Dictionary of Painters and Engravers.* Vol. 3. London: G. Bell and Sons, 1904.

Bubenik, Andrea. "The Shape of Things to Come: Dürer's Polyhedron." In *The Persistence of Melancholia in Arts and Culture*, edited by Andrea Bubenik, 68–93. New York: Routledge, 2019.

Bucher, François. *Architector: The Lodge Books and Sketchbooks of Medieval Architects.* New York: Abaris Books, 1979.

Büchsel, Martin. *Albrecht Dürers Stich Melencolia I: Zeichen und Emotion; Logik einer kunsthistorischen Debatte.* Munich: Wilhelm Fink, 2010.

Burack, Benjamin. *Ivory and Its Uses.* Rutland, VT: Charles E. Tuttle, 1984.

Burt, Justin L., Jose L. Elechiguerra, Jose Reyes-Gasga, J. Martin Montejano-Carrizales, and Miguel Jose-Yacaman. "Beyond Archimedean Solids: Star Polyhedral Gold Nanocrystals." *Journal of Crystal Growth* 285, no. 4 (2005): 681–691.

Buschmann, R., ed. *800 Jahre Leipziger Messe: Festschrift des Leipziger Messeamtes zur Jubiläumsmesse, 1965; 1165–1965*. Leipzig: Seemann, 1965.

Casazza, Ornella. "Gli avori di Mattias de' Medici." In *The Museo degli Argenti: Collections and Collectors*, edited by Ornella Casazza and Marilena Mosco, 120–135. Florence: Giunti, 2004.

Cheney, Liana De Girolami. "Giorgio Vasari's Fine Arts: Neoplatonic Visualization of Invention, Imitation and Beauty." *Figura* 1 (2013): 54–79.

Cheney, Liana De Girolami, and John Hendrix, eds. *Neoplatonic Aesthetics: Music, Literature, and the Visual Arts*. New York: Peter Lang, 2004.

Christian I of Saxony. *Perspectief Buch*. 1576. Dresden, Kupferstichkabinett, Grünes Gewölbe, Ca 61.

Colsman, Edla. *Möbel, Gotik bis Jugendstil: Die Sammlung im Museum für angewandte Kunst Köln*. Stuttgart: Arnoldsche Art Publishers, 1999.

Concinnitas Portfolio. Edited by Dan Rockmore. New York: Harlan & Weaver, 2014.

Connors, Joseph. "*Ars Tornandi*: Baroque Architecture and the Lathe." *Journal of the Warburg and Courtauld Institutes* 53 (1990): 217–236.

Cornford, Francis MacDonald. *Plato's Cosmology: The Timaeus of Plato*. London: Routledge, 2014.

Cosgrove, Denis. "Ptolemy and Vitruvius: Spatial Representations in the Sixteenth-Century Texts and Commentaries." In *Architecture and the Sciences: Exchanging Metaphors*, edited by Antoine Picon and Alessandra Ponte, 20–51. New York: Princeton Architectural Press, 2003.

Cousin, Jean. *Livre de perspective de Iehan Cousin senonois, maistre painctre à Paris*. Paris: Impr. de I. le Royer, 1560.

Cromwell, Peter R. *Polyhedra*. Cambridge: Cambridge University Press, 1997.

Daston, Lorraine. "The Physicalist Tradition in Early Nineteenth Century French Geometry." *Studies in History and Philosophy of Science* 17, no. 3 (1986): 269–295.

Daston, Lorraine, and Katharine Park. *Wonders and the Order of Nature, 1150–1750*. New York: Zone Books, 1998.

Davis, Margaret Daly. *Piero della Francesca's Mathematical Treatises: The Trattato d'abaco and Libellus de quinque corporibus regularibus*. Ravenna: Longo, 1977.

Dee, John. "Mathematicall Praeface." In *The Elements of Geometrie of the Most Auncient Philosopher Evclide of Megara*, translated by Henry Billingsley. London: Iohn Daye, 1570.

Dijksterhuis, Eduard Jan. *Simon Stevin: Science in the Netherlands around 1600*. The Hague: Martinus Nijhoff, 1970.

Distelberger, Rudolf. "The Castrucci and the Miseroni: Prague, Florence, Milan." In *Art of the Royal Court: Treasures in Pietre Dure from the Palaces of Europe*, edited by Wolfram Koeppe, 29–39. New York: Metropolitan Museum of Art, 2008.

Doorly, Patrick. "Dürer's *Melencolia I*: Plato's Abandoned Search for the Beautiful." *Art Bulletin* 86, no. 2 (June 2004): 255–276.

Doppelmayr, Johann Gabriel. *Historische Nachricht von den nürnbergischen Mathematicis und Künstlern....* Nuremberg: P. C. Monath, 1730.

Duhem, Pierre. "Plato's Theory of Space and the Geometrical Composition of the Elements." In *The Concepts of Space and Time: Their Structure and Their Development*, edited by Milič Čapek, 22–25. Dordrecht: Reidel, 1976.

Dupré, Sven, Dedo von Kerssenbrock-Krosigk, and Beat Wismer, eds. *Art and Alchemy: The Mystery of Transformation*. Munich: Hirmer, 2014.

Dürer, Albrecht. *The Human Figure: The Complete "Dresden Sketchbook."* Edited by Walter L. Strauss. New York: Dover, 1972.

Dürer, Albrecht. *The Painter's Manual: A Manual of Measurement of Lines, Areas, and Solids by Means of Compass and Ruler Assembled by Albrecht Dürer for the Use of All Lovers of Art with Appropriate Illustrations Arranged to Be Printed in the Year MDXXV*. Translated and edited by Walter L. Strauss. New York: Abaris Books, 1977.

Dürer, Albrecht. *Records of Journeys to Venice and the Low Countries*. Edited by Roger Fry. Boston: Merrymount Press, 1913.

Dürer, Albrecht. *Underweysung der Messung*. Nuremberg, 1525.

Dürer, Albrecht. *Underweysung der Messung*. Nuremberg, 1538.

Dürer, Albrecht. *Vier Bücher von menschlicher Proportion*. Nuremberg, 1528.

Ebelmann, Hans Jakob. *Gebäude Bekrönungen Decken*. Collection of prints. Germany, 1609.

Eckert, Willehad Paul, and Christoph von Imhoff. *Willibald Pirckheimer: Dürers Freund im Spiegel seines Lebens, seiner Werke, und seiner Umwelt*. Cologne: Wienand, 1971.

Eichholz, Klaus. "Die Polyeder-Sonnenuhr des Ludwig Hohenfeld von 1596." *Jahresschrift* (Deutsche Gesellschaft für Chronometrie) 49 (2010): 169–186.

Eimer, Gerhard. "Abstrakte Figuren in der Kunst der abendländischen Renaissance." *Konsthistorisk Tidskrift* 25, nos. 1–4 (1956): 113–145.

Ekserdjian, David. *Parmigianino*. New Haven: Yale University Press, 2006.

Ekserdjian, David. "Two Drawings by Parmigianino for Prints." *Master Drawings* 46, no. 3 (Autumn 2008): 367–373.

Enyclopaedia Britannica Online. "Wolfgang Ratke." Accessed January 6, 2016. www.britannica.com/biography/Wolfgang-Ratke.

Epple, Moritz, Tinne Hoff Kjeldsen, and Reinhard Siegmund-Schultze. "From 'Mixed' to 'Applied' Mathematics: Tracing an Important Dimension of Mathematics and Its History." *Oberwolfach Reports* 10, no. 1 (2013): 657–660.

Erlande-Brandenburg, Alain, Roland Bechmann, Jean Gimpel, and Régine Pernoud. *Carnet de Villard de Honnecourt*. Paris: Stock, 1986.

Euclid. *Euclidis Megarensis mathematici clarissimi Elementorum geometricorum libri XV*. Basel: per Iohannem Hervagivm, 1546.

Euclid. *Preclarissimu[m] opus Elementor[um] Euclidis Megare[n]sis: Vna cu[m] co[m]mentis Campani p[er]spicacissimi in arte[m] geometria[m] incipit felicit[er]*. Venice: Erhardus Ratdolt, 1482.

Euclid. *The Thirteen Books of Euclid's Elements*. Edited by Sir Thomas Heath. 2nd ed. Chicago: Encyclopaedia Britannica, 1990.

Fagg, William Buller. *Afro-Portuguese Ivories*. London: Batchworth Press, 1959.

Fara, Giovanni Maria. "La ricezione italiana della *Unterweisung der Messung* di Albrecht Dürer fra XVI e XVII secolo *Modelli di lettura*." In *Buchmalerei der Dürerzeit: Dürer und die neue Mathematik: Neues Dürerforschung*, edited by G. Ulrich Grossmann, 139–156. Dürer-Forschungen 2. Nuremberg: Verlag des Germanischen Nationalmuseums, 2009.

Farmer, John David. *The Virtuoso Craftsman: Northern European Design in the Sixteenth Century*. Worcester, MA: Worcester Art Museum, 1969.

Faulhaber, Johann. *Newe geometrische vnd perspektiuische Inuentiones*. Frankfurt am Main, 1610.

Federico, Pasquale Joseph. *Descartes on Polyhedra: A Study of the De solidorum elementis*. New York: Springer, 1982.

Ficino, Marsilio. *All Things Natural: Ficino on Plato's Timaeus*. Translated by Arthur Farndell. London: Shepheard-Walwyn, 2010.

Field, Judith V. "Kepler's Star Polyhedra." *Vistas in Astronomy* 23 (1979): 109–141.

Field, Judith V. *Piero della Francesca: A Mathematician's Art*. New Haven: Yale University Press, 2005.

Field, Judith V. "Rediscovering the Archimedean Polyhedra: Piero della Francesca, Luca Pacioli, Leonardo da Vinci, Albrecht Dürer, Daniele Barbaro, and Johannes Kepler." *Archive for History of Exact Sciences* 50, no. 3/4 (1997): 241–289.

Flade, Helmut. *Intarsia: Europäische Einlegekunst aus sechs Jahrhunderten*. Munich: C. H. Beck, 1986.

Flamingh, Alida de, Ashley Coutu, Judith Sealy, Shadreck Chirikure, Armanda D. S. Bastos, Nzila M. Libanda-Mubusisi, Ripan S. Malhi, and Alfred L. Roca. "Sourcing Elephant Ivory from a Sixteenth-Century Portuguese Shipwreck." *Current Biology* 31, no. 3 (2021): 621–628.

Folkerts, Menso. "Euclid in Medieval Europe." In *The Development of Mathematics in Medieval Europe: The Arabs, Euclid, Regiomontanus*, 1–64. Hampshire: Ashgate, 2006.

Folkerts, Menso. "Die Mathematik im sächsisch-thüringischen Raum im 15. und 16. Jahrhundert." In *Kaufmanns-Rechenbücher und mathematische Schriften der frühen Neuzeit*, edited by Rainer Gebhardt, 1–22. Annaberg-Buchholz: Adam-Ries-Bund, 2011.

Frank, Martin. "Mathematics, Technics, and Courtly Life in Late Renaissance Urbino." *Archive for History of Exact Sciences* 67, no. 3 (2013): 305–330.

Frankenburger, Max, ed. *Beiträge zur Geschichte Wenzel Jamnitzers und seiner Familie.* Strasbourg: J. H. E. Heitz, 1901.

Friedman, Michael. *A History of Folding in Mathematics: Mathematizing the Margins.* Cham: Springer, 2018.

Friese, Friedrich. *Der vornehmsten Künstler und Handwerker....* Leipzig, 1708.

Gabbarelli, Jamie. "A Solid Mistake: An Early State of Caraglio's *Diogenes* after Parmigianino." *Journal of the Warburg and Courtauld Institutes* 80 (2017): 231–241.

Gluch, Sibylle. "Hans Lencker, die Perspektive und der kursächsiche Hof." In *Visier und Rechenbücher der frühen Neuzeit*, edited by Rainer Gebhardt, 379–396. Annaberg-Bucholz: Adam-Ries-Bund, 2008.

Gnann, Achim. *Chiaroscuro: Renaissance Woodcuts from the Collections of Georg Baselitz and the Albertina, Vienna*. London: Royal Academy of Arts, 2014.

Gnann, Achim. *Parmigianino: Die Zeichnungen.* 2 vols. Petersberg: Michael Imhof, 2007.

Gnann, Achim. *Parmigianino und sein Kreis: Druckgraphik aus der Sammlung Baselitz.* Ostfildern: Hatje Cantz, 2007.

Goldthwaite, Richard A. *The Economy of Renaissance Florence*. Baltimore: Johns Hopkins University Press, 2009.

Goldthwaite, Richard A. *Wealth and the Demand for Art in Italy, 1300–1600*. Baltimore: Johns Hopkins University Press, 1993.

Göttler, Christine. "Allegories of Fire and of the Arts." In *Art and Alchemy: The Mystery of Transformation*, edited by Sven Dupré, Dedo von Kerssenbrock-Krosigk, and Beat Wismer, 134–138. Munich: Hirmer, 2014.

Goulding, Robert. *Defending Hypatia: Ramus, Savile, and the Renaissance Rediscovery of Mathematical History.* New York: Springer, 2010.

Grasselli, Luigi. "Piero della Francesca, Mathematics and Perspective." In Massimo Mussini and Luigi Grasselli, *Piero della Francesca: De prospectiva pingendi; Saggio critic / Critical Essay*, 162–261. Sansepolcro: Aboca Museum, 2008.

Grendler, Paul F. "The Universities of the Renaissance and Reformation." *Renaissance Quarterly* 57, no. 1 (2004): 1–42.

Grossman, Wendy A., and Edouard Sebline, eds. *Man Ray: Human Equations: A Journey from Mathematics to Shakespeare*. Ostfildern: Hatje Cantz, 2015.

Gruber, Alain, ed. *The History of Decorative Arts.* [Vol. 1], *The Renaissance and Mannerism in Europe.* New York: Abbeville Press, 1993.

Gruber, Peter M. "History of Convexity." In *Handbook of Convex Geometry*, edited by Peter M. Gruber and Jörg M. Wills, 1–15. Amsterdam: North-Holland, 1993.

Grünbaum, Branko. "Polyhedra with Hollow Faces." In *Proceedings of the NATO Advanced Study Institute on Polytopes: Abstract, Convex and Computational*, edited by Ted Bisztriczky, Peter McMullen, Rolf Schneider, and Asia Ivi´c Weiss, 43–70. Dordrecht: Kluwer.

Gundestrup, Bente. *Det Kongelige danske Kunstkammer 1737 / The Royal Danish Kunstkammer, 1737.* Copenhagen: Nationalmuseet and A. Busc, 1991.

Haag, Sabine. "A Signed and Dated Ivory Goblet by Marcus Heiden." *J. Paul Getty Museum Journal* 24 (1996): 45–59.

Hainhofer, Philipp. *Des augsburger Patriciers Philipp Hainhofer Reisen nach Innsbruck und Dresden*. Edited by Oscar Doering. Vienna: Carl Graeser, 1901.

Hainhofer, Philipp. "Stutgarter Rayss Ao. 1616 auf die Kündtstauf." In "Philipp Hainhofers Bericht über die Stuttgarter Kindtaufe im Jahre 1616," edited by Adolf von Oechelhäuser. *Neue Heidelberger Jahrbücher* 1 (1891), 277–335.

Halt, Peter. *Perspektivische Reiß Kunst.* Augsburg, 1625.

Hampe, Theodor. *Nürnberger Ratsverlässe über Kunst und Kunstler im Zeitalter der Spätgotik und Renaissance (1449) 1474–1618 (1633)*. 2 vols. Quellenschriften für Kunstgeschichte und Kunsttechnik ..., n.s., XI–XIII. Vienna: Graeser & Kie., 1904.

Hankins, James. "The Study of the *Timaeus* in Early Renaissance Italy." In *Natural Particulars: Nature and the Disciplines in Renaissance Europe*, edited by Anthony Grafton and Nancy G. Siraisi, 77–119. Cambridge, MA: MIT Press, 1999.

Hannam, James. "Teaching Natural Philosophy and Mathematics at Oxford and Cambridge 1500–1570." PhD diss., University of Cambridge, 2008.

Hanschmidt, Alwin. "Elementarbildung und Berufsausbildung, 1450 bis 1750: Inhalte und Institutionen." In *Elementarbildung und Berufsausbildung, 1450–1750*, edited by Alwin Hanschmidt and Hans-Ulrich Musolff, 19–46. Cologne: Böhlau, 2005.

Hart, George. "Fra Giovanni's Intarsia Polyhedra." www.georgehart.com/virtual-polyhedra/intarsia.html.

Haug, Henrike. "Artificial Interventions in the Natural Form of Things: Shared Metallogenetical Concepts of Goldsmiths and Alchemists." In *Laboratories of Art: Alchemy and Art Technology from Antiquity to the 18th Century*, edited by Sven Dupré, 79–103. Cham: Springer, 2014.

Hauptmann, Andreas, Gabi Schneider, and Christoph Bartels. "The Shipwreck of *Bom Jesus*, AD 1533: Fugger Copper in Namibia." *Journal of African Archaeology* 14, no. 2 (2016): 184–207. *Die Hausbücher der Nürnberger Zwölfbrüderstiftungen.* Nuremberg, fifteenth-nineteenth centuries. Stadtbibliothek Nürnberg.

Hauschke, Sven. "Albrecht Dürers Perspektivtische und ihre Nachfolger im 16. Jahrhundert." In *Buchmalerei der Dürerzeit: Dürer und die neue Mathematik; Neues Dürerforschung*, edited by G. Ulrich Grossmann and Thomas Eser, 173–189. Nuremberg: Verlag des Germanischen Nationalmuseums, 2009.

Hauschke, Sven. "The Mathematical Instruments of Wenzel Jamnitzer (1508–1585)." In *European Collections of Scientific Instruments, 1550–1750*, edited by Giorgio Strano, Stephen Johnston, Mara Miniati, and Alison D. Morrison-Low, 1–13. Leiden: Brill, 2009.

Hauschke, Sven. "Der soziale Aufstieg eines Handwerkers: Die Grabplatte des Goldschmieds Wenzel Jamnitzer." In *Künstlergrabmäler: Genese, Typologie, Intention, Metamorphosen*, edited by Birgit Ulrike Münch, 97–109. Petersberg: Michael Imhof, 2011.

Hayward, John Forrest. *Virtuoso Goldsmiths and the Triumph of Mannerism, 1540–1620.* London: Sotheby Parke Bernet, 1976.

Heiden, Markus. *Beschreibung eines von Helffenbein gedrehten Kunstücks....* Coburg, 1640.

Hein, Jørgen. *The Treasure Collection at Rosenborg Castle: The Inventories of 1696 and 1718; Royal Heritage and Collecting in Denmark–Norway, 1500–1900.* 3 vols. Copenhagen: Museum Tusculanum Press, University of Copenhagen, 2009.

Hein, Jørgen, and Peter Kristiansen. *Rosenborg: A Guide to the Danish Royal Collections.* Copenhagen: Rosenborg, 2005.

Hellwag, Fritz. *Die Geschichte des deutschen Tischlerhandwerks, vom 12. bis zum 20. Jahrhundert.* Berlin: Verlagsanstalt des Deutschen Holzarbeiter-Verbandes, 1924.

Hendrix, John. *Platonic Architectonics: Platonic Philosophies and the Visual Arts.* New York: Peter Lang, 2004.

Henninger-Voss, Mary. "Working Machines and Noble Mechanics: Guidobaldo del Monte and the Translation of Knowledge." *Isis* 91, no. 2 (2000): 233–259.

Heuer, Christopher P. *The City Rehearsed: Object, Architecture, and Print in the Worlds of Hans Vredeman de Vries.* London: Routledge, 2009.

Heuer, Christopher P. "Dürer's Folds." *RES: Anthropology and Aesthetics*, no. 59/60 (2011): 249–265.

Heuer, Christopher P. "Kinesis and Death in Lautensack." In *The Primacy of the Image in Northern European Art, 1400–1700: Essays in Honor of Larry Silver*, edited by Debra Taylor Cashion, Henry Luttikhuizen, and Ashley D. West, 178–192. Leiden: Brill, 2017.

Hirschvogel, Augustin. *Ein aigentliche vnd grundtliche anweysung, in die Geometria, sonderlich aber, wie alle Regulierte, vnd Vnregulierte Corpora, in den grundt gelegt, vnd in das Perspecktiff gebracht, auch mit jren Linien auffzogen sollen werden.* Nuremberg, 1543.

Hirschvogel, Augustin. *Geometria.* Nuremberg, 1543.

Hoffmann, Rainer. *Im Zwielicht: Zu Albrecht Dürers Meisterstich Melencolia I.* Cologne: Böhlau, 2014.

Holtzapffel, John Jacob. *Hand or Simple Turning: Principles and Practice* [1881]. New York: Dover, 1976.

Hooykaas, Reijer. *Humanisme, science et réforme: Pierre de la Ramée (1515–1572).* Leiden: E. J. Brill, 1958.

Jamnitzer, Wenzel. *Perspectiva corporum regularium.* Nuremberg, 1568.

Jamnitzer, Wenzel. *Perspectiva corporum regularium* [1568]. Edited by Albert Flocon. Paris: Gutenberg Reprints, 1981.

Jeanneret, Michel. *Perpetual Motion: Transforming Shapes in the Renaissance from Da Vinci to Montaigne.* Baltimore: Johns Hopkins University Press, 2001.

Jervis, Simon. *Printed Furniture Designs before 1650.* Leeds: Furniture History Society, 1974.

Johnston, Stephen. "The Identity of the Mathematical Practitioner in 16th-Century England." In *Der "Mathematicus": Zur Entwicklung und Bedeutung einer neuen Berufsgruppe in der Zeit Gerhard Mercators,* edited by Irmgarde Hantsche, 93–120. Bochum: Brockmeyer, 1996.

Kappel, Jutta. "Der drechselnde Herrscher: Fürstliche Drechselkunst." In *Fürst und Fürstin als Künstler: Herrschaftliches Künstlertum zwischen Habitus, Norm und Neigung,* edited by Annette C. Cremer, Matthias Müller, and Klaus Pietschmann, 161–177. Berlin: Lukas, 2018.

Kappel, Jutta. *Elfenbeinkunst im Grünen Gewölbe zu Dresden: Geschichte einer Sammlung; Wissenschaftlicher Bestandskatalog—Statuetten, Figurengruppen, Reliefs, Gefäße, Varia.* Staatliche Kunstsammlungen Dresden. Dresden: Sandstein, 2017.

Kappel, Jutta. "Turned Ivory Works." In *Princely Splendor: The Dresden Court, 1580–1620,* edited by Dirk Syndram and Antje Scherner, 176–179. Dresden: Staatliche Kunstsammlungen Dresden; Milan: Mondadori Electa, 2004.

Katalog der Gedenkschau Augustin Hirschvogel (1503–1553). Vienna: Historisches Museum der Stadt Wien, 1953.

Kavaler, Ethan Matt. *Renaissance Gothic: Architecture and the Arts in Northern Europe, 1470–1540.* New Haven: Yale University Press, 2012.

Kayser, Petra. "The Intellectual and the Artisan: Wenzel Jamnitzer and Bernard Palissy Uncover the Secrets of Nature." *Australian and New Zealand Journal of Art* 7, no. 2 (2006): 45–61.

Keller, James R. "The Science of Salvation: Spiritual Alchemy in Donne's Final Sermon." *Sixteenth Century Journal* 23, no. 3 (Autumn 1992): 486–493.

Kemp, Martin. *Leonardo da Vinci: Experience, Experiment and Design.* London: V&A Publications, 2006.

Kemp, Martin. *The Science of Art: Optical Themes in Western Art from Brunelleschi to Seurat.* New Haven: Yale University Press, 1990.

Kepler, Johannes. *Johannes Keplers gesammelte Werke.* Edited by Walther von Dyck, Max Caspar, and Franz Hammer. 22 vols. Munich: C. H. Beck, 1938–1998.

Klibansky, Raymond, Erwin Panofsky, and Fritz Saxl. *Saturn and Melancholy: Studies in the History of Natural Philosophy, Religion and Art.* London: Nelson, 1964.

Koeppe, Wolfram, ed. *Art of the Royal Court: Treasures in Pietre Dure from the Palaces of Europe.* New York: Metropolitan Museum of Art, 2008.

Koeppe, Wolfram, ed. *Making Marvels: Science and Splendor at the Courts of Europe.* New York: Metropolitan Museum of Art, 2019.

Koerner, Joseph Leo. *The Reformation of the Image.* Chicago: University of Chicago Press, 2004.

Korey, Michael. *Die Geometrie der Macht, die Macht der Geometrie: Mathematische Instrumente und fürstliche Mechanik um 1600 aus dem Mathematisch-Physikalischen Salon.* Munich: Deutscher Kunstverlag, 2007.

Kotrč, Ronald F. "The Dodecahedron in Plato's *Timaeus*." *Rheinisches Museum für Philologie* 124, no. 3–4 (1981): 212–222.
Kraus, Hans P. *A Rare Book Saga: The Autobiography of H. P. Kraus*. New York: Putnam, 1978.
Kremer, Richard L. "Experimenting with Paper Instruments in Fifteenth- and Sixteenth-Century Astronomy: Computing Syzygies with Isotemporal Lines and Salt Dishes." *Journal for the History of Astronomy* 42, no. 2 (May 2011): 223–258.

Landolt, Elisabeth. *Das Amerbach-Kabinett: Beiträge zu Basilius Amerbach*. Basel: Kunstsammlung, 1991.
Laue, Georg, ed. *Perspectiva: A Nuremberg Renaissance Casket for the Marquesses of Lothan*. London: Kunstkammer Ltd., 2018.
Lautensack, Heinrich. *Des Circkels vnnd Richtscheyts, auch der Perspectiua, vnd Proportion der Menschen und Rosse, kurtze, doch gründliche vnderweisung* .... Frankfurt am Main, 1564.
Lefèvre, Wolfgang, ed. *Picturing Machines, 1400–1700*. Cambridge, MA: MIT Press, 2004.
Lencker, Johannes. *Perspectiva*. Nuremberg, 1571.
Lencker, Johannes. *Perspectiva literaria*. Nuremberg, 1567.
Livy. *The History of Rome*. London: Henry G. Bohn, 1850–1856.
Long, Pamela O. "Power, Patronage, and the Authorship of *Ars*: From Mechanical Know-How to Mechanical Knowledge in the Last Scribal Age." *Isis* 88, no. 1 (1997): 1–41.
Lucretius. *De rerum natura*. Edited and translated by Cyril Bailey. 3 vols. Oxford: Clarendon Press, 1947.
Luminet, Jean-Pierre. "Science, Art and Geometrical Imagination." *Proceedings of the International Astronomical Union* 5, no. S260 (2009): 248–273.
Lynch, Terence. "The Geometric Body in Dürer's Engraving *Melencholia I*." *Journal of the Warburg and Courtauld Institutes* 45 (1982): 226–232.

Mackinnon, Nick. "The Portrait of Fra Luca Pacioli." *Mathematical Gazette* 77, no. 479 (1993): 130–219.
Maes, Frans W. "Entzerrung der Sonnenuhr des Ludwig Hohenfeld von 1596." *Jahresschrift* Deutsche Gesellschaft für Chronometrie 49 (2010): 163–168.
Mager, Raymond. "Zur Technik der 'Basler Goldschmiedrisse.'" *Jahresberichte der Oeffentlichen Kunstsammlung Basel*, 1963: 35–37.
Malkevitch, Joseph. "Milestones in the History of Polyhedra." In *Shaping Space: Exploring Polyhedra in Nature, Art, and the Geometrical Imagination*, edited by Marjorie Senechal, 53–63. New York: Springer, 2013.
Manders, Kenneth. "Algebra in Roth, Faulhaber, and Descartes." *Historia Mathematica* 33, no. 2 (2006): 184–209.
Mark, Peter. "Towards a Reassessment of the Dating and the Geographical Origins of the Luso-African Ivories, Fifteenth to Seventeenth Centuries." *History in Africa* 34 (2007): 189–211.
Mark, Peter, and José da Silva Horta. *The Forgotten Diaspora: Jewish Communities in West Africa and the Making of the Atlantic World*. New York: Cambridge University Press, 2011.
Marr, Alexander. "The Production and Distribution of Mutio Oddi's *Dello squadro* (1625)." In *Transmitting Knowledge: Words, Images, and Instruments in Early Modern Europe*, edited by Sachiko Kusukawa and Ian Maclean, 165–192. Oxford: Oxford University, 2006.
Marr, Alexander. "Walther Ryff, Plagiarism and Imitation in Sixteenth-Century Germany." *Print Quarterly* 31, no. 2 (2014): 131–143.
Massing, Jean Michel. "The Quest for the Exotic: Albrecht Dürer in the Netherlands." *Circa 1492: Art in the Age of Exploration*, edited by Jay Levenson, 115–119. Washington, DC: National Gallery of Art; New Haven: Yale University Press, 1991.
Maurice, Klaus. *Der drechselnde Souverän: Materialen zu einer fürstlichen Maschinenkunst / Sovereigns as Turners: Materials on a Machine Art by Princes*. Zurich: Ineichen, 1985.
Maurice, Klaus. "Fürstliche Drechselkunst zur äesthetischen Bedeutung von Naturwissenschaft und Technologie" / "The Princely

Art of Turning: On the Aesthetic Significance of the Natural Sciences and Technology." In *Gedrehte Kostbarkeiten / Turned Treasures*, edited by Georg Laue, 8–23. Munich: Kunstkammer Georg Laue, 2004.

McNamee, Megan C. "Imaging and Imagining Solidity." In *After the Carolingians: Re-defining Manuscript Illumination in the 10th and 11th Centuries*, edited by Beatrice Kitzinger and Joshua O'Driscoll, 86–117. Berlin: Walter de Gruyter, 2019.

"Medici Splendour: Pope Leo X and Florence." Press release, 2013.

Melanchthon, Philip. *Orations on Philosophy and Education*. Edited by Sachiko Kusukawa. Translated by Christine F. Salazar. Cambridge: Cambridge University Press, 1999.

Merback, Mitchell B. *Perfection's Therapy: An Essay on Albrecht Dürer's Melencolia I*. New York: Zone Books, 2017.

Methuen, Charlotte. "The Role of the Heavens in the Thought of Philip Melanchthon." *Journal of the History of Ideas* 57, no. 3 (1996): 385–403.

Möller, Lieselotte. *Der Wrangelschrank und die verwandten süddeutschen Intarsienmöbel des 16. Jahrhunderts*. Berlin: Deutschen Verein für Kunstwissenschaft, 1956.

Moore, Ernst D. *Ivory: Scourge of Africa*. New York: Harper & Brothers, 1931.

Moran, Bruce T. "German Prince-Practitioners: Aspects in the Development of Courtly Science, Technology, and Procedures in the Renaissance." *Technology and Culture* 22, no. 2 (April 1981): 253–274.

Moran, Bruce T. "Princes, Machines and the Valuation of Precision in the 16th Century." *Sudhoffs Archiv* 61, no. 3 (1977): 209–228.

Morrall, Andrew. "Art, Geometry, and the Landscape of Ruin in the Sixteenth-Century German *Kunstkabinett*." In *Imagery and Ingenuity in Early Modern Europe: Essays in Honor of Jeffrey Chipps Smith*, edited by Catharine Ingersoll, Alisa McCusker, and Jessica Weiss, 201–212. Turnhout: Brepols, 2018.

Morrall, Andrew. "Defining the Beautiful in Early Renaissance Germany." In *Concepts of Beauty in Renaissance Art*, edited by Francis Ames-Lewis and Mary Rogers, 80–92. Brookfield, VT: Ashgate, 1998.

Moxon, Joseph. *Mechanick Exercises: Or The Doctrine of Handy-Works* [1703]. Dedham, MA: Toolemera Press, 2009.

Müller, Christian. "Holbein oder Holbein-Werkstatt? Zu einem Pokalentwurf der Gottfried Keller-Stiftung im Kupferstichkabinett Basel." *Zeitschrift für schweizerische Archäologie und Kunstgeschichte* 47 (1990): 33–42.

Mulzer, Erich. "Das Jamnitzerhaus in Nürnberg und der Goldschmied Wenzel Jamnitzer." *Mitteilungen des Vereins für Geschichte der Stadt Nürnberg* 61 (1974): 48–89.

Neudörffer, Johann. *Des Johann Neudörfer Schreib- und Rechenmeisters zu Nürnberg Nachrichten von Künstlern und Werkleuten daselbst aus dem Jahre 1547*. Stadtarchiv Nürnberg. Vienna: W. Braumüller, 1875.

Newman, William R. *Promethean Ambitions: Alchemy and the Quest to Perfect Nature*. Chicago: University of Chicago Press, 2004.

Niceron, Jean François. *La perspective cvrievse*. Paris: Langlois, 1652.

Nilson, Christoph Andreas. *Anleitung zur Linearperspective*. Augsburg, 1812.

Oosterhoff, Richard. *Making Mathematical Culture: University and Print in the Circle of Lefèvre d'Étaples*. Oxford: Oxford University Press, 2018.

Pacioli, Luca. *De divina proportione*. 1496–1498. Milan, Biblioteca Ambrosiana.

Pacioli, Luca. *De divina proportione*. Venice: Paganinus de Paganinus, 1509.

Pacioli, Luca. *De divina proportione*. Sansepolcro: Aboca Museum Edizioni, 2009.

Pacioli, Luca, Duilio Contin, Piergiorgio Odifreddi, and Antonio Pieretti. *Antologia della divina proporzione di Luca Pacioli, Piero della Francesca e Leonardo da Vinci*. Sansepolcro: Aboca Museum, 2010.

Palmer, Ada. *Reading Lucretius in the Renaissance*. Cambridge, MA: Harvard University Press, 2014.

Panofsky, Erwin. *Dürers Kunsttheorie*. Berlin: Georg Reimer, 1915.

Panofsky, Erwin. *The Life and Art of Albrecht Dürer*. Princeton University Press, 1955.

Paracelsus. *Paracelsus (Theophrastus Bombastus von Hohenheim, 1493–1541): Essential Theoretical Writings*. Edited and translated by Andrew Weeks. Leiden: Brill, 2008.

Park, Katharine. "Allegories of Knowledge." In *Prints and the Pursuit of Knowledge in Early Modern Europe*, edited by Susan Dackerman, 358–365. Cambridge, MA: Harvard Art Museums, 2011.

Park, Katharine. "The Organic Soul." In *The Cambridge History of Renaissance Philosophy*, edited by Charles B. Schmitt, Quentin Skinner, Eckhard Kessler, and Jill Kraye, 464–484. Cambridge: Cambridge University Press, 1988.

Park, Mungo. *Travels in the Interior Districts of Africa: Performed under the Direction and Patronage of the African Association, in the Years 1795, 1796, and 1797*. London: printed by W. Bulmer for the author, 1799.

Payne, Alina. "Order and Disorder in Renaissance Architecture." Lecture. University of Chicago, November 2007.

Pedersen, Olaf. "Tradition and Innovation." In *Universities in Early Modern Europe (1500–1800)*, edited by Hilde de Ridder-Symoens, 451–488. Vol. 2 of *A History of the University in Europe*. Cambridge: Cambridge University Press, 1996.

Petz, H. "Urkundliche Nachrichten über den literarischen Nachlass Regiomontans und B. Walters 1478–1522." *Mitteilungen des Vereins für Geschichte der Stadt Nürnberg* 7 (1888): 237–262.

Pfinzing von Henfenfeld, Paul. *Ein schöner kurtzer Extract der Geometriae vnnd Perspectiuae*. Nuremberg: Valentin Fuhrmann, 1599.

Phillips, Christopher J. *The New Math: A Political History*. Chicago: University of Chicago Press, 2015.

Piero della Francesca. *L'opera "De corporibus regularibus" di Pietro Franceschi detto della Francesca, usurpata da fra Luca Pacioli*, edited by Girolamo Mancini. Rome: R. Accademia dei Lincei, 1915.

Plato. *The Dialogues of Plato*. Translated by Benjamin Jowett. 3rd ed. 5 vols. New York: Oxford University Press, 1892.

Plumier, Charles. *L'art de tourner....* Paris: C.-A. Jombert, 1749.

Pont, Jean-Claude. *La topologie algébrique: Des origines à Poincaré*. Paris: Presses Universitaires de France, 1974.

Popham, Arthur E. *Catalogue of the Drawings of Parmigianino*. 3 vols. New Haven: Yale University Press, 1971.

Prass, Reiner. "Ausbildung und Schriftkenntnisse von Thüringer Bauhandwerken, 1600–1750." In *Elementarbildung und Berufsausbildung, 1450–1750*, edited by Alwin Hanschmidt and Hans-Ulrich Musolff, 145–166. Cologne: Böhlau, 2005.

Price, David H. *Albrecht Dürer's Renaissance: Humanism, Reformation, and the Art of Faith*. Ann Arbor: University of Michigan Press, 2003.

Principe, Lawrence M. *The Secrets of Alchemy*. Chicago: University of Chicago Press, 2013.

Quiccheberg, Samuel. *The First Treatise on Museums: Samuel Quiccheberg's Inscriptiones, 1565*. Edited and translated by Mark A. Meadow and Bruce Robertson. Los Angeles: Getty Research Institute, 2013.

Raggio, Olga, and Antoine M. Wilmering. *The Gubbio Studiolo and Its Conservation*. 2 vols. New York: Metropolitan Museum of Art, 1999.

Ramond, Pierre. *Masterpieces of Marquetry*. Los Angeles: J. Paul Getty Museum, 2000.

Ramus, Petrus. *Arithmeticae libri duo: Geometriae septem et viginti*. Basel: per Eusebium Episcopium & Nicolai fratris haeredes, 1569.

Ramus, Petrus. *Prooemium mathematicum. Ad Catharainam Mediceam, reginam, matrem regis*. Paris, 1567.

Ramus, Petrus. *Via regia ad geometriam. The Vvay to Geometry. Being Necessary and Usefull, for Astronomers. Geographers. Landmeaters. Sea-men. Engineres. Architecks. Carpenters. Paynters. Carvers, &c.* Translated by William Beadvvell. London: printed by Thomas Cotes, 1636.

Rebel, Ernst. *Albrecht Dürer: Maler und Humanist*. Munich: C. Bertelsmann, 1996.

Reisch, Gregor. *Margarita philosophica*. Strasbourg, 1503.

Reitman, Nimrod. "Allegory's Lament from Benjamin to Dante." *MLN* 133, no. 5 (2018): 1362–1389.

Richeson, David S. *Euler's Gem: The Polyhedron Formula and the Birth of Topology*. Princeton: Princeton University Press, 2008.

Richter, Fleur. *Die Ästhetik geometrischer Körper in der Renaissance*. Stuttgart: G. Hatje, 1995.

Roriczer, Mathes, and Ferdinand Geldner. *Das Büchlein von der Fialen Gerechtigkeit. Faksimile der Originalausg, Regensburg 1486*. Wiesbaden: Pressler, 1965.

Rosenthal, Erwin. "Dürers Buchmalereien für Pirckheimers Bibliothek." *Jahrbuch der Preussischen Kunstsammlungen* 49 (1928): 1–54.

Rostenberg, Leona. "The Libraries of Three Nuremberg Patricians, 1491–1568." *Library Quarterly* 13, no. 1 (January 1943): 21–33.

Rothstein, Bret. "Making Trouble: Strange Wooden Objects and the Early Modern Pursuit of Difficulty." *Journal for Early Modern Cultural Studies* 13, no. 1 (Winter 2013): 96–129.

Rowe, David E. "Euclidean Geometry and Physical Space." *Mathematical Intelligencer* 28, no. 2 (2006): 51–59.

Rowe, David E. "Mathematical Models as Artefacts for Research: Felix Klein and the Case of Kummer Surfaces." *Mathematische Semesterberichte* 60 (2013): 1–24.

Rowe, David E. "On Models and Visualizations of Some Special Quartic Surfaces." *Mathematical Intelligencer* 40, no. 1 (2018): 59–67.

Rowlands, John. *Drawings by German Artists and Artists from German-Speaking Regions of Europe in the Department of Prints and Drawings in the British Museum*. London: British Museum Press, 1993.

Rubin, Patricia Lee. *Giorgio Vasari: Art and History*. New Haven: Yale University Press, 1995.

Rupprich, Hans, ed. *Dürer: Schriftlicher Nachlass*. 3 vols. Berlin: Deutscher Verein für Kunstwissenschaft, 1956–1969.

Rupprich, Hans. "Die kunsttheoretischen Schriften L. B. Albertis und ihre Nachwirkung bei Dürer." *Schweizer Beiträge zur Allgemeinen Geschichte* 18–19 (1960–1961): 219–239.

Rutkin, H. Darrel. "Astrology." In *The Cambridge History of Science*, vol. 3, *Early Modern Science*, edited by Katharine Park and Lorraine Daston, 541–561. Cambridge: Cambridge University Press, 2006.

Ryder, A. F. C. *Benin and the Europeans, 1485–1897*. New York: Humanities Press, 1969.

Ryder, A. F. C. "A Note on the Afro-Portuguese Ivories." *Journal of African History* 5, no. 3 (1964): 363–365.

Sanders, Philip Morris. "Charles de Bovelles's Treatise on the Regular Polyhedra (Paris, 1511)." *Annals of Science* 41, no. 6 (1984): 513–566.

Sasaki, Chikara. *Descartes's Mathematical Thought*. Dordrecht: Kluwer, 2003.

Sattelmacher, Anja. "Geordnete Verhältnisse: Mathematische Anschauungsmodelle im frühen 20. Jahrhundert." *Berichte zur Wissenschaftsgeschichte* 36, no. 4 (2013): 294–312.

Schaper, Christa. *Die Hirschvogel von Nürnberg und ihr Handelshaus*. Nuremberg: Verein für Geschichte der Stadt Nürnberg, 1973.

Schätze der Astronomie: Arabische und deutsche Instrumente aus dem Germanischen Nationalmuseum / Treasures of Astronomy: Arabic and German Instruments of the German National Museum. [In German, English, and Arabic.] Nuremberg: Germanisches Nationalmuseum, 1983.

Scheible, Heinz. *Philipp Melanchthon: Leben und Werk in Bildern*. Karlsruhe: Landesbildstelle Baden, 1998.

Schmidt, Eike D., and Maria Sframeli. *Diafane passioni: Avori barocchi dalle corti europee*. Livorno: Sillabe, 2013.

Schmidt, Suzanne Karr. *Altered and Adorned: Using Renaissance Prints in Daily Life*. Chicago: Art Institute of Chicago, 2011.

Schmidt, Suzanne Karr. "Georg Hartmann and the Development of Printed Instruments in Nuremberg." In *Prints and the Pursuit of Knowledge in Early Modern Europe*, edited by Susan Dackerman, 268–279. Cambridge, MA: Harvard Art Museums, 2011.

Schmidt, Suzanne Karr. "Georg Hartmann as Interactive Printmaker." In *Interactive and Sculptural Printmaking in the Renaissance*, 205–240. Leiden: Brill, 2017.

Schmitt, Lothar. "Education and Learning among Sixteenth-Century German Artists." *Studies in the History of Art* 60 (2001): 72–81.

Schmuttermayer, Hans. *Fialenbüchlein*. Nuremberg, 1486.

Schön, Erhard. *Unterweisung der Proportion und Stellung der Possen...* Nuremberg, 1538.

Schöner, Andreas. *Gnomonice*. Nuremberg, 1562.

Schönherr, David von. "Wenzel Jamnitzers Arbeiten für Erzherzog Ferdinand." *Mittheilungen des Instituts für Oesterreichische Geschichtsforschung* 9 (1888): 289–305.

Schreiber, Peter. "A New Hypothesis on Dürer's Enigmatic Polyhedron in His Copper Engraving 'Melencolia I.'" *Historia Mathematica* 26, no. 4 (1999): 369–377.

Schreiber, Peter, Gisela Fischer, and Maria Luise Sternath. "New Light on the Rediscovery of the Archimedean Solids during the Renaissance." *Archive for History of Exact Sciences* 62, no. 4 (2008): 457–467.

Schwarz, Karl. *Augustin Hirschvogel: Ein deutscher Meister der Renaissance*. Berlin: Julius Bard, 1917.

Schwarz, Karl. *Augustin Hirschvogel: I. Lebensbeschreibung und Zeichnungen*. Heidelberg: Rößler & Herbert, 1915.

Schwenter, Daniel. *Geometriae practicae novae et auctae tractatus*. Nuremberg: bey Jeremia Dümlern, 1641.

Seeliger-Zeiss, Anneliese. *Lorenz Lechler von Heidelberg und sein Umkreis: Studien zur Geschichte der spätgotischen Zierarchitektur und Skulptur in der Kurpfalz und in Schwaben*. Heidelberg: C. Winter, 1967.

Seidenfuß, Birgit. *"Daß wirdt also die Geometrische Perspektiv genand": Deutschsprachige Perspektivtraktate des 16. Jahrhunderts*. Weimar: VDG, 2006.

Seipel, Wilfried, ed. *Spielwelten der Kunst: Kunstkammerspiele*. Vienna: Kunsthistorisches Museum Wien, 1998.

Shalem, Avinoam. *The Oliphant: Islamic Objects in Historical Context*. Leiden: Brill, 2004.

Shelby, Lonnie Royce, ed. and trans. *Gothic Design Techniques: The Fifteenth-Century Design Booklets of Mathes Roriczer and Hanns Schmuttermayer*. Carbondale: Southern Illinois University Press, 1977.

Skinner, Quentin. *The Foundations of Modern Political Thought*. Vol. 2, *The Age of Reformation*. Cambridge: Cambridge University Press, 1978.

Smith, Jeffrey Chipps. "Albrecht Dürer as Collector." *Renaissance Quarterly* 64, no. 1 (2011): 1–49.

Smith, Pamela H. *The Body of the Artisan: Art and Experience in the Scientific Revolution*. Chicago: University of Chicago Press, 2004.

Smith, Pamela H. "Vermilion, Mercury, Blood, and Lizards: Matter and Meaning in Metalworking." In *Materials and Expertise in Early Modern Europe: Between Market and Laboratory*, edited by Ursula Klein and E. C. Spary, 29–49. Chicago: University of Chicago Press, 2009.

Smith, Pamela H., and T. Beentjes. "Nature and Art, Making and Knowing: Reconstructing Sixteenth-Century Life-Casting Techniques." *Renaissance Quarterly* 63, no. 1 (2010): 128–179.

Söll-Tauchert, Sabine, Raphael Beuing, and Burkard von Roda. *Die grosse Kunstkammer: Bürgerliche Sammler und Sammlungen in Basel*. Basel: Historisches Museum; C. Merian, 2011.

Springett, David. *Woodturning Wizardry*. East Sussex: Guild of Master Craftsman Publications, 2005.

Steck, Max. *Dürers Gestaltlehre der Mathematik und der bildenden Künste*. Halle: Niemeyer, 1948.

Stephenson, Bruce, Marvin Bolt, and Anna Felicity Friedman. *The Universe Unveiled: Instruments and Images through History*. Cambridge: Cambridge University Press, 2000.

Stetten, Paul von. *Kunst-, Gewerb- und Handwerks Geschichte der Reichs-Stadt Augsburg*. 2 vols. Augsburg, 1779–1788.

Stevin, Simon. *The Principal Works of Simon Stevin*. Vol. 2. Edited by Dirk J. Struik. 2 vols. Amsterdam: C. V. Swets and Zeitlinger, 1958.

Stevin, Simon. *Problemata geometrica*. Antwerp: J. Bellerus, 1583.

Stöer, Lorentz. *Geometria et perspectiva*. Augsburg, 1567.

Stöer, Lorentz. *Geometria et perspectiva corpora regulata et irregulata*. 1562–1584. Universitätsbibliothek München, Handschrift Cim 103.

Stöer, Lorentz. *Optica*. Late 16th century. Universitätsbibliothek Erlangen-Nürnberg, MS 2774.

Stöer, Lorentz. *Polyhedra and Other Solid-Geometric Figures*. 1585, 1598, and undated. Houghton Library, Harvard University, Cambridge, MA, Typ 520.67.810.

Stöer, Lorentz. Sketchbook. 1595. Staats- und Stadtbibliothek Augsburg, 4 Cod. Aug. 247.

Strano, Giorgio, Stephen Johnston, Mara Miniati, and Alison D. Morrison-Low, eds. *European Collections of Scientific Instruments, 1550–1750*. Leiden: Brill, 2009.

Strauss, Walter L. *The Complete Drawings of Albrecht Dürer*. 6 vols. New York: Abaris Books, 1974.

Syndram, Dirk. "Amassing Magnificence: The Role of the Kunstkammer in Princely Self-Representation." In *Making Marvels: Science and Splendor at the Courts of Europe*, edited by Wolfram Koeppe, 35–40, 263. New York: Metropolitan Museum of Art, 2019.

Syndram, Dirk. "Princely Diversion and Courtly Display: The *Kunstkammer* and Dresden's Renaissance Collections." In *Princely Splendor: The Dresden Court, 1580–1620*, edited by Dirk Syndram and Antje Scherner, 54–69. Dresden: Staatliche Kunstsammlungen Dresden; Milan: Mondadori Electa, 2004.

Syndram, Dirk. *Das Schloss zu Dresden: Von der Residenz zum Museum*. Munich: Koehler & Amelang, 2001.

Syndram, Dirk. "Three Turned-Ivory Pieces (cat. 55)." In *Making Marvels: Science and Splendor at the Courts of Europe*, edited by Wolfram Koeppe, 142. New York: Metropolitan Museum of Art, 2019.

Syndram, Dirk, Jutta Kappel, and Ulrike Weinhold. *The Baroque Treasury at the Grünes Gewölbe Dresden*. Munich: Deutscher Kunstverlag; Dresden: Staatliche Kunstsammlungen Dresden, 2006.

Syndram, Dirk, and Martina Minning, eds. *Die Inventare der kurfürstlich-sächsischen Kunstkammer in Dresden*. 4 vols. Staatliche Kunstsammlungen Dresden. Dresden: Sandstein, 2010–2012.

Syndram, Dirk, and Antje Scherner, eds. *Princely Splendor: The Dresden Court, 1580–1620*. Dresden: Staatliche Kunstsammlungen Dresden; Milan: Mondadori Electa, 2004.

Teuber, Johann Martin. *Vollständiger Unterricht der gemeinen und höheren Dreh-Kunst*. Regensburg: J. P. Lang, 1740.

Thieme, Ulrich, Felix Becker, et al., eds. *Allgemeines Lexikon der bildenden Künstler von der Antike bis zur Gegenwart*. Vol. 32. Leipzig: E. A. Seemann, 1938.

Thorndike, Lynn. *The Sphere of Sacrobosco and Its Commentators*. Chicago: University of Chicago Press, 1949.

Timann, Ursula. *Untersuchungen zu Nürnberger Holzschnitt und Briefmalerei in der ersten Hälfte des 16. Jahrhunderts*. Münster: Lit, 1993.

Tormey, Alan, and Judith Farr Tormey. "Renaissance Intarsia: The Art of Geometry." *Scientific American* 247, no. 1 (July 1982): 136–143.

Tunstall, Cuthbert. *De arte supputandi*. Paris, 1538.

Tzschimmer, Gabriel. *Durchtigsten Zusammenkunfft*. Nuremberg, 1680.

Ueberwasser, Wilhelm. "Spätgotische Bau-Geometrie: Untersuchungen an den 'Basler Goldschmiedrissen.'" *Jahresberichte der Oeffentlichen Kunstsammlung Basel*, 1928–1930: 79–123.

Vasari, Giorgio. *Lives of the Most Eminent Painters, Sculptors, and Architects*. Translated by Gaston du C. De Vere. 10 vols. London: Macmillan and the Medici Society, 1912–1915.

Vasari, Giorgio. *Vasari on Technique*. New York: Dover, 2012.

Vasari, Giorgio. *Le vite dei più eccellenti pittori, scultori e architetti*. Edited by Maurizio Marini. 3rd ed. Rome: Newton & Compton, 1997.

Vasari, Giorgio. *Le vite de' più eccellenti pittori, scultori e architettori: Nelle redazioni del*

1550 e 1568. Edited by Rosanna Bettarini and Paola Barocchi. Florence: Sansoni, 1966–1987.

Veltman, Kim H. "Geometric Games: A Brief History of the Not So Regular Solids." www.sumscorp.com/perspective/geometric_games/ (site discontinued).

Vickers, Michael, *Ivory: A History and Collectors Guide*. London: Thames & Hudson, 1987.

Vincent, Clare. "Prince Karl I of Liechtenstein's Pietre Dure Tabletop." *Metropolitan Museum Journal* 22 (1987): 157–178.

Voltelini, Hans von, ed. "Urkunden und Regesten aus dem K. u. K. Haus-, Hof- und Staatsarchiv in Wien." *Jahrbuch der kunsthistorischen Sammlungen in Wien* 11 (1890): i–lxviii.

Waddington, Charles. *Ramus (Pierre de la Ramée): Sa vie, ses écrits et ses opinions*. Paris: C. Meyrueis, 1855.

Wamberg, Jacob, ed. *Art and Alchemy*. Copenhagen: Museum Tusculanum Press, University of Copenhagen, 2006.

Warburg, Aby. *Der Bilderatlas Mnemosyne*. Edited by Martin Warnke, with Claudia Brink. Berlin: Akademie Verlag, 2008.

Warburg, Aby. *Bilderatlas Mnemosyne: The Original*. Edited by Robert Ohrt and Axel Heil. Berlin: Hatje Cantz, 2020.

Warncke, Carsten-Peter. *Die ornamentale Groteske in Deutschland, 1500–1650*. 2 vols. Berlin: Volker Spiess, 1979.

Weitzel, Hans. "A Further Hypothesis on the Polyhedron of A. Dürer's Engraving Melencolia I." *Historia Mathematica* 31, no. 1 (2004): 11–14.

Weisstein, Eric W. "Regular Polygon." From MathWorld—A Wolfram Web Resource. www.mathworld.wolfram.com/RegularPolygon.html.

Welper, Eberhard. *Neu-vermehrte Welperische Gnomonica....* Nuremberg, 1708.

Wesoly, Kurt. "Berufsausbildung von Handwerkslehrlingen und Elementarbildung vornehmlich am Mittelrhein bis ins 17. Jahrhundert." In *Elementarbildung und Berufsausbildung, 1450–1750*, edited by Alwin Hanschmidt and Hans-Ulrich Musolff, 109–124. Cologne: Böhlau, 2005.

Winckelmann, Johann Joachim. *Winckelmann: Writings on Art*. Edited by David Irwin. London: Phaidon, 1972.

Witcombe, Christopher L. C. E. *Copyright in the Renaissance: Prints and the* Privilegio *in Sixteenth-Century Venice and Rome*. Leiden: Brill, 2004.

Wolff, Christian. *Elementa matheseos universae*. Vol. 1. New ed. Geneva, 1743.

Wright, William J. "Evaluating the Results of Sixteenth Century Educational Policy: Some Hessian Data." *Sixteenth Century Journal* 18, no. 3 (Autumn 1987): 411–426.

Zimmermann, Konrad. "Vorgeschichte und Anfänge der Dresdener Skulpturensammlung." In *Die Dresdener Antiken und Winckelmann*, edited by Konrad Zimmermann, 9–32. Berlin: Akademie-Verlag, 1977.

Zinner, Ernst. *Regiomontanus: His Life and Work*. Translated by Ezra Brown. Amsterdam: Elsevier Science & Technology, 1990.

Zorach, Rebecca. "Meditation, Idolatry, Mathematics: The Printed Image in Europe around 1500." In *The Idol in the Age of Art: Objects, Devotions and the Early Modern World*, edited by Michael W. Cole and Rebecca Zorach, 317–342. Burlington, VT: Ashgate, 2009.

INDEX

Adelard of Bath, 72, 76
Alberti, Leon Battista, 72, 76, 155
Albrecht V (duke of Bavaria), 211, 221
Albrecht VII (archduke of Austria), 175
alchemy, 163
Aldegrever, Heinrich, 17
Ambras Castle, 155, 175, 221, 261
Amerbach, Basilius, 145
Amerbach-Kabinett, 145, 148. *See also* Basler Goldschmiedrisse
Amman, Jost, 122, 123, 155, 157, 166–167
Andreae, Hieronymus, 89–92
Aquinas, Thomas, 61
Archimedean solids, 14, 19, 31, 36–37, 79, 89, 95, 116, 119, 247, 256
Archimedes, 30, 31, 72
Arcimboldo, Giuseppe, 229
Arfe, Juan de, 114
Aristotle, 17, 44, 106, 161, 188
Augsburg, 18, 171–172, 174, 179, 180, 185–186, 201–202
Augsburg Confession, 104
August (elector of Saxony), 209–212, 215, 217, 220, 223–229, 239
August the Younger (duke of Brunswick-Lüneburg), 114, 209. *See also* Herzog August Bibliothek

Barbaro, Daniele, 14, 37, 38, 44, 47, 86, 155, 174, 261
Basler Goldschmiedrisse, 148–154
Beham, Hans Sebald, 58, 59
Beltrami, Eugenio, 246, 251
Benjamin, Walter, 202
Besson, Jacques, 209, 214
Bill, Max, 245
Billingsley, Henry, 24, 25, 247
Bloemaert, Abraham, 59
bodies (*corpora*), regular and irregular (*regulata et irregulata*), 13–14, 19, 27, 36–37. *See also* Archimedean solids; Platonic solids
- as emblems of knowledge, 17, 24, 27, 54, 143, 178, 189, 201
- as physical models, 12–13, 14, 17, 19–23, 41, 47–48, 52–54, 98, 116–128, 131, 196, 210, 218, 229–230, 239, 246–247 (*see also* paper: models of polyhedra)
- in relation to the human body, 32, 75, 128
- as subjects of perspective, 14, 17, 23, 36–41, 43–60, 86–87, 196, 210
- unfolding of, 19–20, 82–89, 93–97, 98, 109, 116–119, 125, 131, 141, 160, 252
- as workshop tools, 14, 17, 41, 47–48, 67, 120, 196–200, 220–221

Bonnemer, Marin, 59–60, 62–63
Bovelles, Charles de, 14, 83
Brahe, Tycho, 248
Braun, Hieronymus, 24
bravura. *See* virtuosity, artistic

Calcidius, 43, 83, 86, 96, 120, 189, 257
Cambiaso, Luca, 76
Camerarius, Joachim, 69, 78, 259
Campanus of Novara, 33
Caraglio, Giovanni Jacopo, 41, 48–51, 54, 60
Caron, Joseph, 245
Carpaccio, Vittore, 14, 41, 48, 52, 210
Casa de Guiné, 221
Casa de Índia, 221
Castiglione, Giovanni Benedetto, 59
Castrucci, Cosimo di Giovanni, 170, 175
Cellini, Benvenuto, 141
Cesariano, Cesare, 102
Charles V (emperor), 141, 172
Christian I (elector of Saxony), 194, 196–197, 199, 211, 218
Christian II (elector of Saxony), 211
Cicero, Marcus Tullius, 106
Cock, Hieronymus, 180
contrefait spheres, 211, 217, 218, 233, 235, 236, 239
copying, of polyhedra, 17, 18, 93, 94, 134, 180, 185, 195–200
- techniques, 76, 94, 196–199

Cosimo II de' Medici, 102
Cranach, Lucas, the Elder, 227
Crespi, Daniele, 234, 236, 241

de' Barbari, Jacopo, 44, 45, 52, 54, 61, 70, 120
de Bruyn, Nicolaes, 180
de Bry, Théodore, 180
Dee, John, 247
Delsenbach, Johann Adam, 66
Descartes, René, 48, 246–247, 249, 250
Diderot, Denis, 116
Dietterlin, Wendel, 180
Diogenes, 49, 51
Donatello (Donato di Niccolò di Betto Bardi), 174
Donne, John, 163
Doppelmayr, Johann Gabriel, 112, 114, 141

drawing. *See also* paper: designing on
tools and apparatus for, 120–127, 210
training in, 14, 32, 75, 116, 120, 132, 136, 143, 144, 199
Dresden, court and castle, 136, 155, 209–211, 217–233, 236, 237, 240. *See also* August (elector of Saxony); Christian I (elector of Saxony); Christian II (elector of Saxony); Friedrich III (elector of Saxony)
Grünes Gewölbe, 47, 196
Staatliche Kunstsammlungen, 229–232
du Cerceau, Jacques Androuet, 180, 200
Dürer, Albrecht, 14, 19, 37, 44, 47, 59, 61, 69–99, 102, 108–110, 112, 116, 119, 120, 123, 125, 128, 132, 134, 160, 189, 196, 218, 221–222, 229, 230, 246, 250
Melencolia I, 14, 55, 58–59, 258
Underweysung der Messung, 19, 37, 47, 55, 70, 75, 78–96, 98, 109, 116, 123, 142, 160, 246
Vier Bücher von menschlicher Proportion, 74–79, 128, 259
Dutch East India Company, 222

Ebelmann, Hans Jakob, 196
Eck, Oswald von, 13, 14
Eckher von Kapfing, Johann Franz, 186
Einstein, Albert, 245
elements (earth, air, fire, water), associated with Platonic solids, 18, 42–44, 98, 156, 160, 161, 188
Escher, M. C., 245
etching, 114, 202
Euclid, *Elements*, 13, 14, 17, 23, 27, 32, 36, 44, 60, 61, 120, 132, 155, 163, 178, 235, 246, 248, 251. *See also* non-Euclidean geometry
Dürer and, 70, 72–73, 75–79, 82, 86, 94, 96, 109
educational use of, 13, 102, 105–106, 108, 111–112, 125, 134, 188, 195
limits of visualization in, 34, 47–48
polyhedra used as aids in understanding, 54–55, 109
printed editions of, 24–25, 27, 32–36, 51, 70, 72, 76, 78, 133, 209, 247
Euler, Leonhard, 143, 246, 247, 250, 265

Faulhaber, Johannes, 46, 48, 52, 210, 247
Ferdinand I (emperor), 114, 141, 186
Ferdinand II (archduke of Tyrol), 151, 155, 172, 175, 221
Ferdinand III (emperor), 217
Ficino, Marsilio, 42–43, 51
Fleischer, Hieronymus, 172
Flötner, Peter, 22, 23
Fludd, Robert, 248
Foix-Candale, François de, 246–247
Francesco I de' Medici, 217
Franz Josef II (prince of Lichtenstein), 188
Frederik III (king of Denmark), 217
Friedrich I (duke of Württemberg), 18, 20, 209, 248, 250, 251, 252
Friedrich III (elector of Saxony), 44, 69, 104
Frisius, Gemma, 110
Fugger, family, 171
Fugger, Jakob, 171
Fuller, R. Buckminster, 245

Galileo Galilei, 102, 248
Gauss, Carl Friedrich, 245
geometria, 98, 109, 112, 258
geometry, study of, 13–19, 24, 26, 36. *See also* Euclid, *Elements*; perspective, study of
allegorical representation of, 59, 60
experimentation with, 17, 18, 19, 23, 26–27, 37, 76, 82, 94, 98–99, 109, 112, 123, 125, 128, 132, 143, 148, 189, 193, 196, 202, 220, 245, 250–251
in workshops, 17, 18–19, 23, 27, 128
Germanisches Nationalmuseum, 52–53, 192–193
Gessner, Conrad, 34
Giovanni da Verona, Fra, 41, 55
Giuliano da Maiano, 41
goldsmiths, 18, 23, 48, 67, 79, 102, 114, 128, 141–143, 144, 161–165
drawings produced by, 143, 144–154, 162, 195
Gottlieb, Conrad, 175

Haiden, Hans, 120, 127
Hainhofer, Philipp, 209–210, 217, 265
Halt, Peter, 120, 121, 160, 189
Hannibal, 221
Hartmann, Georg, 83, 84–85
Harvard University, Houghton Library, 186, 188, 193, 269

Hauslab, Frank, Ritter von, 188, 269
Heiden, Markus, 218, 222
Herzog August Bibliothek, 102, 114, 136, 196
Hirschvogel, Augustin, 26, 109, 112–120, 123, 125, 128, 131, 132, 133, 134, 136, 247
 Geometria, 114–119, 128, 209
Hofer, Philip, 188
Hohenfeld, Ludwig, 18, 20–21, 249
Holbein, Hans, the Younger, 149, 151, 152
Homer, *Iliad*, 164
Honter, Johannes, 110
Hörislamb, Adam, 244

Ibn Al-haitan, 72
intarsia, 18, 41, 55, 144, 171, 172, 174–175, 180, 185, 199, 201
inventories, of libraries and collections, 47, 72, 75, 136, 175, 209–210, 217, 218, 220, 227, 233, 250
Italy, artistic and artisanal exchanges with Germany, 44–47, 70, 72, 217
ivory, 18, 175, 208–211, 217–240
 procurement of, 222–227, 237
 properties of, 222
 value of, 217, 223

Jacquard loom, 215
Jamnitzer, Albrecht, 141
Jamnitzer, Wenzel, 67–68, 109, 120–123, 126, 133, 142, 145, 148, 151, 161–162, 185, 186, 193, 196, 204, 209, 218, 246, 250
 Perspectiva corporum regularium, 23, 26, 44, 123, 141–145, 154–167, 178, 189, 196, 202
Jean de Meung, 95
Jeweller's Pocket Book, 151–152
Joachim (elector of Brandenburg), 44
Johann Casimir (duke of Saxony-Coburg), 218
Jordan, Camille, 247
Judd, Donald, 245

Kant, Immanuel, 80
Karl I (prince of Liechtenstein), 170, 175
Kepler, Johannes, 14, 102, 209, 248–253
Kieser, Eberhard, 66, 68
Klein, Felix, 245, 246, 250, 251
Koberger, Anton, 72
Kratzer, Nikolaus, 78, 255
Kraus, H. P., 188, 269
Kunstkammern, 18, 155, 209–211, 212, 217–221, 227–229, 250
Kuster, Peter, 148

lathe turning, 211–217, 220, 222, 227, 230–236
 technology of, 212, 214–215, 222, 230–231, 234–235, 271
Lautensack, Hanns, 12, 13
Lautensack, Heinrich, 109, 112, 120, 128, 131–136, 209
Lautensack, Paul, 128
Lebesgue, Henri, 247
Legendre, Adrien-Marie, 247
Lehrbücher, geometrical, 18, 19, 23, 98, 103, 108–116, 119, 125, 128, 133–136, 144, 156, 160, 195, 209, 250, 262
Leibniz, Gottfried Wilhelm, 247
Lencker, Johannes (Hans), 109, 120, 134, 161, 196, 209, 247
Leonardo da Vinci, 14, 17, 34, 37, 40, 41, 44, 54, 61
Listing, Johann Benedict, 247
Livy (Titus Livius), 221
Lobenigk, Egidius, 218, 219, 245, 247
Lucretius, 160
Luther, Martin, 104, 105

Maggiore, Giovanni Ambrogio, 217, 245
Manger, Michael, 186
Man Ray, 245
Maria (archduchess of Austria), 172
marquetry, 171–172, 175, 180, 185, 200
Master of Perspective (Nuremberg, sixteenth century), 16, 17, 175
Mästlin, Michael, 249, 250
materiality, geometry and, 18, 26, 136, 239
mathematics, modern, 26, 98, 134, 143, 245–246, 247
Mathesius, Johannes, 163
Matthias I (king of Hungary), 68
Maximilian I (emperor), 44, 78, 89, 141, 217
Maximilian II (emperor), 145, 155, 265
measurement, 60, 68, 75, 86, 116, 209, 212, 236, 258. *See also* Dürer, Albrecht: *Underweysung der Messung*
 instruments, 48, 60, 120, 141, 156
 skills, 14, 17, 23, 41, 70, 79, 93, 98, 103, 133, 144
Meisner, Daniel, 67
Melanchthon, Philip, 104–106, 110, 111, 125, 262
metamorphosis, in design, 145, 154, 163, 165, 178, 196. *See also* transformation, of geometrical figures
Metropolitan Museum of Art, New York, 175
Michelangelo Buonarroti, 55
Miseroni, Dionisio, 170
mixed mathematics, 17, 79, 102, 106, 110–112, 125, 133, 136, 201, 229, 247, 255

Möbius, August Ferdinand, 247
Mocenigo, Giovanni (doge), 33
Monge, Gaspard, 134, 245, 251
Moxon, Joseph, 236, 273
Mulhouse, Hôtel de Ville, 171, 172, 174, 200
Museum Angewandte Kunst, Frankfurt, 88, 175
Museum für Angewandte Kunst, Cologne, 175–178

Neptune, 141
Neudörffer, Johann, 52, 53, 66, 67, 141, 172
Niceron, Jean François, 111
Nicolas de Neufchâtel, 52, 53, 61, 67, 140, 210
Nilson, Christoph Andreas, 111–112, 113, 128
non-Euclidean geometry, 236, 245–246, 251
Nuremberg, 23–24, 67–72, 89, 108, 110, 123, 128, 133, 141–145, 148, 156, 164, 175, 186, 217, 239

Olivier, Théodore, 134
Örtel, Anton, 229
Ovid (Publius Ovidius Naso), 18, 248

Pacioli, Luca, 14, 34, 45, 52, 60, 70, 78–79, 83, 95, 189, 218, 234, 256
    De divina proportione, 17, 34, 36–42, 47, 48, 50, 51, 54–57, 60, 82, 86, 96, 160, 203, 210, 250, 256, 261
Pandolfini, Giuliano di Piero, 170
Panofsky, Erwin, 70, 87, 95
paper
    designing on, 23, 80, 112, 143–154, 162, 185–195, 197, 229, 235–236, 239
    models of polyhedra, 14, 24, 47, 86–87, 93, 96, 98, 109, 116, 119–120, 128, 131, 164, 199, 210, 218, 229, 246, 251
    tracing/copying with, 94, 196–197, 199–200
Pappus of Alexandria, 36
Parmigianino (Girolamo Francesco Maria Mazzola), 14, 15, 30, 41, 48, 49, 51
perspective, study of, 14, 17, 23, 26, 27, 32, 36–41, 43–55, 60–61, 70, 72, 75, 78–79, 86–87
    polyhedral models as tools for, 14, 17, 41, 48, 54, 196, 210
Pfinzing von Henfenfeld, Paul, 120, 123–127, 128, 133, 263
Philip I (landgrave of Hesse), 104
Philip II (king of Spain), 172, 180
Philipp II (duke of Pomerania), 209
physicalist tradition of geometry, 134
Piero della Francesca, 14, 36, 70, 155
pietre dure, 174–175, 229
Pincker, Christof, 210
Pirckheimer, Willibald, 69, 70, 72, 76, 78–79, 93, 95
Plato, 13, 42–44, 51, 54, 83, 99, 155, 160, 163, 251
Platonic solids (*corpora regulata*), 13–14, 18, 19, 23, 27, 31, 36, 37, 41, 47, 95, 119, 125, 128–129, 142, 160, 178, 188–189, 221, 235, 239, 249, 251
    Dürer and, 82, 86, 89, 95, 96–99, 110, 125, 160
    experiments with, 36, 37, 43, 119, 193, 245, 246
    Jamnitzer and, 141, 155–160, 162, 188–189, 193
    Kepler and, 249–250
    physical models of, 41, 47–48, 52, 54, 98, 116, 119–120, 123, 128, 131, 196, 210, 218, 230, 239, 247
    Stöer and, 188–189, 193–196
Pliny (Gaius Plinius Secundus), 106, 110
Plumier, Charles, 216
polyhedral nets, 19, 83, 86, 89, 95–96. *See also* bodies (*corpora*): unfolding of
Ptolemy (Claudius Ptolemaeus), 19, 48, 72, 105, 106, 110, 111

Quiccheberg, Samuel, 47, 211, 229

Raimondi, Marcantonio, 44
Ramus, Petrus (Pierre de la Ramée), 102, 125, 133, 264
Raphael (Raffaello Sanzio), 44
Ratdolt, Erhard, 33–36, 47, 51, 52, 72, 76
Regiomontanus, Johannes, 67, 68–69, 70, 72, 76, 78, 95
Reisch, Gregor, 106, 107
Reisner, Frédéric, 133
Reverdy, Georges, 137
Riemann, Bernhard, 245
Riese, Abraham, 210, 211
Riese, Adam, 210
Rivius. *See* Ryff
Robert of Chester, 33
Rodler, Hieronymus, 110, 134, 262
Rogel, Hans (Johann), 186, 197
Rohn, Karl, 245
Roriczer, Mathes, 72, 259
Rudolf II (emperor), 102, 141, 217, 218, 220, 248
Ryff (Rivius), Walther Hermann, 136, 180

Sachs, Hans, 155
Sacrobosco, Johannes de, 72, 106
Sadeler, Raphael I, 212, 213
Saxony, electorate of. *See* Dresden, court and castle
Schmuttermayer, Hans, 72
Schön, Erhard, 76, 110, 238, 239
Schöner, Andreas, 255
Schöner, Johannes, 69, 82, 98, 110
Schwenter, Daniel, 47, 114
Scipio, Lucius Cornelius (Scipio Asiaticus), 221
seriality, 145, 152, 196
Serlio, Sebastiano, 180
Settala, Manfredo, 234, 236, 241
silver, 141, 143
Sirigatti, Lorenzo, 174
Solis, Virgil, 18, 59, 152, 180, 200
Staats- und Stadtbibliothek Augsburg, 189
stereometry, 14, 54, 75–77, 110, 112, 148, 210, 238–239, 259
Stetten, Paul von, 185
Stevin, Simon, 155, 246
Stimmer, Christoph, 102
Stöer, Lorentz, 14, 110, 178–197, 199, 200–203, 230, 233, 247, 248, 263
  Geometria et perspectiva, 3, 178–188, 197, 199, 202, 209
Strada, Jakob, 151
Strohmeir, Lorenz, 172

tactility, in geometry, 17, 44, 80, 96, 103, 109, 133, 230
textbooks. See Euclid, *Elements*; *Lehrbücher*, geometrical
Thiry, Léonard, 180
transformation, of geometrical figures, 43, 89, 94, 98, 109–110, 112, 155–156, 162–164, 185, 188–189, 195–196, 203, 245, 247. *See also* metamorphosis, in design
  in artisanal work, 145, 152, 154, 162, 164–165
  of materials, 43, 162–163, 165, 220–221, 237
Tunstall, Cuthbert, 105
Tzschimmer, Gabriel, 226, 227

Uccello, Paolo, 24, 36, 59, 172
Ugo da Carpi, 30, 31, 41, 54, 60, 96
Universitätsbibliothek Erlangen, 186, 197, 199
Universitätsbibliothek München, 186, 188, 193, 202
University of Bologna, 108
University of Cambridge, 105
University of Heidelberg, 104, 105
University of Helmstedt, 110
University of Leipzig, 105
University of Marburg, 104
University of Oxford, 105
University of Tübingen, 105, 209, 248
University of Vienna, 108
University of Wittenberg, 104, 105, 110

Vasari, Giorgio, 32, 36, 55, 172, 174, 257
Virgil (Publius Vergilius Maro), *Aeneid*, 116
virtuosity, artistic, 18, 142, 144, 189, 220–222, 229
Visconti, Prospero, 217
Vitruvius Pollio, Marcus, 33, 37, 48, 69–70, 102, 245
Vredeman de Vries, Hans, 180, 209

Waldseemüller, Martin, 82
Walther, Bernhard, 68, 69, 70, 72, 76, 95
Warburg, Aby, 95–96
Wecker, Georg, 217, 218, 228, 229, 245
Wecker, Hans (Johann), 217, 233
Weishaupt, Bartholmä, 172
Werner, Johannes, 69
Wierix, Johannes, 59
Wilhelm V (duke of Bavaria), 217
Winckelmann, Johann Joachim, 189
Wolff, Christian, 125, 129
woodblock printing, 33–34, 37, 89–92, 119, 261

Zamberti, Bartolomeo, 34, 51, 70, 76, 78, 256
Zamberti, Giovanni, 261
Zeller, Jakob, 218
Zick, Lorenz, 208, 217
Zick, Peter, 217
Zick, Stephen, 208

ACKNOWLEDGMENTS

The research and writing of this book over the last decade could not have been completed without the many conversations, anecdotes, leads, and references I had the good fortune to collect along the way in museums, conferences, and archives. A particular acknowledgment is due to my dissertation committee at the Department of the History of Science at Harvard University, Ann Blair, Peter Galison, Alina Payne, and especially Katharine Park, for their crucial encouragement in the development of my work. Lorraine Daston (MPIWG) provided many productive insights. Wolfram Koeppe and Nadine Orenstein from the Metropolitan Museum of Art have shown immense intellectual generosity.

My gratitude goes to Bernard Cache, Jimena Canales, Susan Dackerman, Sven Dupré, Thomas Eser, Owen Gingerich, Jørgen Hein, Myles W. Jackson, Koenraad Jonckheere, Jutta Kappel, Michael Korey, Kerstin Losert, Peter Mark, Christof Metzger, Martina Minning, Bruce T. Moran, Laura Moretti, Christopher J. Phillips, Jennifer L. Roberts, Paolo Savoia, Peter Michael Schenkel, Konrad Schlegel, Friedrich Seck, Pamela Smith, Sabine Söll-Tauchert, Virginie Spenlé, Freyda Spira, Zoran Stankovski, Matt Stanley, and Joanna Woodall.

At MIT Press, I could not have wished for a better team than that of my editor, Thomas Weaver, whose stewardship along with the acumen of Matthew Abbate were crucial at the more advanced stages of this project. Mathias Clottu's beautiful design and Jayne Kuchna's scrupulous assistance with the bibliography and referencing each deserve special mention.

Support for this book has been graciously provided by The Graham Foundation for Advanced Studies in the Fine Arts; The Tavolozzo Foundation; The Lila Wallace–Reader's Digest Publications Subsidy at Villa I Tatti, The Harvard University Center for Italian Renaissance Studies; Research Foundation–Flanders; and Jane and Morgan Whitney Fellowships, Department of European Sculpture and Decorative Arts (2016–2017) and Department of Drawings and Prints (2014–2015), The Metropolitan Museum of Art. Additional research was conducted at the Department of Art, Music, and Theatre Sciences, Universiteit Gent.

Selections from earlier drafts of *The Polyhedrists* have appeared in *Journal for the History of Astronomy, Making Marvels: Science and Splendor at the Courts of Europe* (2019), and *Word & Image: A Journal of Verbal/Visual Enquiry*.

I also thank my parents and brother, Naomi, Eugene, and Alon, for their support and affection. My love for Malte, as well as to Cosimo, Mariusz, Celeste, Amedeo, and Tomaso, goes beyond the scope of any singular project. Writing alongside them has made the work undoubtedly more enjoyable. And finally, to Nimrod, whose criticism and companionship have accompanied this project nearly from its inception, and to whom this book is dedicated.

The Polyhedrists: Art and Geometry in the Long Sixteenth Century
Noam Andrews

The MIT Press would like to thank the anonymous peer reviewers who provided comments on drafts of this book. The generous work of academic experts is essential for establishing the authority and quality of our publications. We acknowledge with gratitude the contributions of these otherwise uncredited readers.

Support for this book has been provided by:

The Graham Foundation for Advanced Studies in the Fine Arts

The Tavolozzo Foundation; The Lila Wallace–Reader's Digest Publications Subsidy at Villa I Tatti, The Harvard University Center for Italian Renaissance Studies; Research Foundation–Flanders; Jane and Morgan Whitney Fellowships, Department of European Sculpture and Decorative Arts (2016–2017) and Department of Drawings and Prints (2014–2015), The Metropolitan Museum of Art.

Edited by Thomas Weaver
Designed by Studio Mathias Clottu

Printed and bound in Germany

Library of Congress Cataloging-in-Publication Data is available.
ISBN 978-0-262-04664-0

10 9 8 7 6 5 4 3 2 1

frontispiece Lorentz Stöer (active 1557–1620/21). Drawing of a model with imperial privilege, from *Geometria et perspectiva corpora regulata et irregulata*. 1562–1584. Universitätsbibliothek der LMU München (HS Cim 103, p. 333).

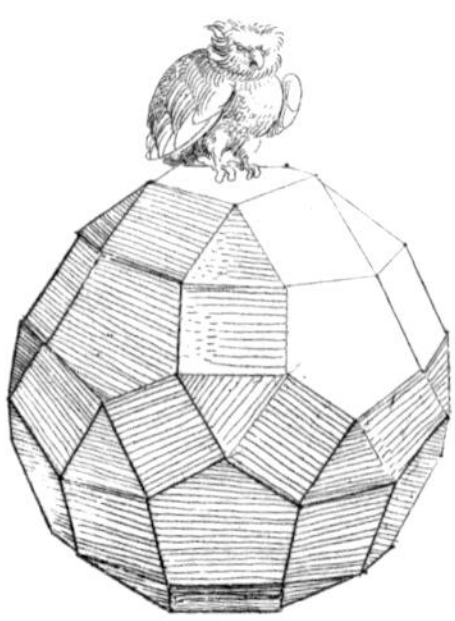